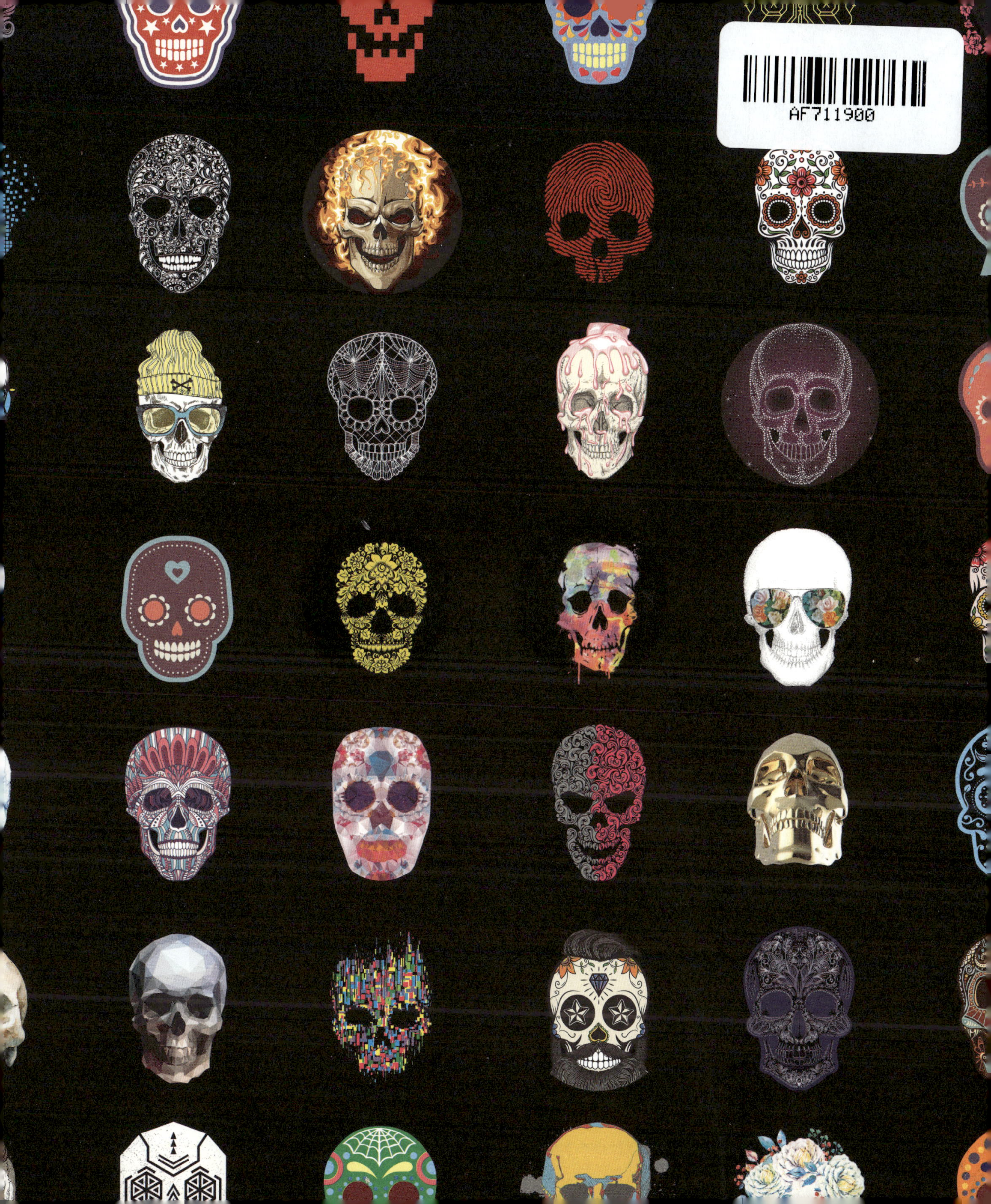

Skull

SOURCEBOOK

Skull

SOURCEBOOK

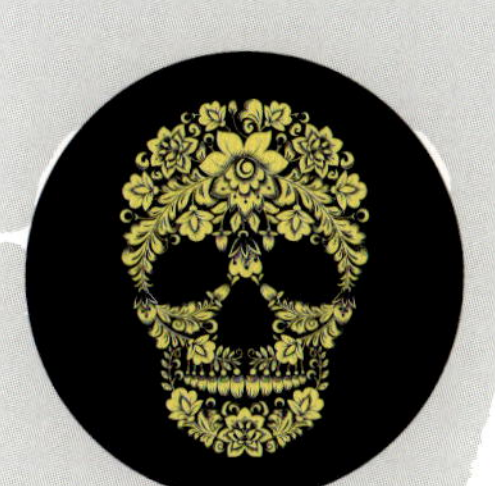

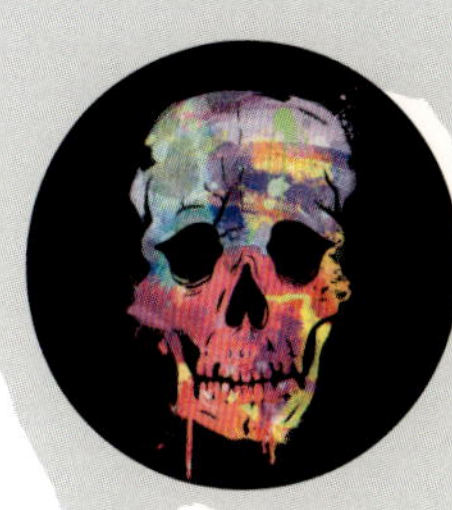

OVER 500 SKULLS IN ART & CULTURE

ADELE NOZEDAR

First published in the United States of America in 2016 by
Race Point Publishing, a member of
Quarto Publishing Group USA Inc.
142 West 36th Street, 4th Floor
New York, New York 10018
quartoknows.com

10 9 8 7 6 5 4 3 2 1

ISBN 978-1-63106-158-5

Library of Congress Cataloging-in-Publication Data

Names: Nozedar, Adele, author.
Title: Skull sourcebook : over 500 skulls in art & culture / by Adele Nozedar.
Description: New York : Race Point Publishing, 2016.
Identifiers: LCCN 2016001488 | ISBN 9781631061585 (paperback)
Subjects: LCSH: Skull in art. | BISAC: ART / Subjects & Themes / Human Figure. | ART / Popular Culture. | ART / Graffiti & Street Art.
Classification: LCC N8243.S57 N69 2016 | DDC 700/.4561--dc23
LC record available at http://lccn.loc.gov/2016001488

Author: Adele Nozedar
Editorial Director: Jeannine Dillon
Managing Editor: Erin Canning
Project Editor: Jason Chappell
Senior Design Manager: Heidi North
Cover and Interior Design: Leah Lococo

Printed in China

CONTENTS

INTRODUCTION

THERE'S SOMETHING BOTH GHOULISH AND FASCINATING ABOUT THE HUMAN SKULL. TRY SOMETHING BEFORE YOU READ ANY FURTHER—IT MIGHT LOOK WEIRD IF ANYONE'S WATCHING! FEEL YOUR FACE WITH THE TIPS OF YOUR fingers. Press into the crevices, noting the lumps, bumps, and accentuations. Smooth over the plane of your forehead and trace around your eye sockets. Touching every part of your head in this way, you quickly come to the realization of your own mortality, our place in the physical world, and our clear belonging to the animal kingdom in spite of any lofty notions otherwise.

Unlike other bones which, unearthed in archaeological digs, might initially be mistaken for stones, or pebbles, or perhaps animals bones, we immediately recognize the skull as the underlying foundation of the face and, as such, the most recognizable remnant of the living human being to whom it once belonged. Those huge blank eye sockets and grinning teeth are full of character and imbued with a sort of comical horror. In spite of the fact that, to see a skull means life is long-gone, it still appears animated to a cartoon-like degree.

Therefore, it's no surprise that the skull occupies such an important symbolic place in our psyche, used—as we shall see—across the globe in just about every walk of life. The skull has been used for millennia as a reminder of mortality, in art, in religion, in films, and in theatre; of late, it has even been used in fashion. The skull is used as both a warning of potential death (for example, on poison bottles) and as a reminder of actual death (on tombstones, where the combination of a skull with wings is a hopeful reminder of the transformative nature of death and the survival of the soul). Use of the skull icon is also an act of defiance. It represents a refusal to fear the inevitability of our own limited time on this planet, as we embrace its imagery as part of our everyday life.

There are inevitable cross-over points in the skull iconography described in the sections of this book; for example, one person's religion may well be the another person's occult practice. Skulls used in literature may also encompass those used in an artistic or literary context. However, let us journey together across time and place, through music, art, and mystery, to discover the place that the skull holds in society.

HIDDEN SKULL ART AND ILLUSION

SOMETIMES, RATHER THAN DISPLAYING THE IMAGE OF A SKULL DIRECTLY, PAINTERS AND ARTISTS CHOOSE TO OBFUSCATE THE PICTURE, LEAVING BOTH THE SKULL AND ITS SYMBOLIC MEANING AS A FEATURE TO BE DISCOVERED (or maybe not) by the observer. In some cases, the skull is revealed only by looking at the painting or artwork from a specific angle. Such a device is called anamorphosis.

One of the best-known examples of anamorphosis is a work of art called *The Ambassadors*, by Hans Holbein the Younger, which is dated 1533 and which is part of the collection at the National Gallery in London. This is a painting that many people are familiar with, although they might not know the artist who painted it. *The Ambassadors* is a masterpiece of anamorphosis, no easy technical feat. The picture shows a pair of finely-dressed gentlemen, placed amidst a display of great wealth and riches. The only anomaly in this scene of prosperity is what looks like a weird smudge in the bottom center of the painting. This is the skull, which is only visible if the picture is viewed from the side of the canvas. The ability to perform such artistic gymnastics, as you can imagine, gave Holbein a chance to show off his impressive skills as well as providing a "surprise" element which continues to intrigue. It's possible that the skull is used by Holbein as a reminder that even the very wealthy are mortal, a common theme of artists of Vanitas paintings.

The 1852 painting of Shakespeare's tragic heroine, *Ophelia*, by pre-Raphaelite artist John Everett Millais, is believed by some to include a hidden skull. Although it is indeed possible to make out dark hollow eyes and a gaping mouth in the undergrowth, this resemblance to a skull may be purely accidental. It's likely that we'll never know for sure.

In contrast to the hidden skulls, is a painting by American artist and illustrator Charles Allan Gilbert. The skull featured in *All is Vanity* is plain to see, not at all disguised; here, what we're looking at is more of a skull illusion. The "eyes" of the skull are cleverly rendered by Gilbert as a beautiful woman staring back at her own reflection. Again, the skull is used as a reminder of mortality and the folly of vanity. The title of the painting itself is a Biblical reference. Gilbert produced this piece of work in 1892 when he was just 18 years old.

Hidden or ambiguous skull imagery became something of a fashion in the late nineteenth and early twentieth centuries; the advent of cheaper methods of printing and the popularity of picture postcards certainly helped. There's a slew of such imagery available, cleverly contriving to combine images of war, nudity, and debauchery into an overarching skull icon. Alas, the artists of these pieces are often anonymous. Despite the tendency toward anonymity, some of the artists who experimented with illusory and metamorphic skull art during this period are very well-known. And the Hungarian artist István Orosz produced a series of etchings, *Ship of Fools*, each featuring a skull.

One of the more dramatic and poignant uses of the hidden skull motif was employed in an advertising campaign by United States advocacy group, Mothers Against Drunk Driving (also known by their initials, MADD). Photographed upwards into a circle of young graduates who celebrate by throwing their caps into the air, the caps themselves form the eyes and nose of the skull, with the waving hands beneath outlining the teeth. The message speaks for itself: a salutary reminder that all the expectation and hope of youth can be dashed in an instant.

Other artists have also hidden skulls in their work. Christina Burch's series, *Garland of Skulls,* uses familiar motifs like flowers to create skulls. Gianni Sarcone, on the other hand, creates a skull by cleverly using a bridge and some rocks.

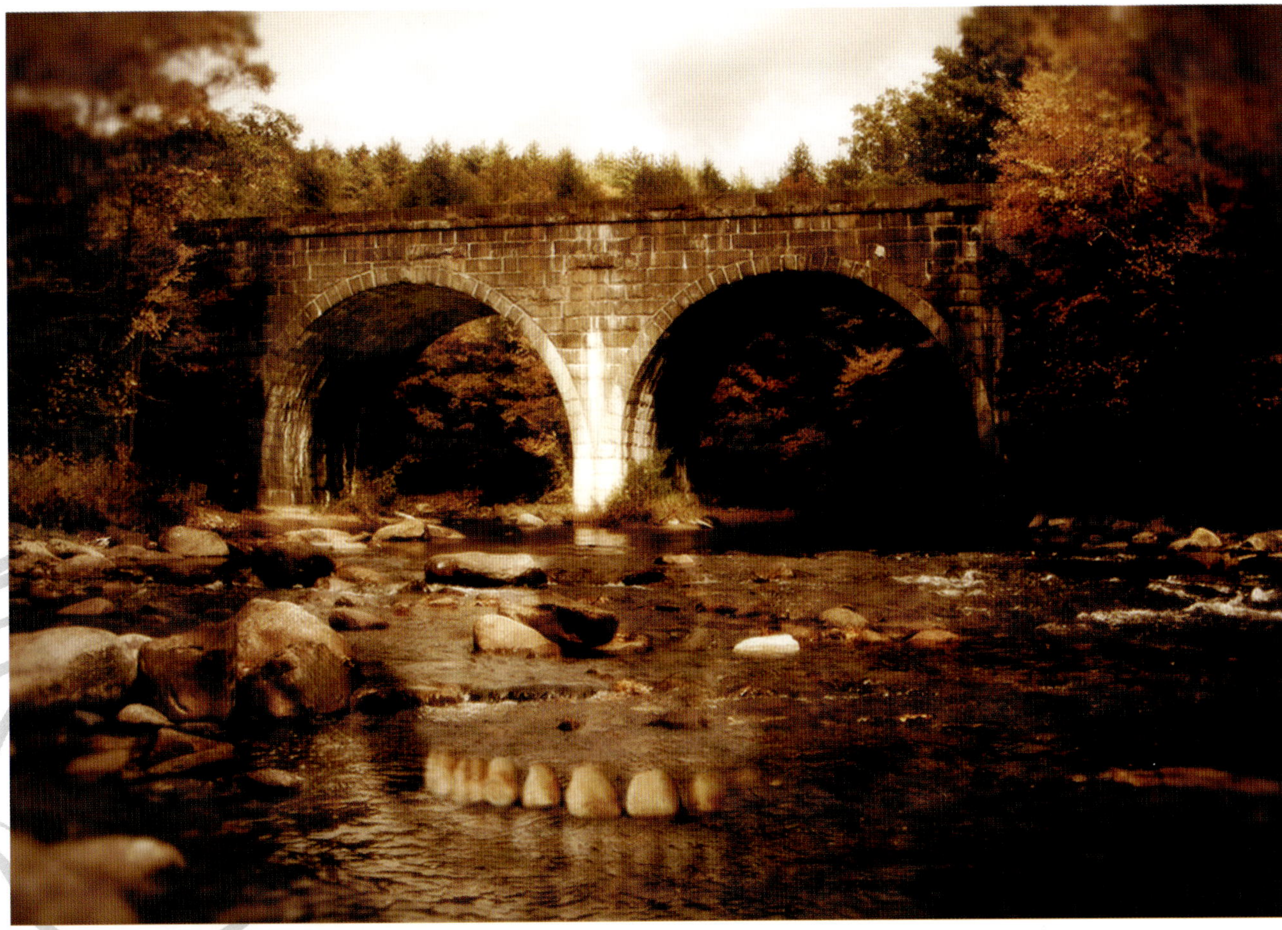

Gianni Sarcone's *Bridge to Eternity* hides a skull inside this peaceful scene of a bridge over water.

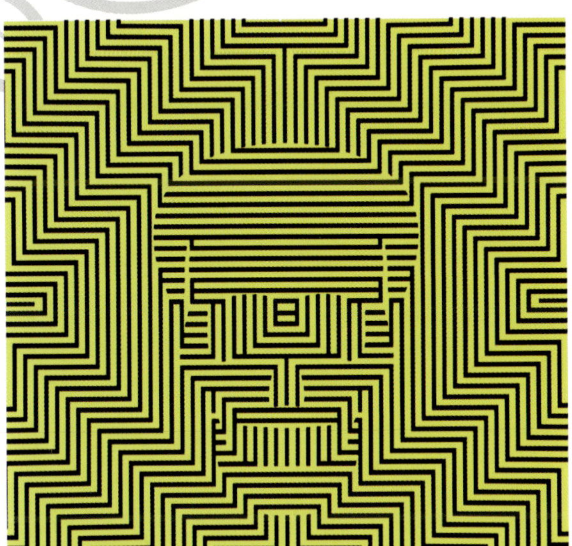

This pattern, dubbed *Vibrating Skull* by creator Gianni Sarcone, hides a skull that appears to move.

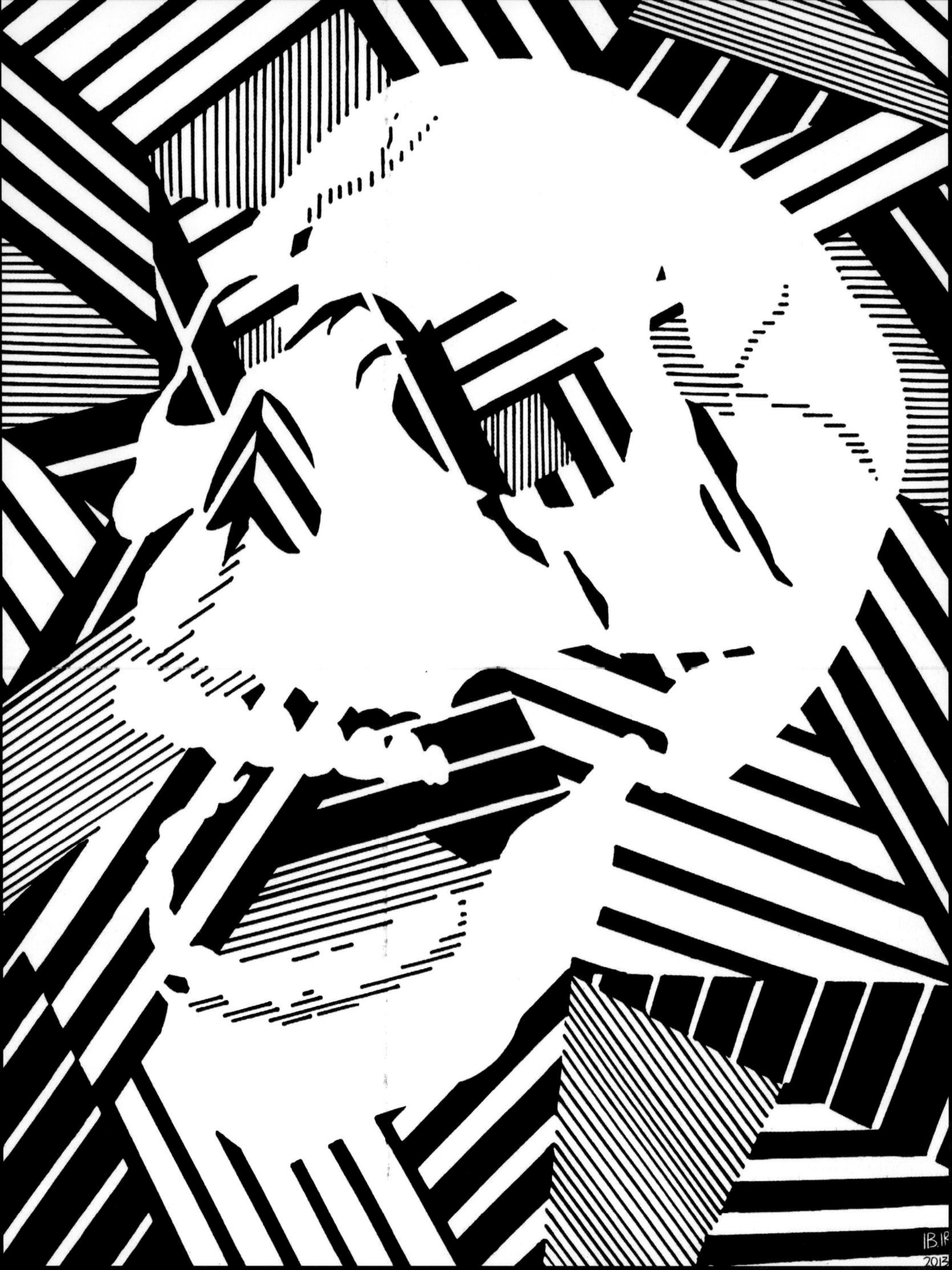

Hidden Between the Lines, an implied line design, was created using illustration marker on bristol board.

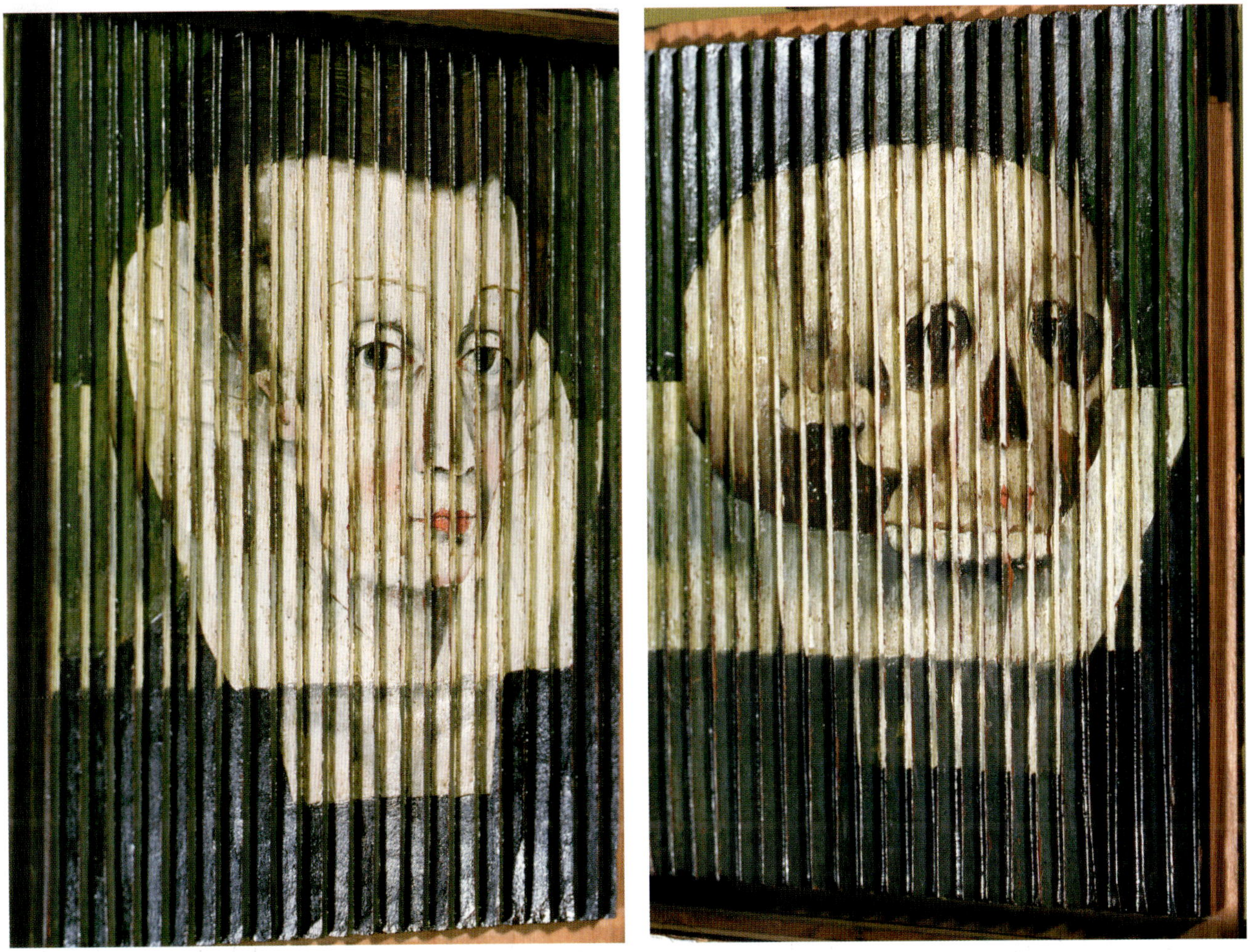

Anamorphosis of a skull: alternative views reveal an unknown portrait of a late 16th century woman formally identified as Mary Queen of Scots.

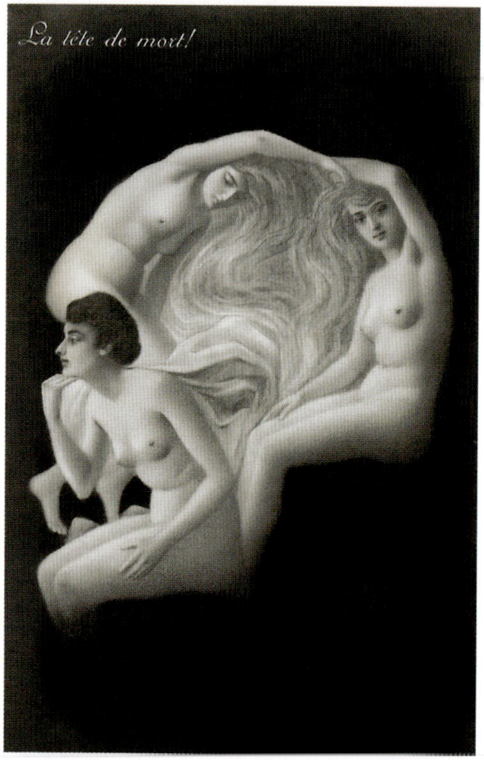
La tête de mort!

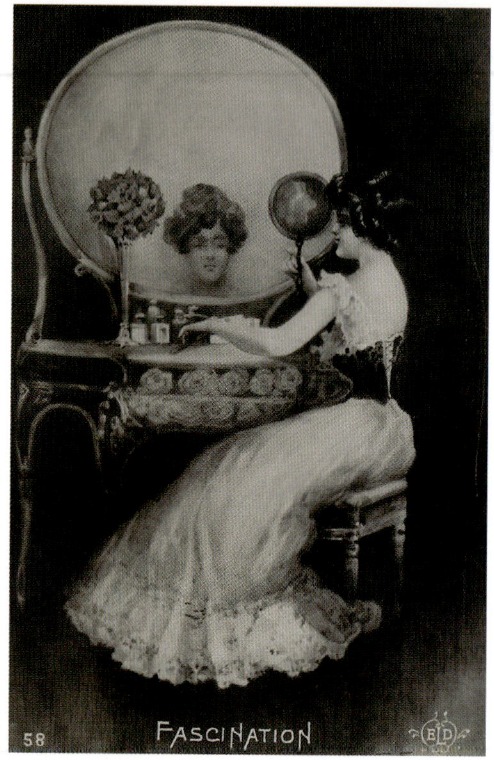
58
FASCINATION

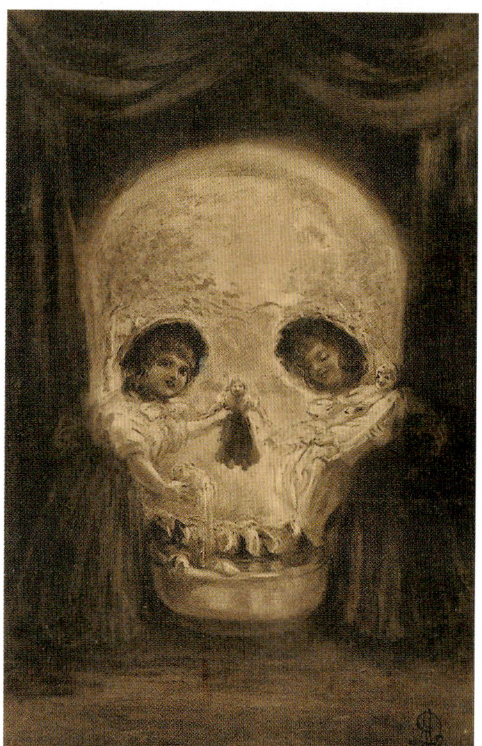

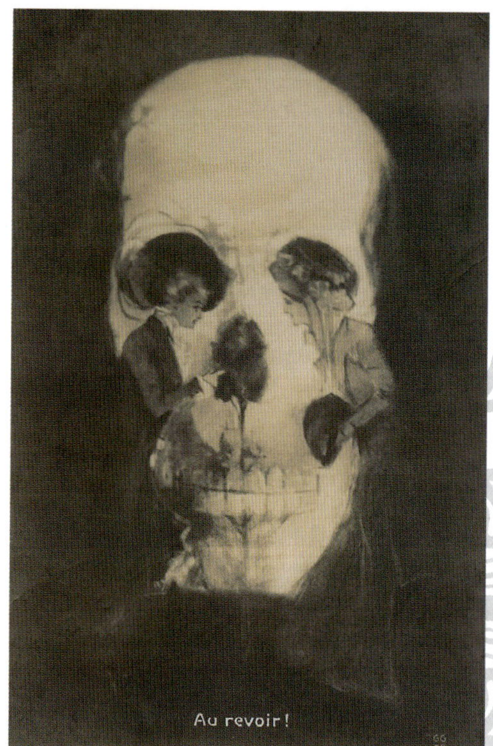
Au revoir!

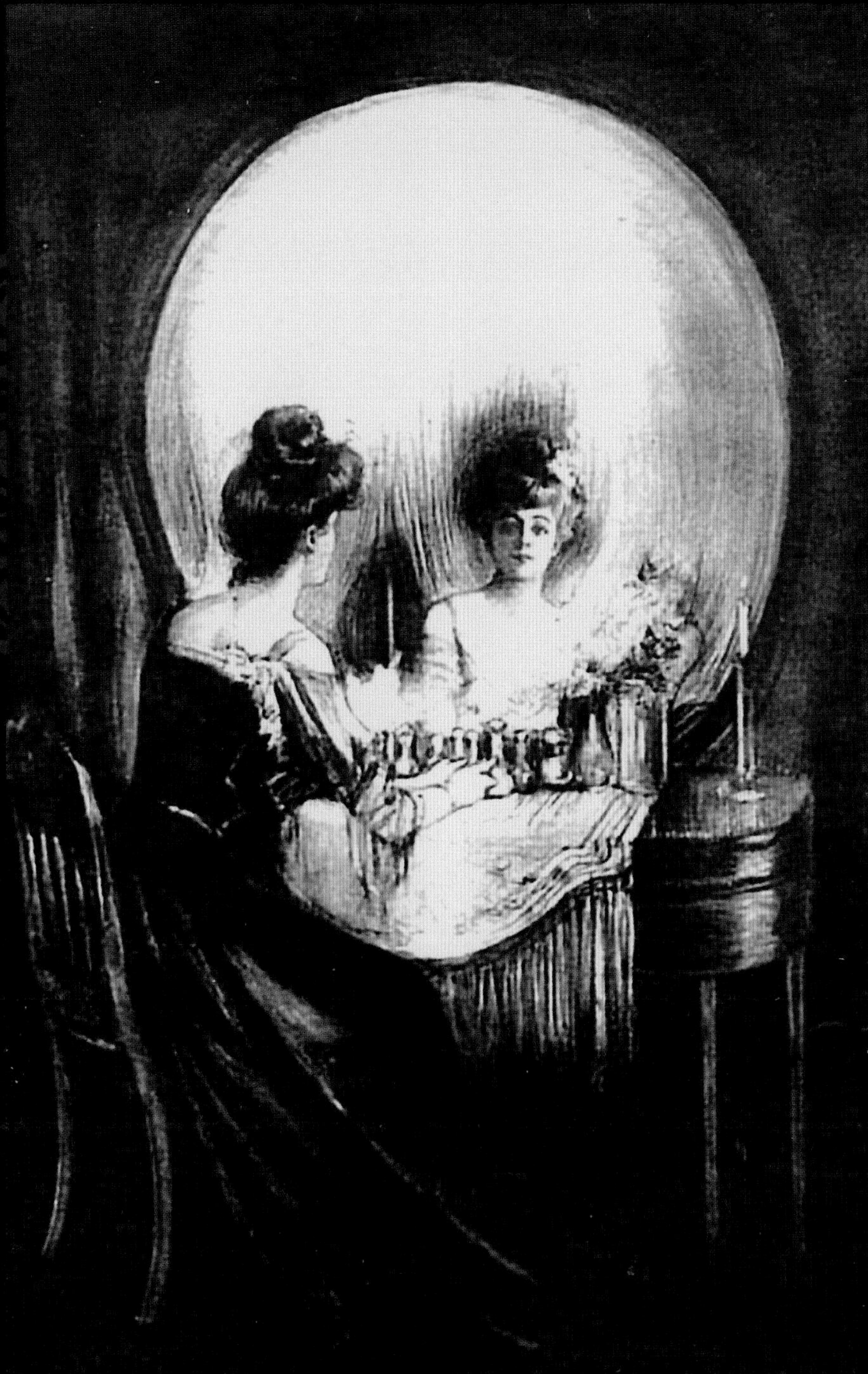

All is Vanity, an 1892 painting by Charles Allan Gilbert.

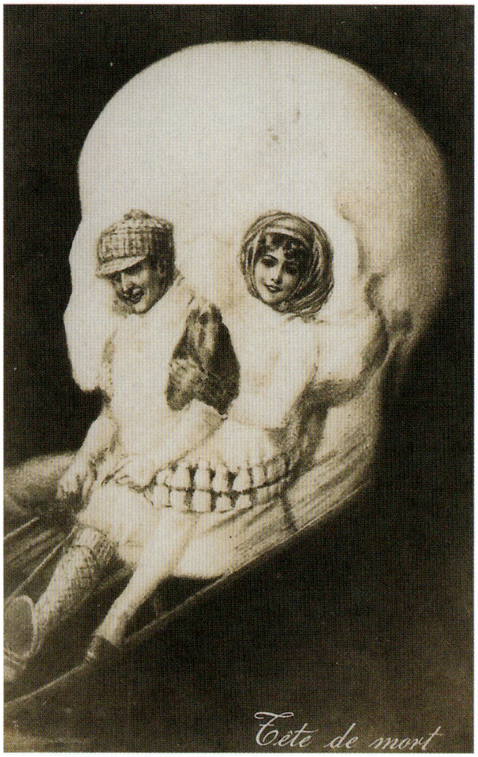
Tête de mort

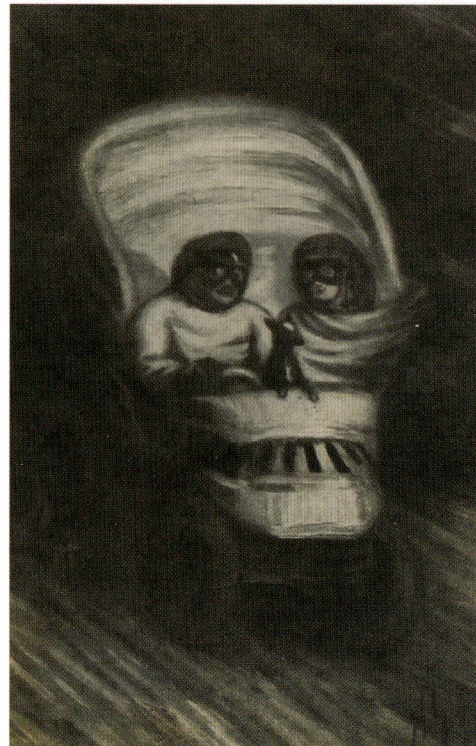

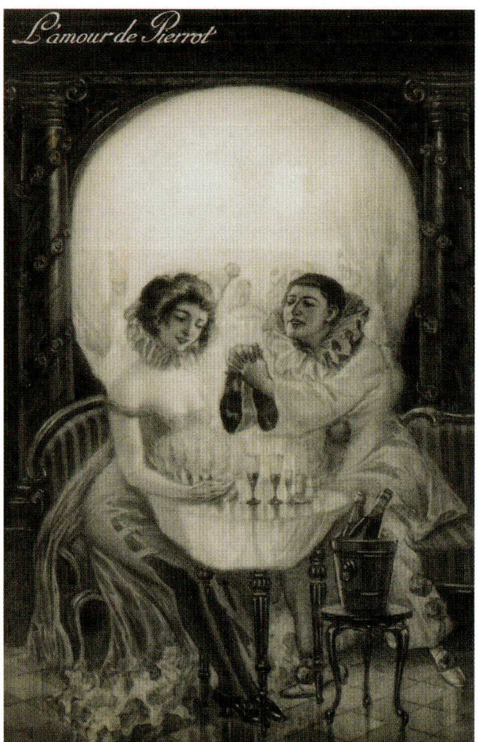
L'amour de Pierrot

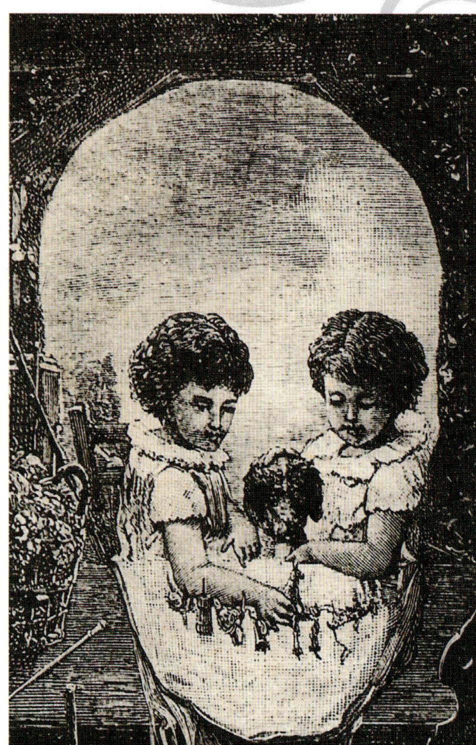

Illustrated by Kerby Rosanes (www.kerbyrosanes.com) using pen and ink on paper.

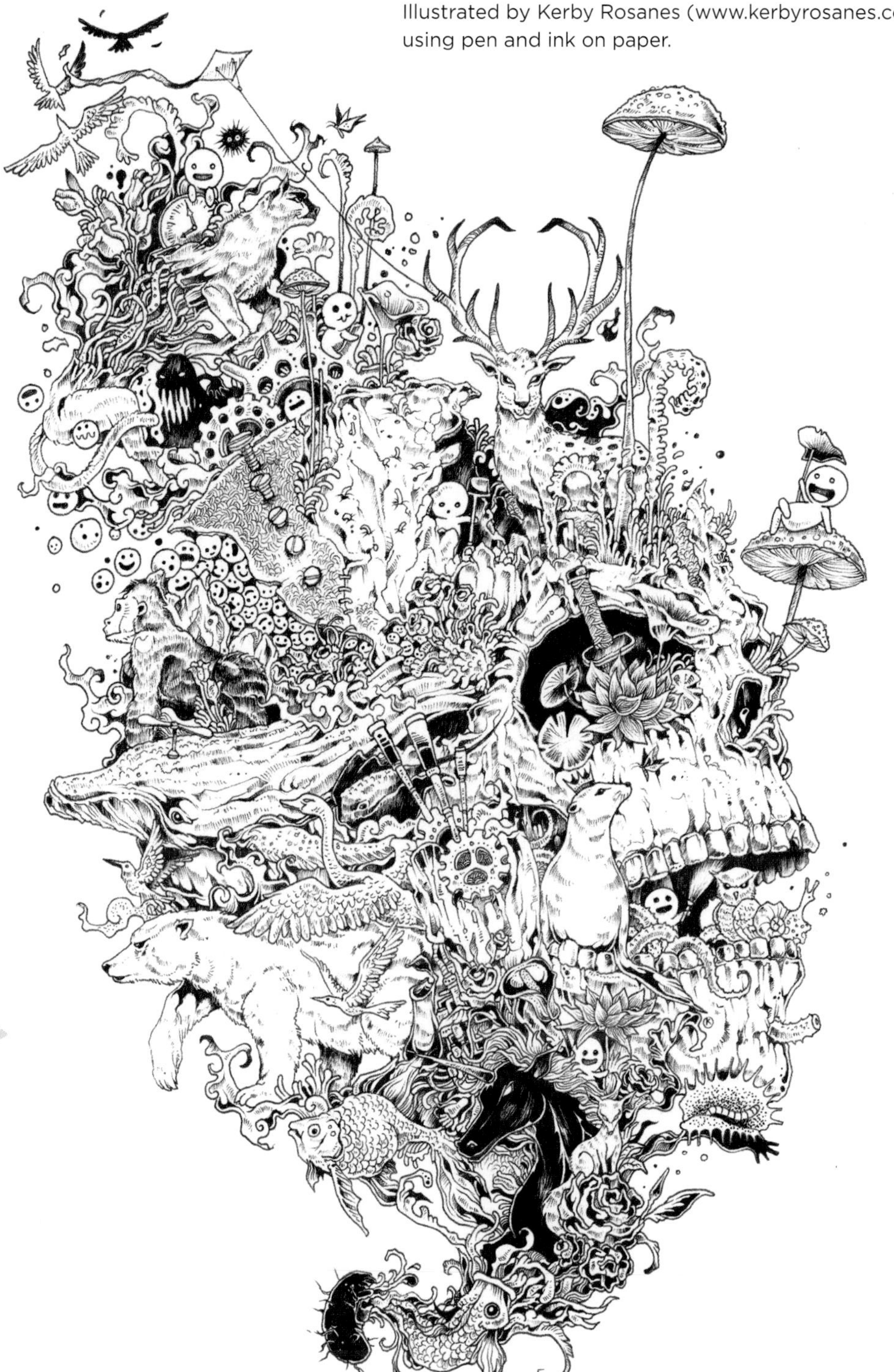

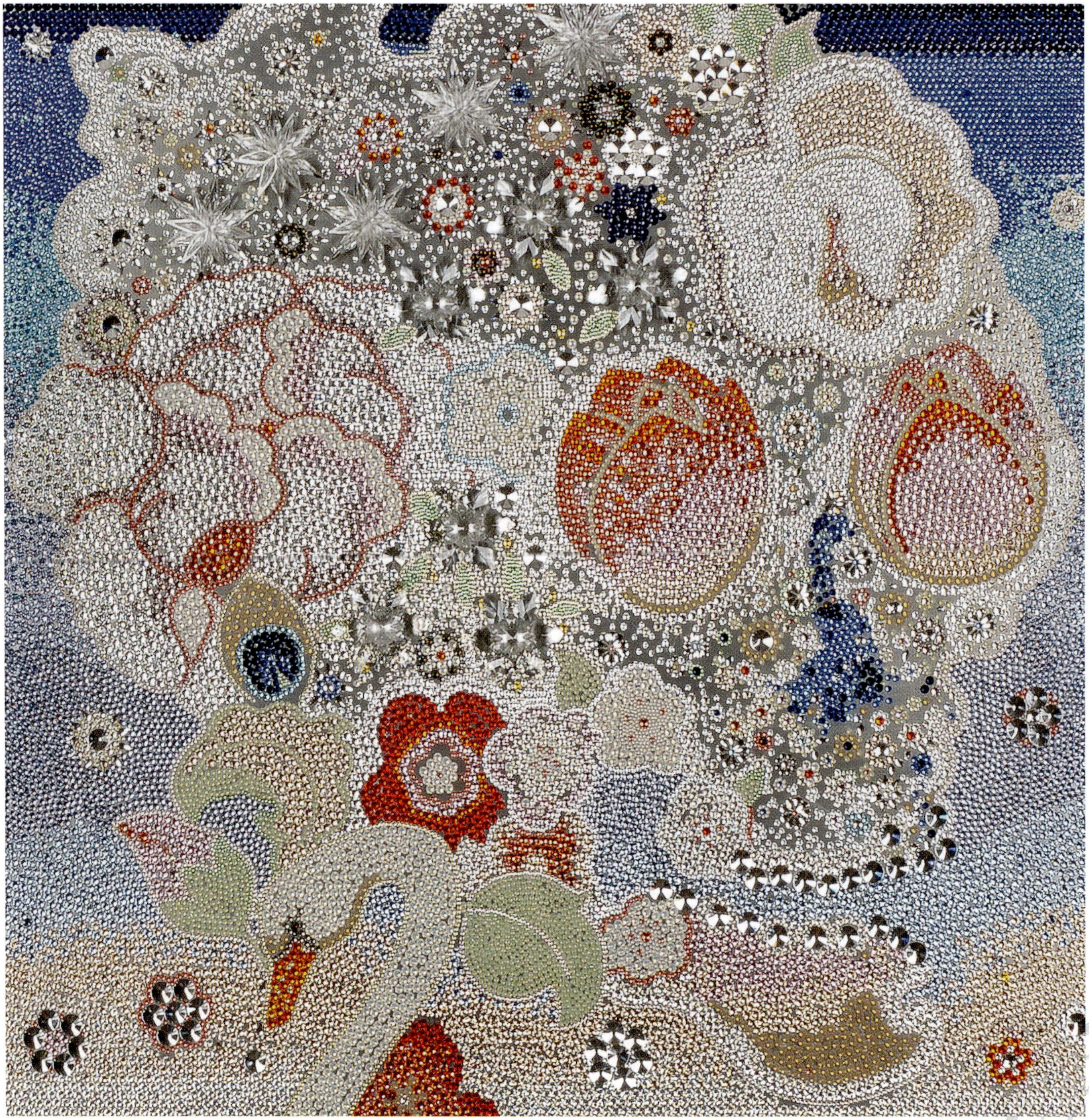

This spread and pages 20–22: Selections from the series, *Garland of Skulls* by artist Christina Burch.

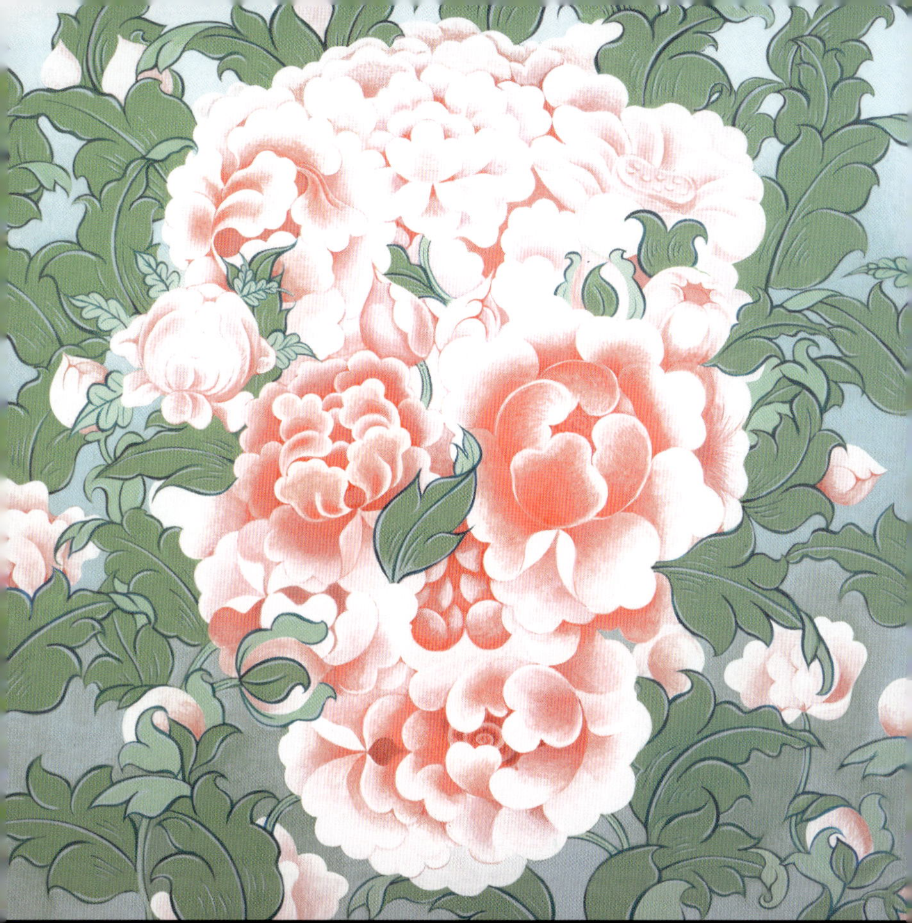

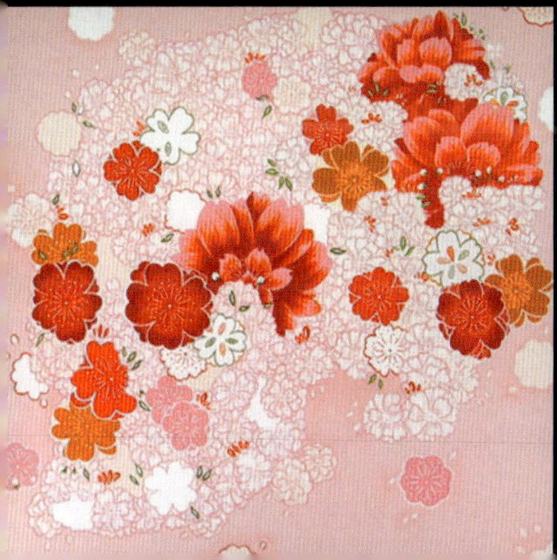

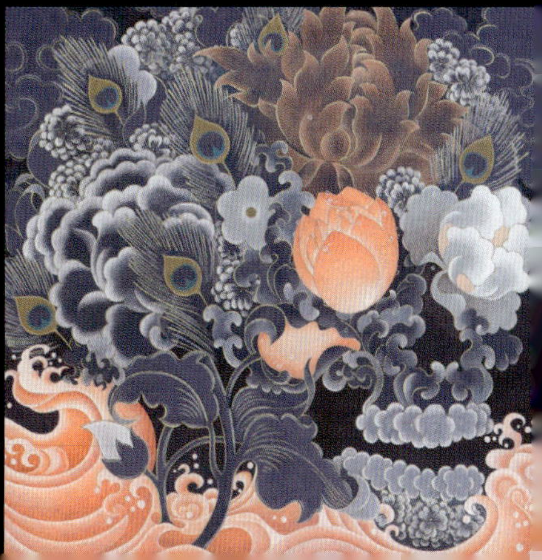

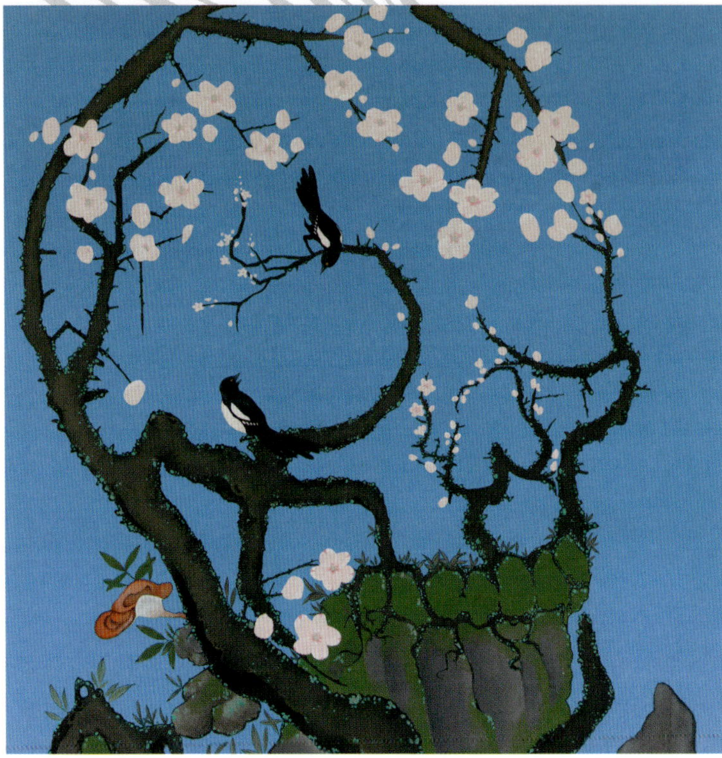

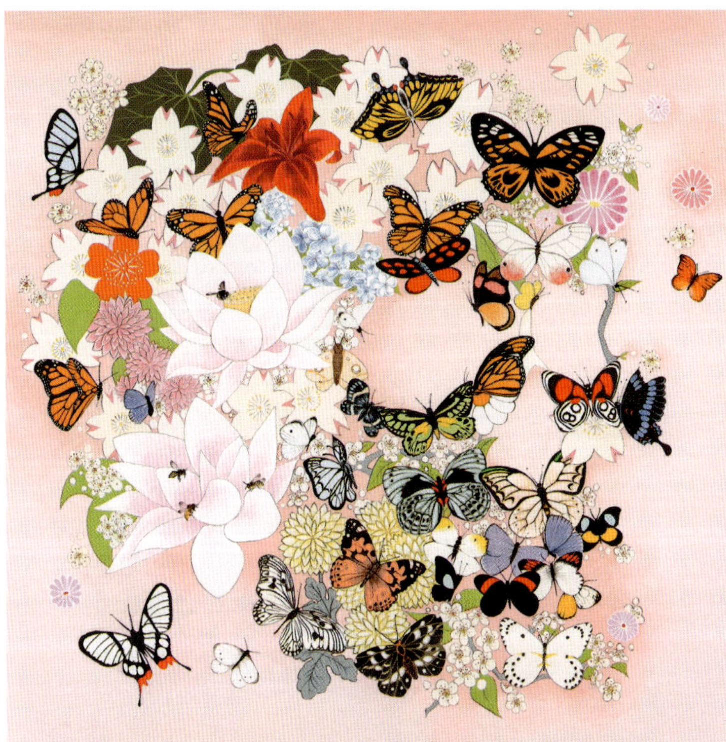

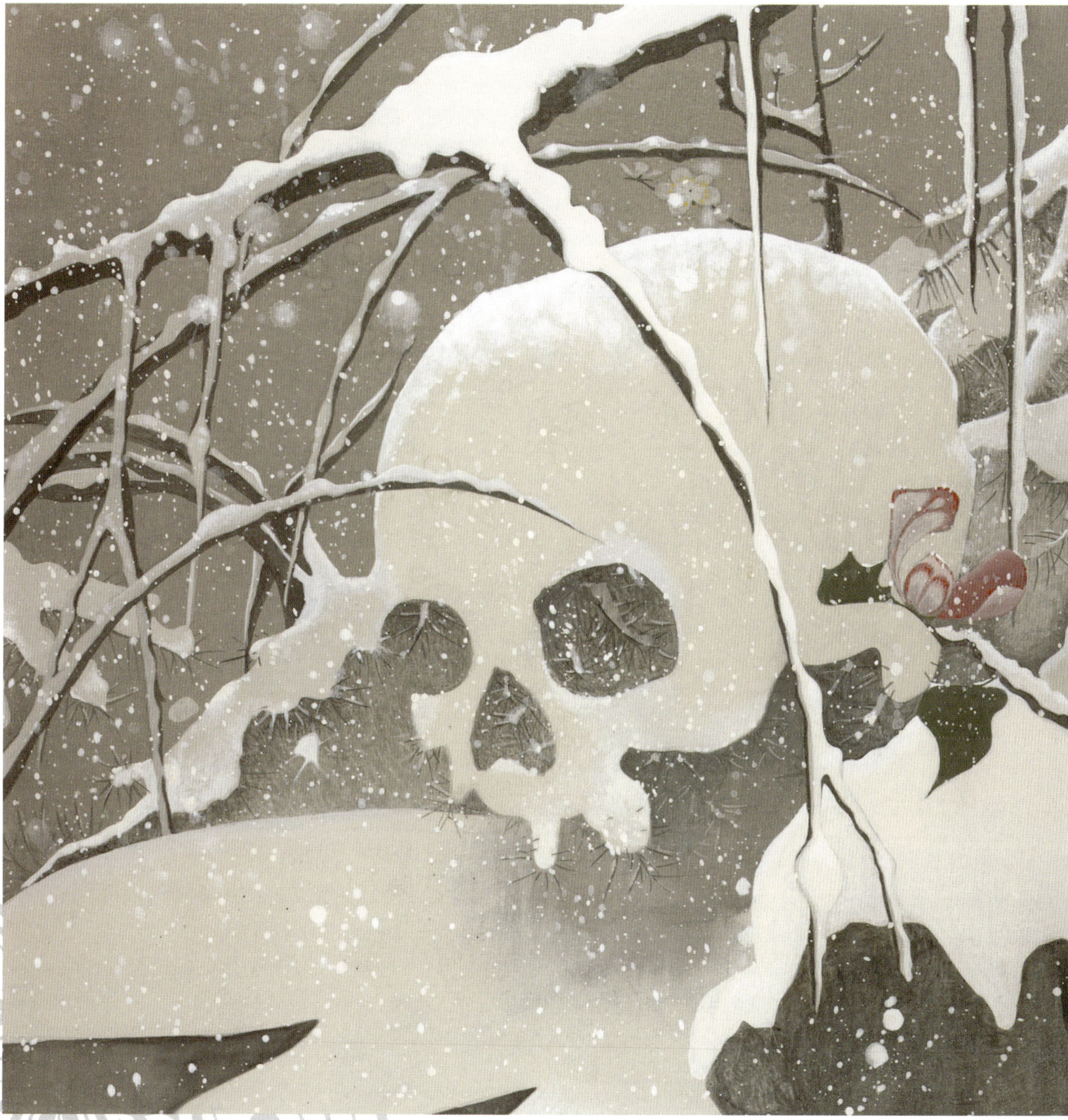

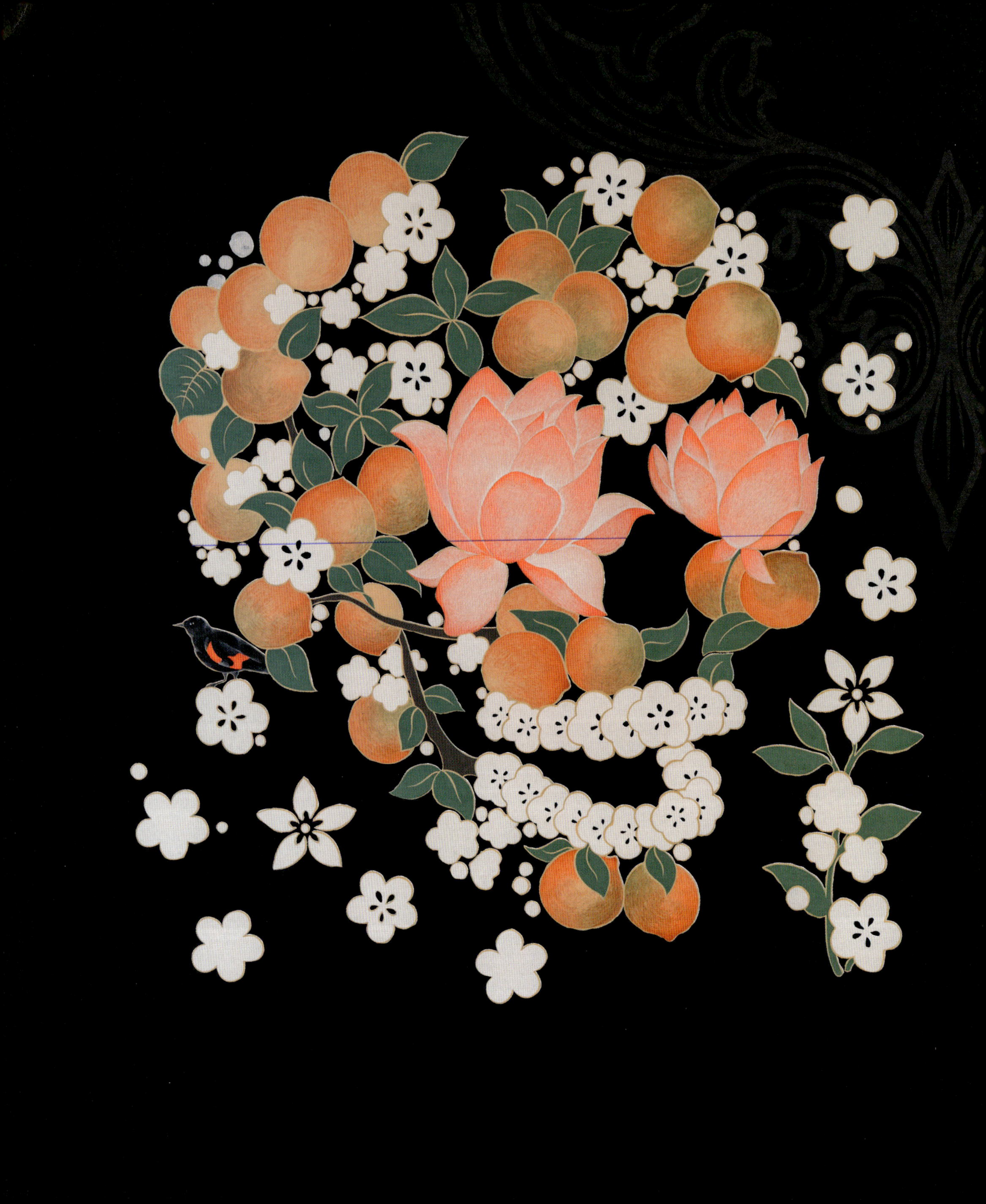

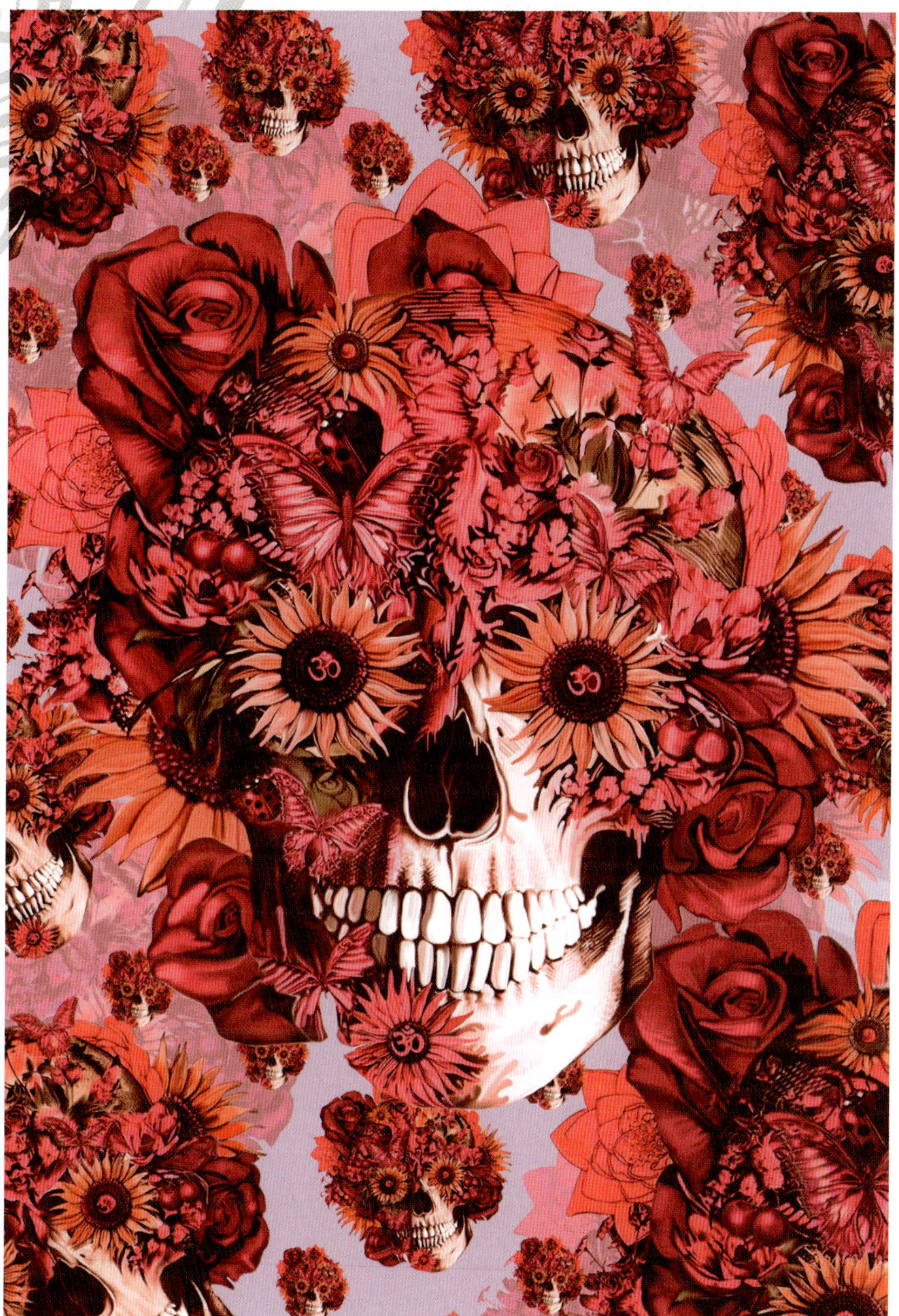

Girly By Nature, a selection from the series *Skull Collection*—an exploration of the subconcscious based on the evolution of a damaged psyche.

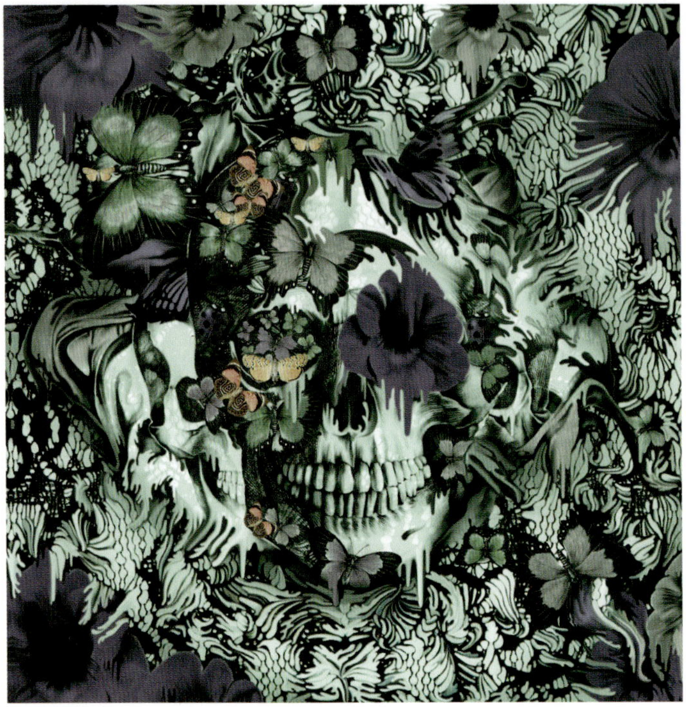

Anatomically Incorrect from the series *Skull Collection.*

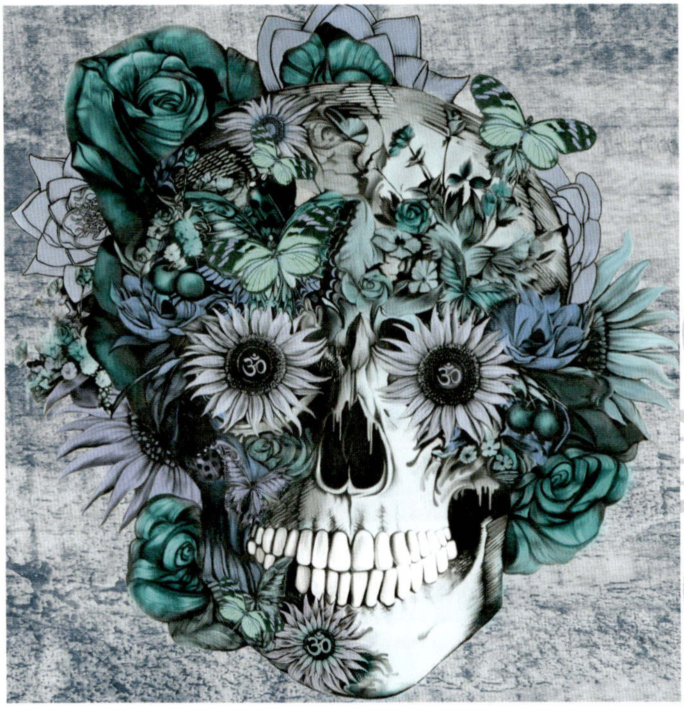

Blue Grunge Ohm from the series *Skull Collection.*

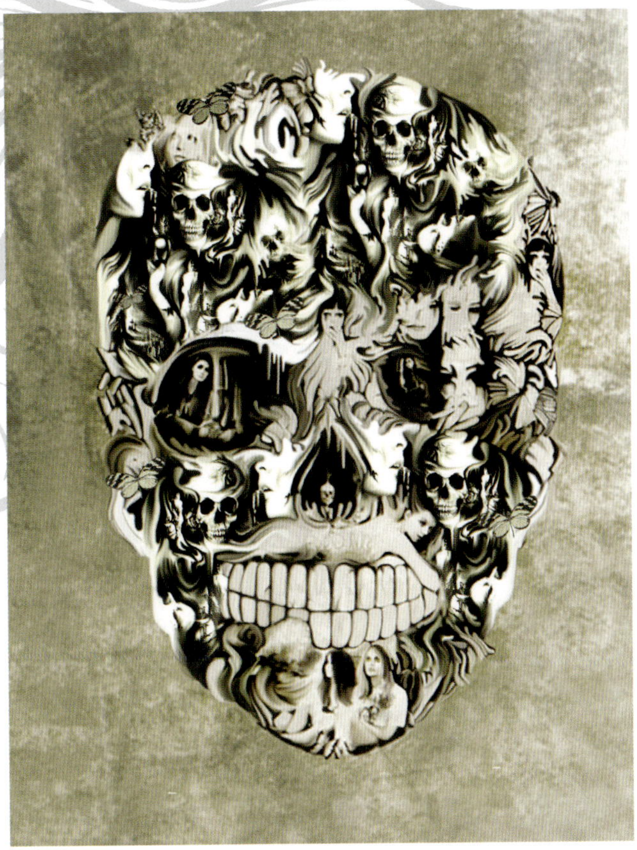

Feed My Soul from the series *Skull Collection.*

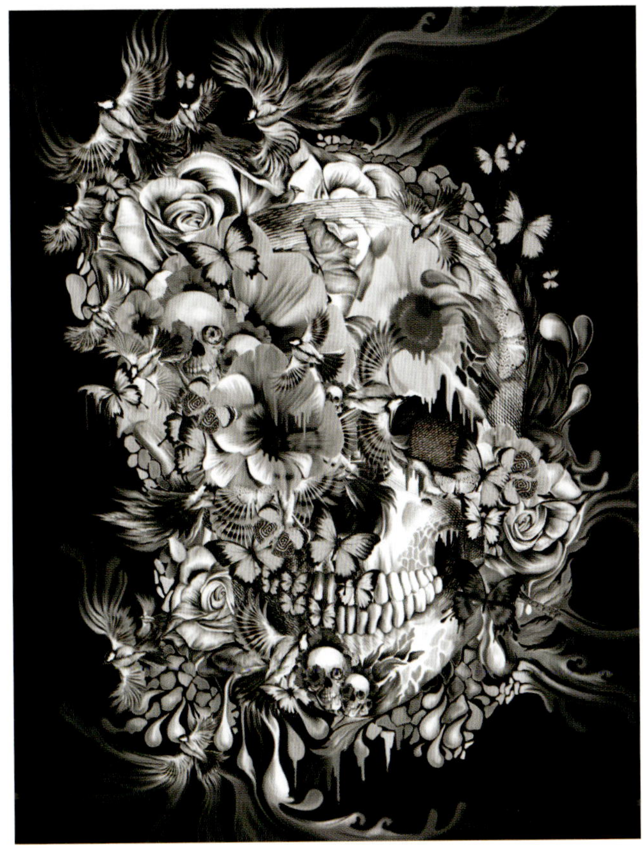

Flight from the series *Skull Collection.*

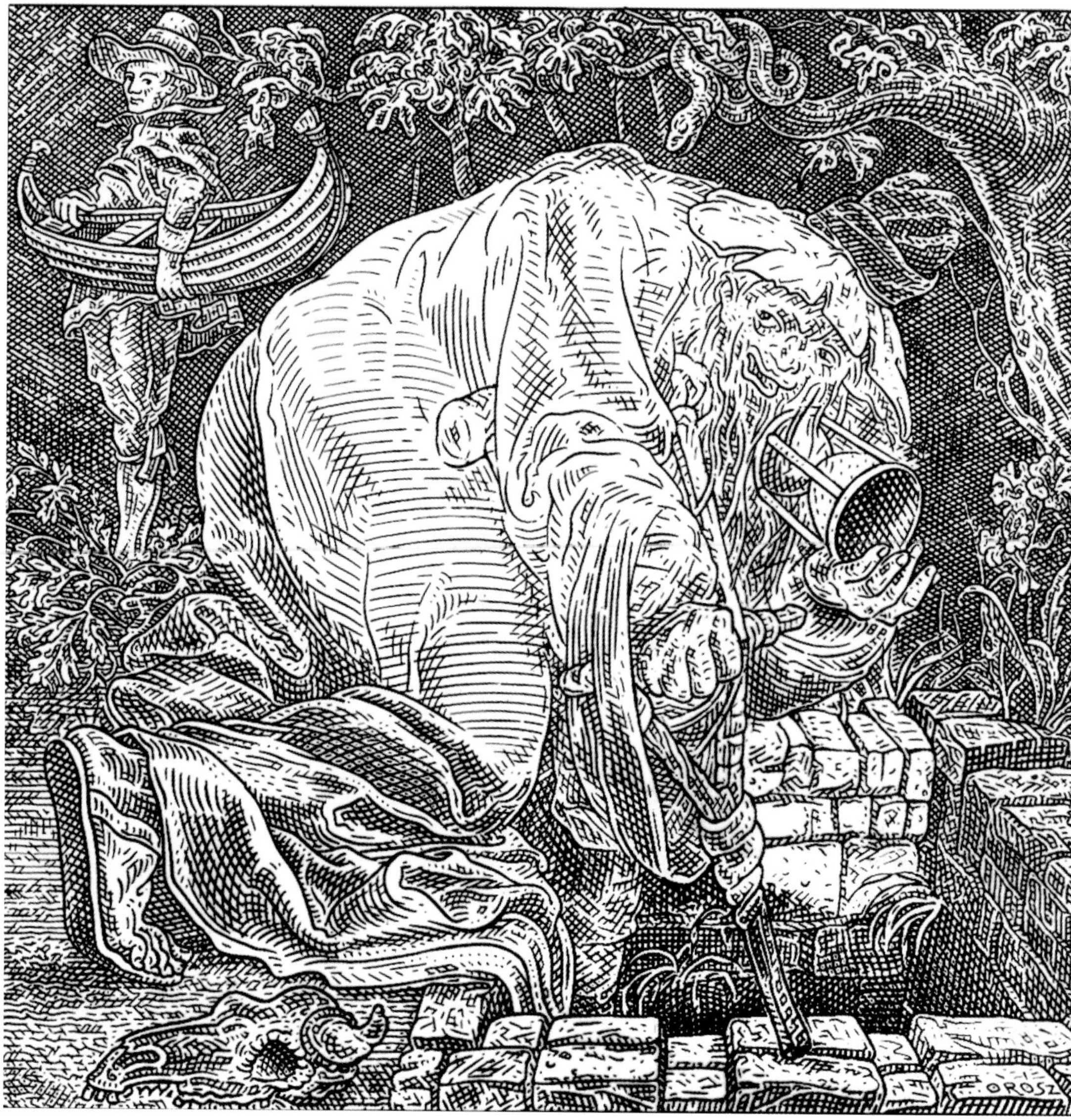

This spread and page 28: From the series *Ship of Fools,* a 20th century take on a 15th century tale, by István Orosz.

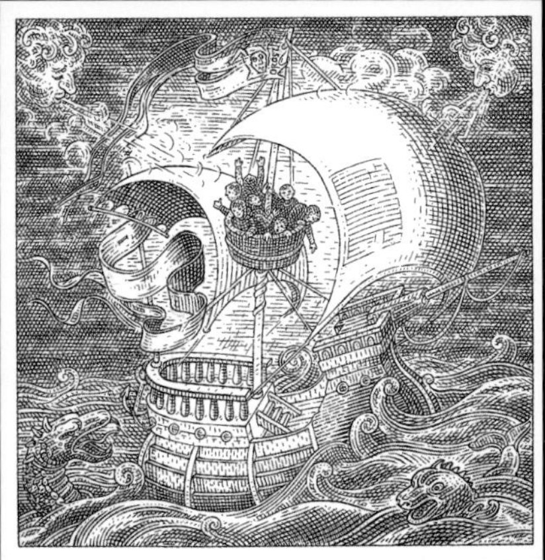

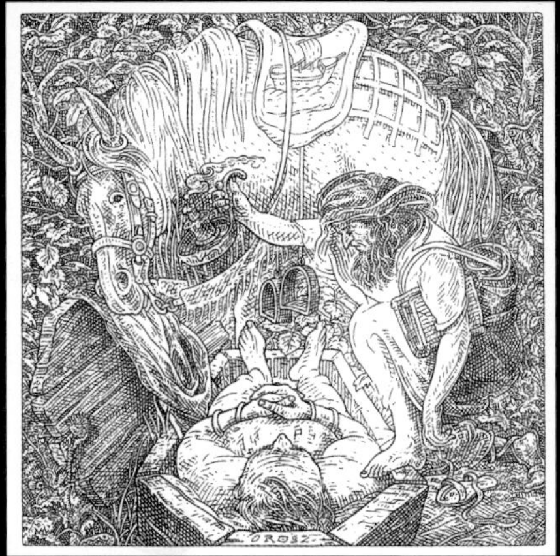
OROSZ

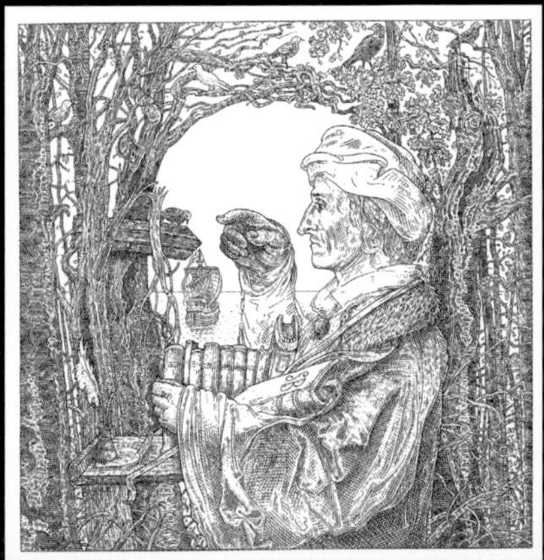

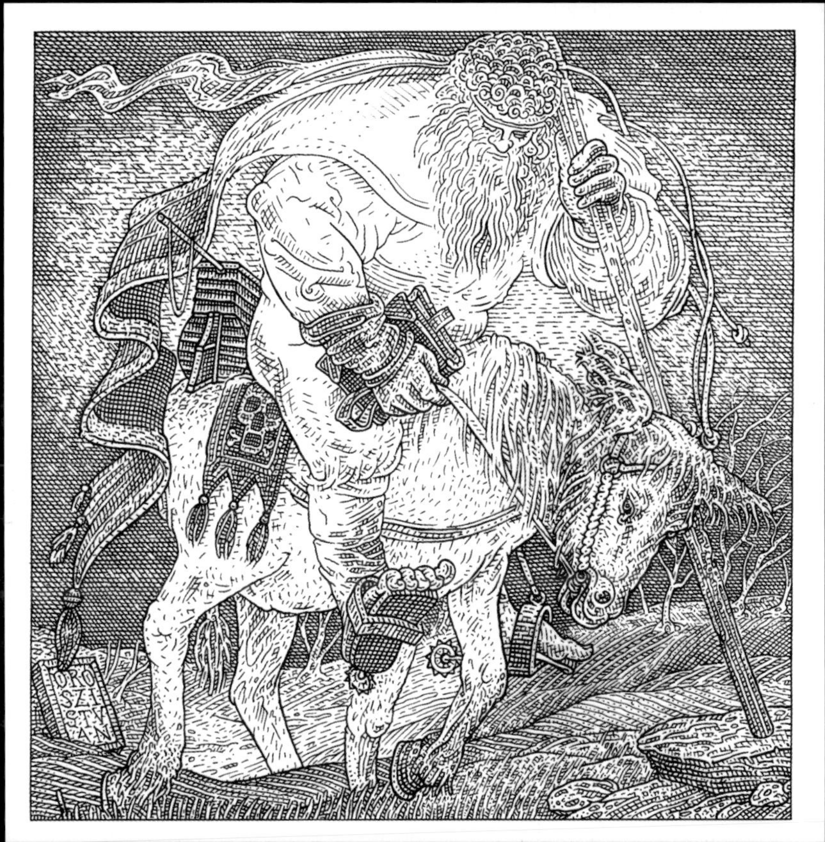

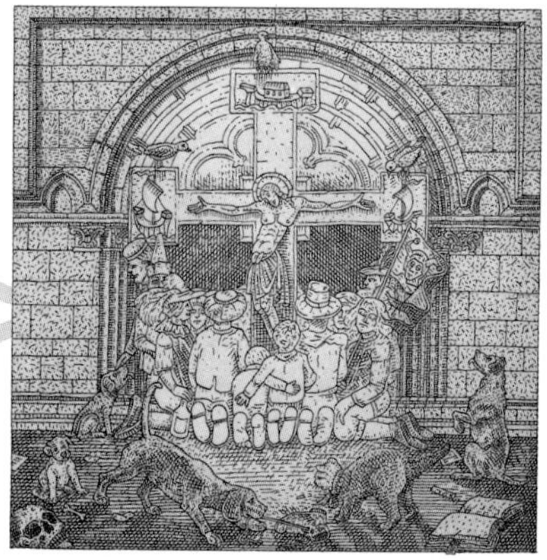

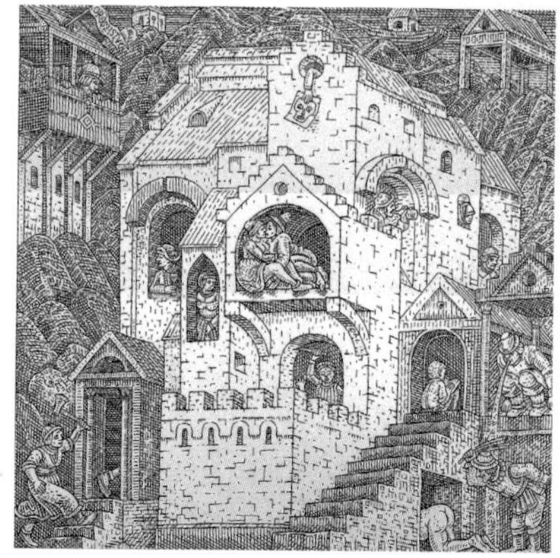

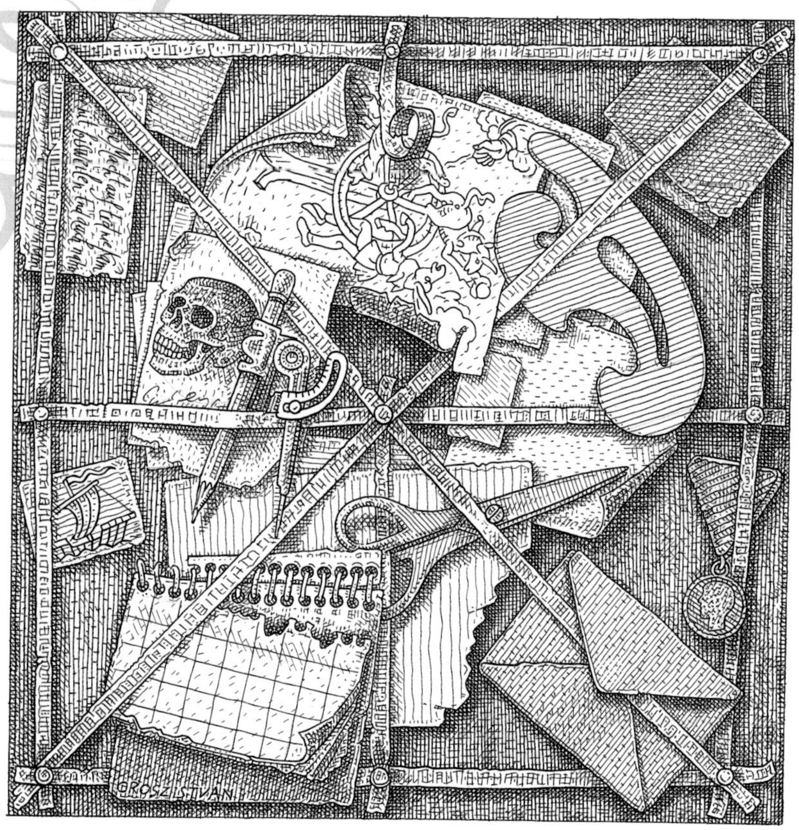

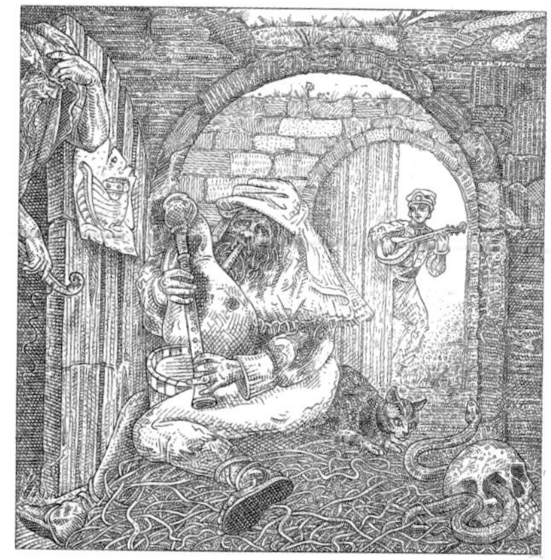

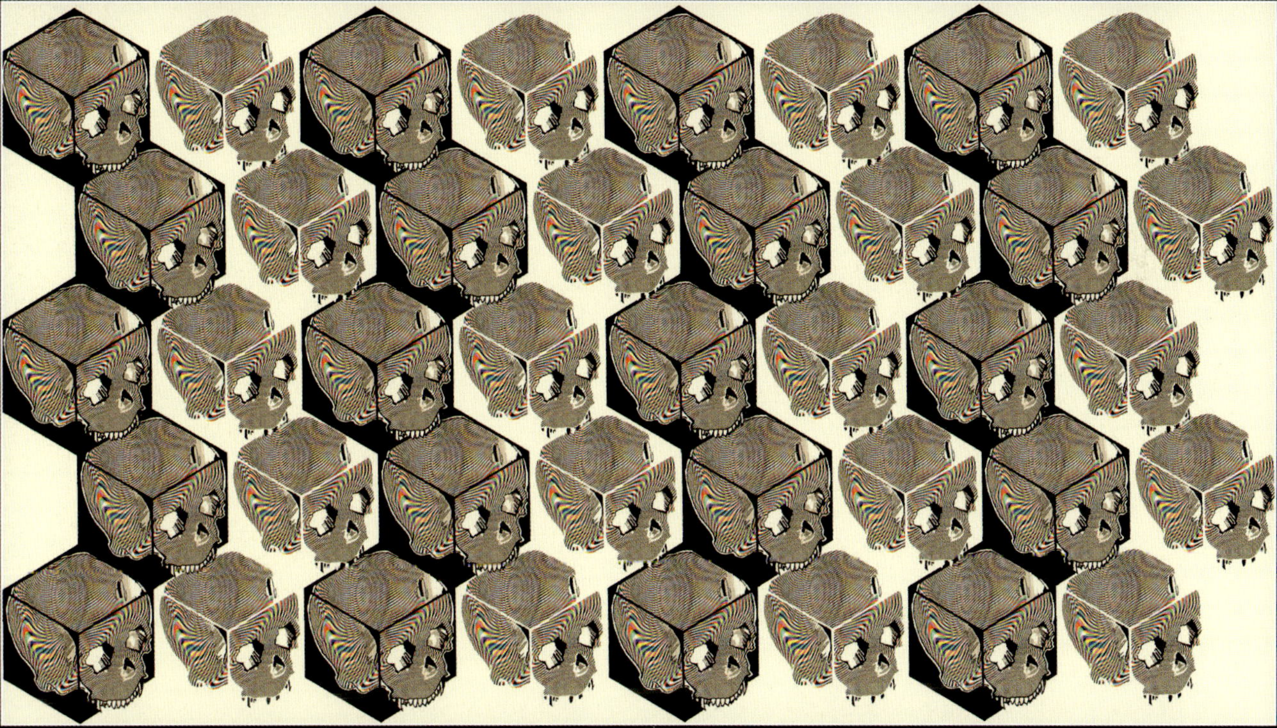

Pattern Illusion

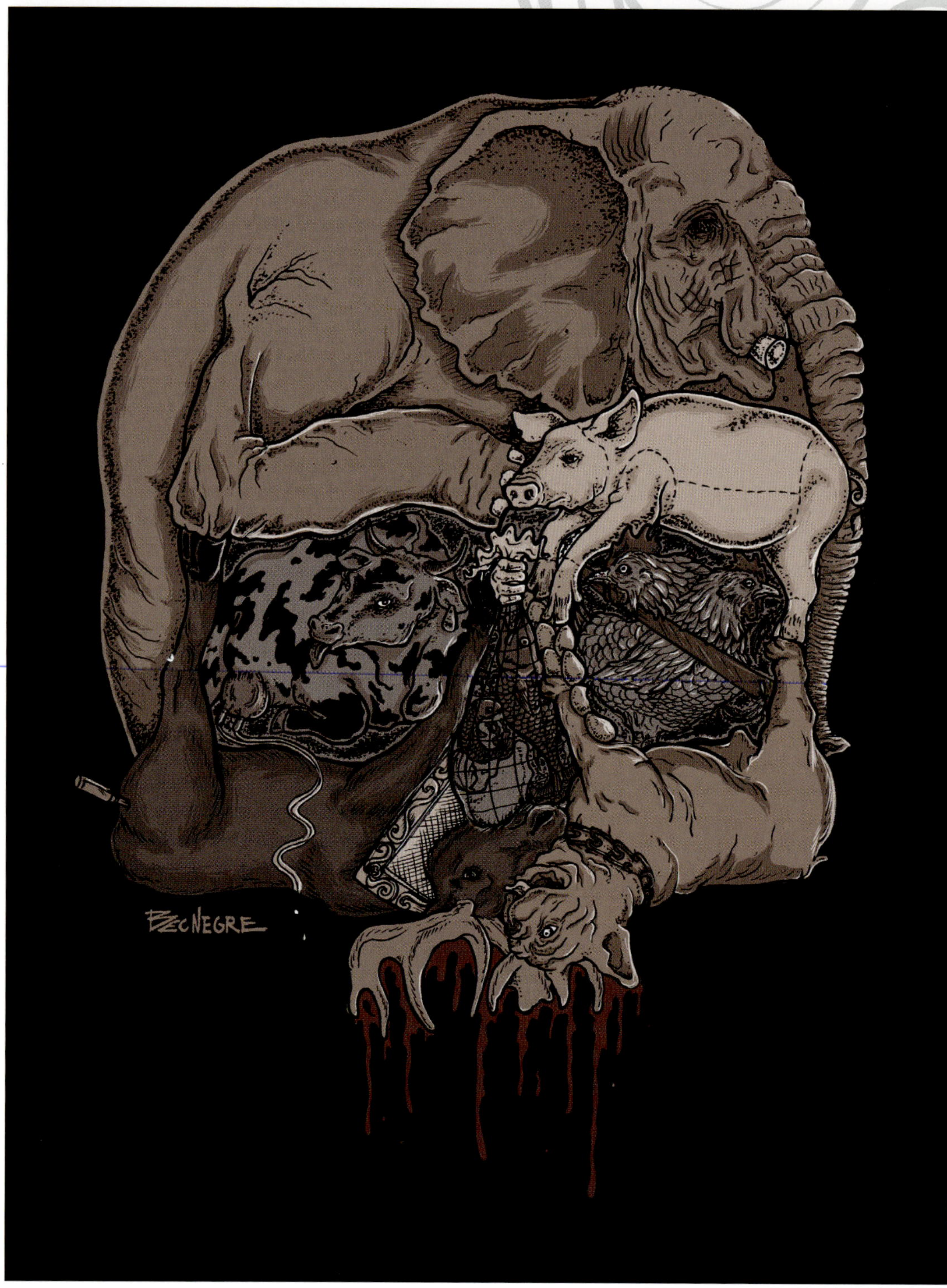

His Death or Exploitation: Your Decision, a painting by Joan Alturo.

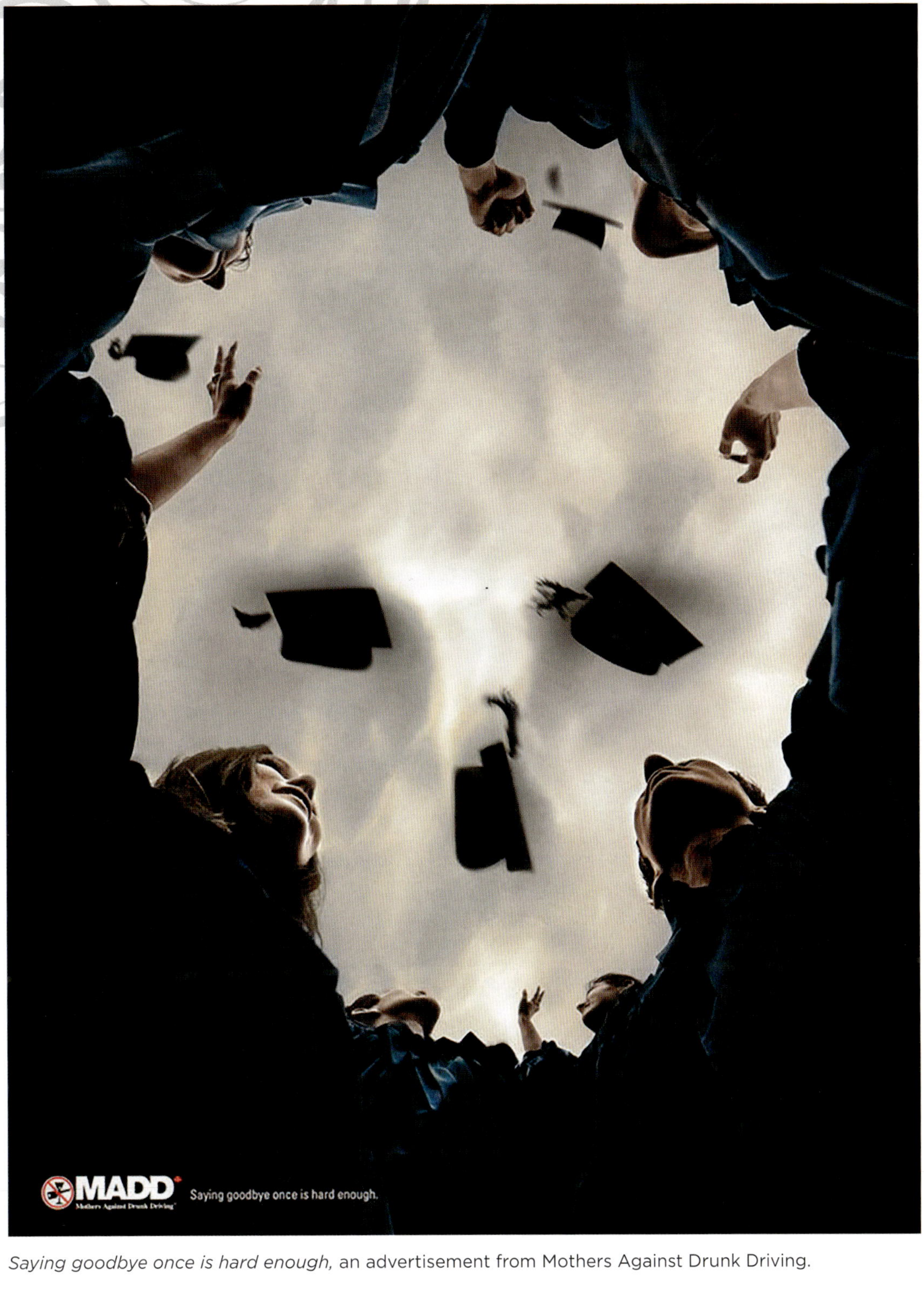

Saying goodbye once is hard enough, an advertisement from Mothers Against Drunk Driving.

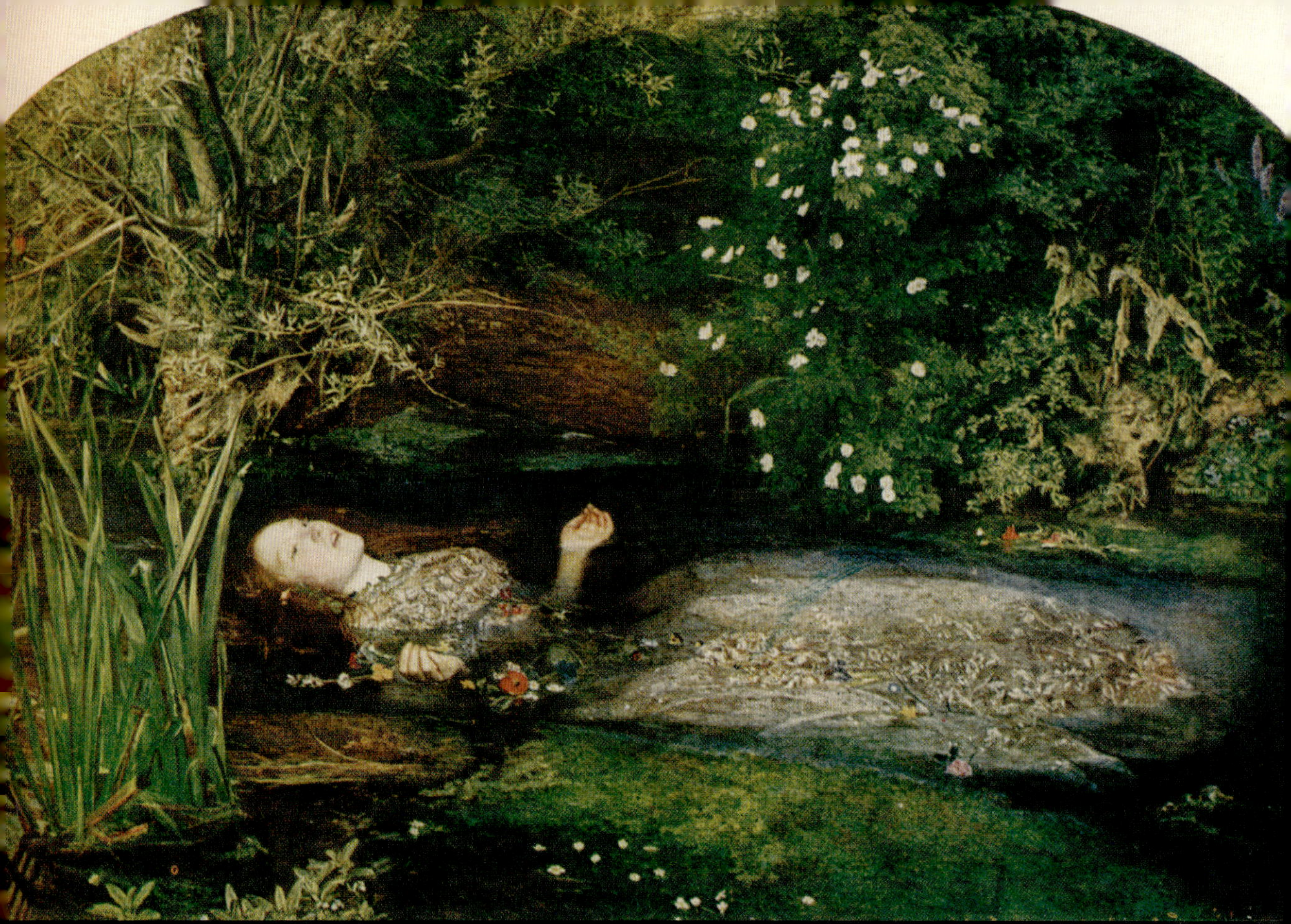

Ophelia, an 1852 painting by Sir John Everett Millais

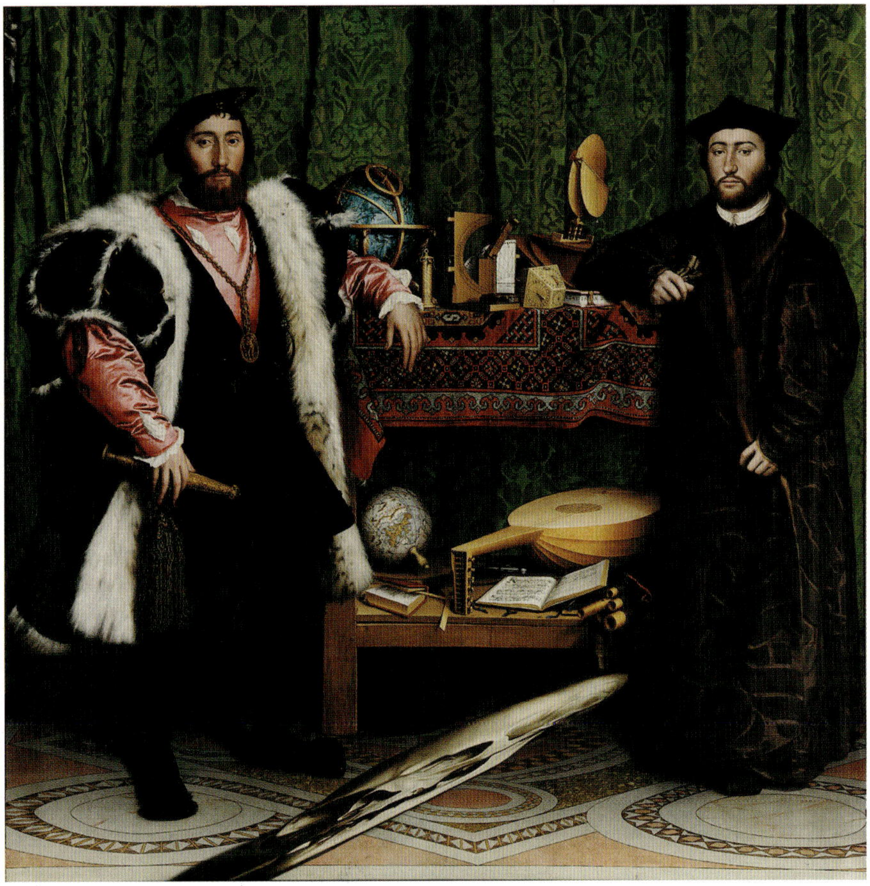

The Ambassadors, a 1533 painting by Hans Holbein the Younger.

SKULLS AND MUSIC

FOR THE REBELLIOUS IN SOCIETY THAT SCORN HUMAN MORTALITY AND CONVENTIONAL LIFESTYLES, THE SKULL IS THE ULTIMATE ACCESSORY. IT SUITS THE "LIVE FAST; DIE YOUNG" MENTALITY ADOPTED BY A MULTITUDE OF ROCK bands, particularly those belonging to the heavy rock/punk/thrash metal fraternity. The skull is a dangerous and edgy image; because it is so closely connected to death, it is also subliminally sexy. It is no surprise, then, that so many of these bands have opted for imagery which, to a casual observer, could be interchangeable with one another. An album sleeve should give the potential buyer all the clues he or she needs to make a judgement about the actual music, and a stylized skull can speak volumes in this instance.

A recent poll by *TIME* magazine listed the Grateful Dead as the musical leader in the use of skull imagery. The image is appropriate not only because of the band's name, but for their followers for whom the band were more of a lifestyle statement than a mere musical ensemble. Skeletons, and skulls in particular, decorated the vast majority of the Grateful Dead's albums.

An early punk outfit, the Misfits, used an unusual skull with eyes in the sockets, which was inspired by the film *The Crimson Ghost*. This particular skull has been hailed as one of the most enduring punk icons, and as such has become a very valuable commodity. In fact, it was the subject of a 2014 lawsuit between the band's founder, Glenn Danzig, and the rest of the band. Danzig alleged that one of his former band-mates was promoting merchandise with the image without his approval and without sharing any of the profits. The case was ultimately dismissed.

English rock band Def Leppard used a hidden skull device for the cover of their 1993 album, *Retro Active*. Close up, the cover appears to feature a beautiful young woman gazing at her reflection in a mirror. Viewed at arm's length, however, the image of the skull becomes apparent. If this sounds familiar, that's because the architects of this album's design, Hugh Syme and Nels Israelson, were directly inspired by the earlier work of Charles Allan Gilbert.

The skull mascot of thrash metal band Megadeth is named Vic Rattlehead. Devised in 1986 by the band's guitarist Dave Mustaine, Vic appears in different forms but his eyes and ears are always covered over with a visor and a clamp respectively, and his mouth is clamped shut, epitomizing the proverb "see no evil, hear no evil, speak no evil." The "Vic"' part of his name is a shortened version of "victim," while "Rattlehead" was inspired by a comment made by Mustaine's mother after she told her son off for headbanging, in case he rattled something loose in his brain. Vic even has his own song, "Skull beneath the Skull," from the album entitled *Killing Is My Business...and Business Is Good!*

Eddie the Head is the skull-like image belonging to Iron Maiden. Eddie has assumed something of superhero status, having been an active part of the band's imagery, and merchandising, since at least 1980. In true cartoon style, Eddie appears in different guises appropriate to the band's material. Although over the years he's made appearances as a zombie, a cyborg and a mummy, Eddie's skull-like nature is consistently recognizable.

Metal band Guns N' Roses, too, use the iconography of the skull to share a message of sexy menace. The original cover for their 1987 album *Appetite for Destruction* had a cover whose design was so extreme that retailers refused to stock it. The design was radically altered, with each of the five members of the band—Axl Rose, Izzy Stradlin, Steven Adler, Duff McKagan and Slash—portrayed as cartoon-like skulls, an idea which grabbed peoples' imagination immediately. In a compromise, the original artwork was used on the inner sleeve. The album itself is a perennial hit, still selling consistently across all formats. Arguably the most recognizable of the Guns N' Roses' skull image is that of Slash; this particular skull wears his owner's trademark top hat. The band has made good use of the skull icon, even using hidden skull imagery in some of their logos.

The highly-influential heavy metal band Metallica relies on the skull to convey an instant understanding of what they're about. A flaming skull, portrayed from the side with fire blazing out from behind, is particularly prominent. Over the years, the band have used skulls in combination with the Anarchy sign; as a stylized skull and crossbones logo; as a skull with one eye obscured, and as a skull dressed as a mummy. Like Guns N' Roses, Metallica also depicting the band members as skulls.

Slayer, who are one of the "big four" thrash metal bands listed alongside Metallica, Megadeth, and Anthrax, have used various interpretations of the skull over the years as a suitable way to express the content of their songs, which cover subjects like Satanism, necrophilia, serial killing, and concentration camps.

Skulls in album art are not exclusive to heavy rock, punk, and thrash metal bands. Cypress Hill is an American hip hop and rap group, who have decorated several album covers with skulls. Their fifth studio album, *Skull & Bones*, has two CDs with different musical styles. *Skull*, the first CD, has a hip-hop style, while *Bones* has a rap-metal style.

In addition to decorating album covers, skulls have also been used to decorate musical instruments themselves. This violin is decorated with skulls on both the front and the back and was auctioned off to benefit the Suzuki Music Program at the Canterbury School.

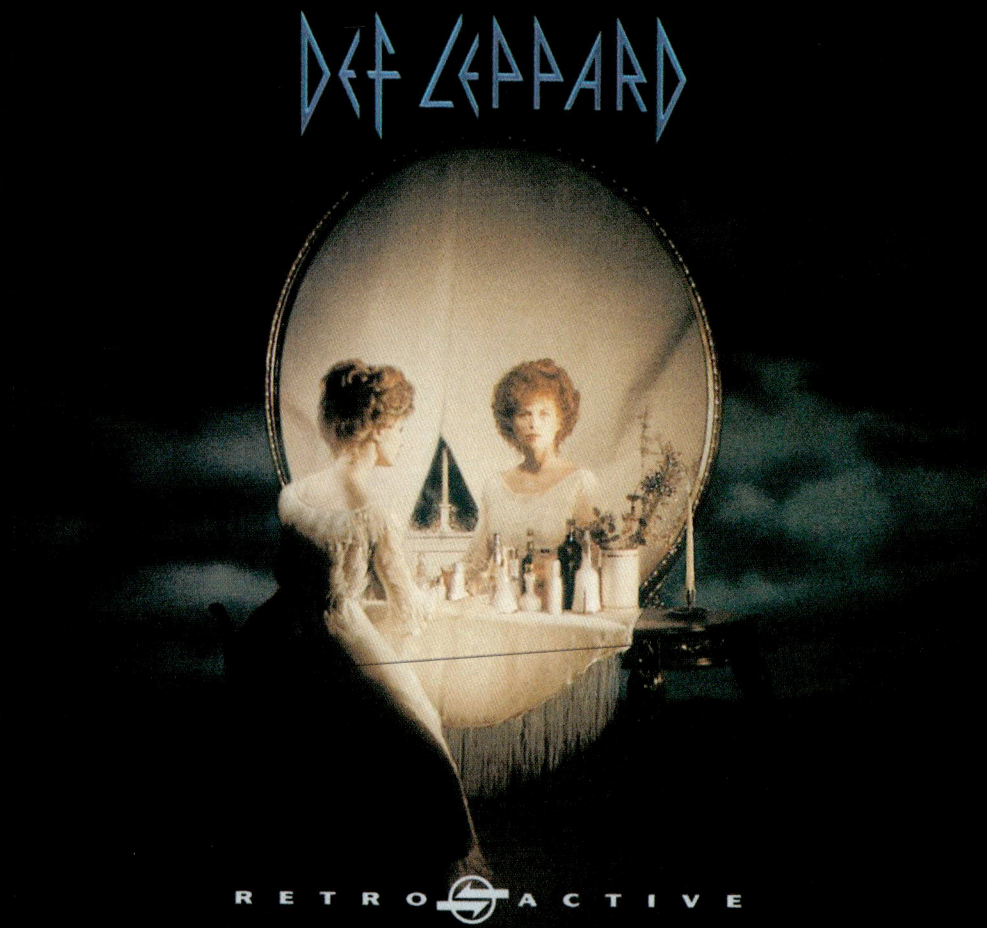

Top: Album cover of 1993's *Retro Active*, from English rock band Def Leppard.

Right: The *Retro Active* album cover above drew inspiration from Charles Allan Gilbert's *All is Vanity.*

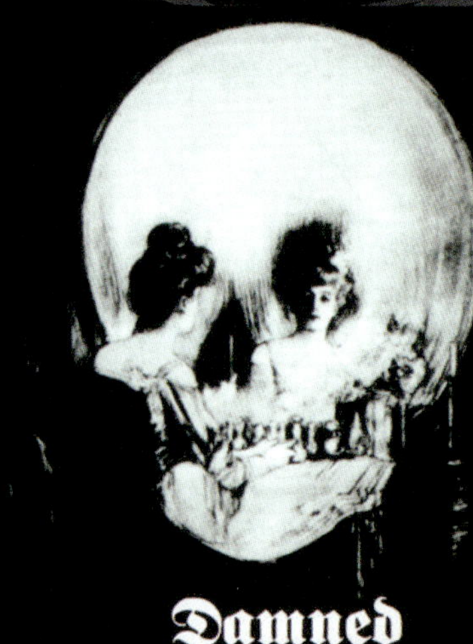

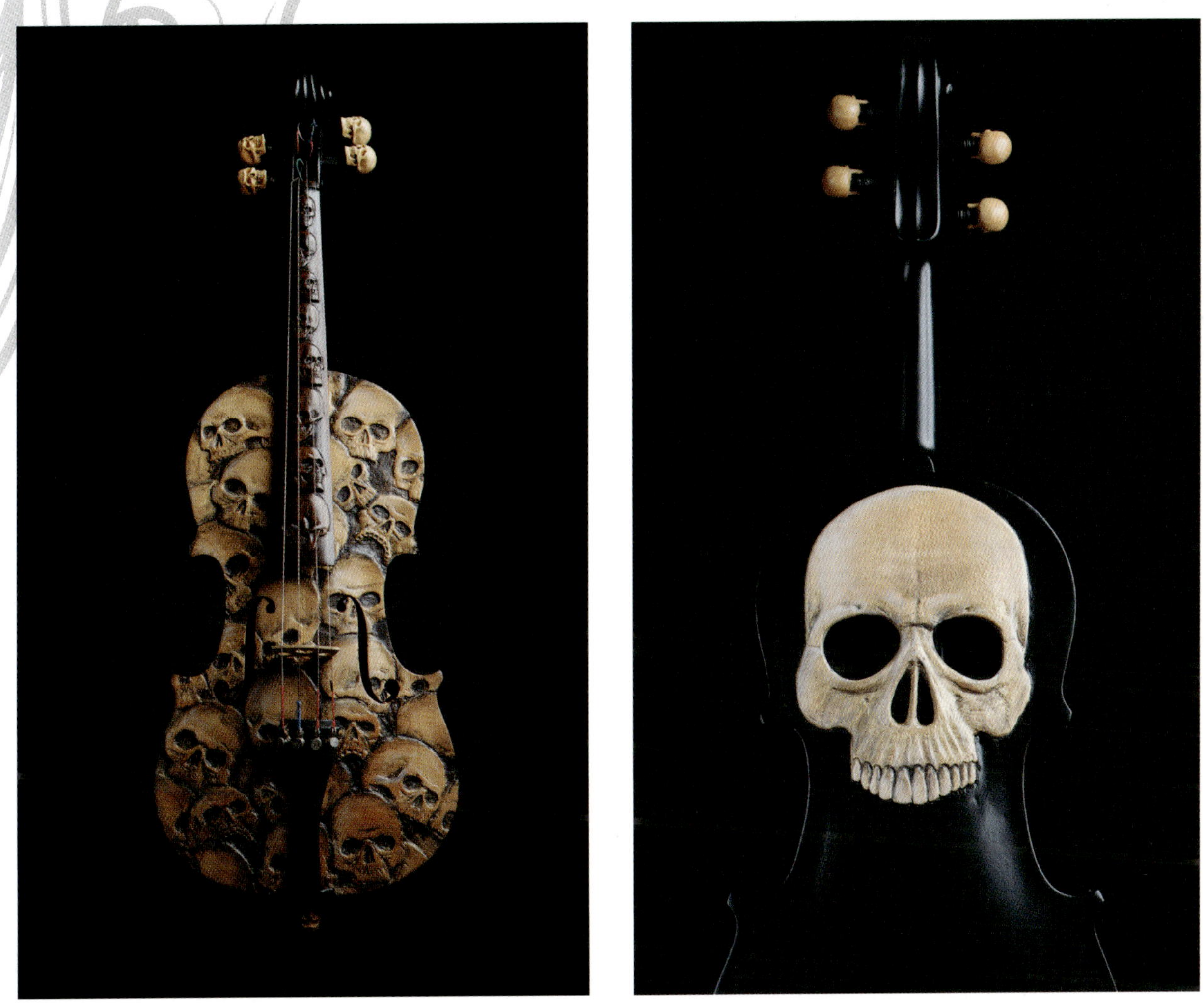

Front and back of a skull-decorated violin, created by American artist Mark Noll and then auctioned off to benefit the Suzuki Music Program at the Canterbury School. The jumble of skulls on the front of the violin is reminiscent of the stacks of skulls in an ossuary.

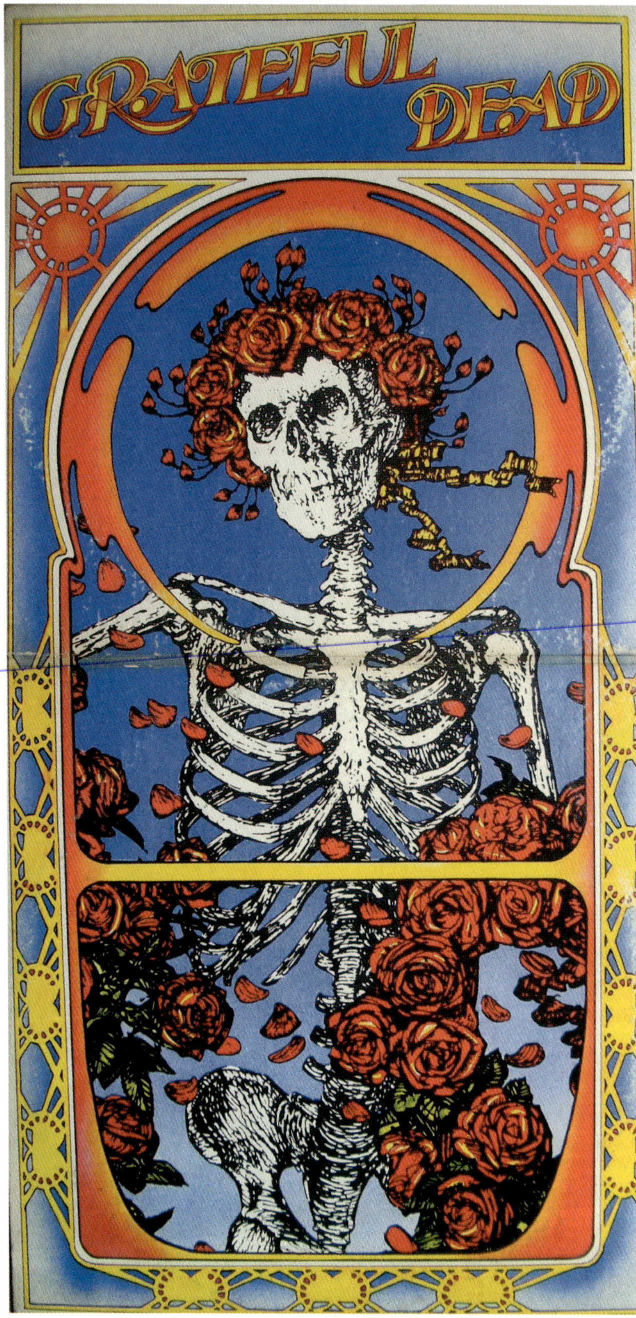

Album cover of 1971's *Grateful Dead*, from the Grateful Dead.

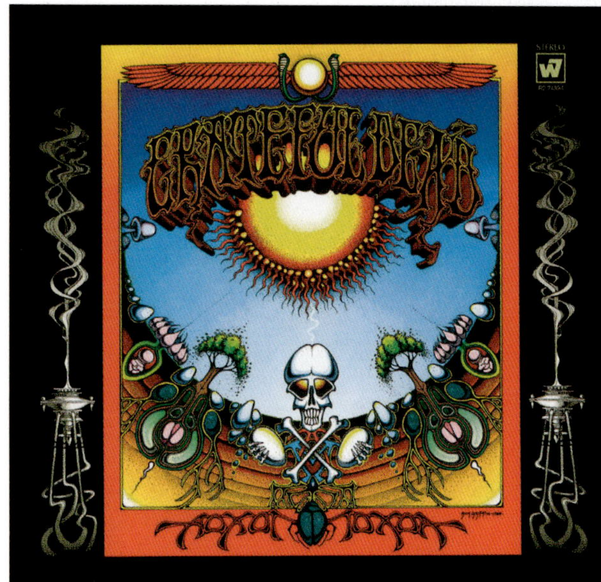

The cover of the Grateful Dead's 1969 album, *Aoxomoxoa*.

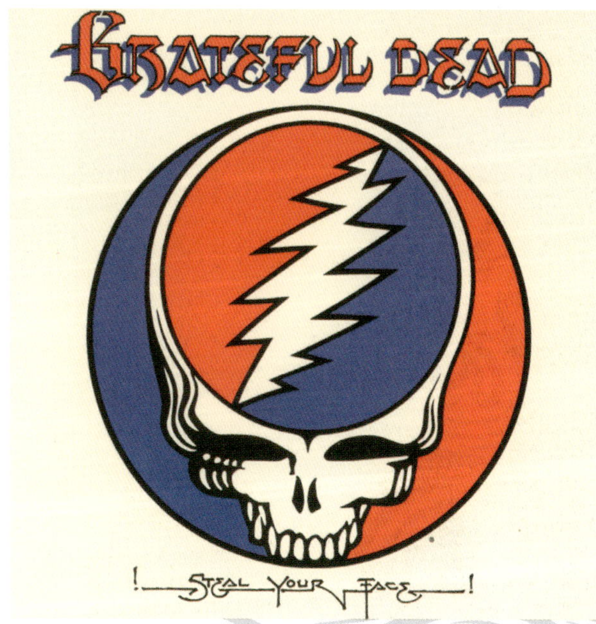

Album cover from 1976's *Steal Your Face*, from the Grateful Dead.

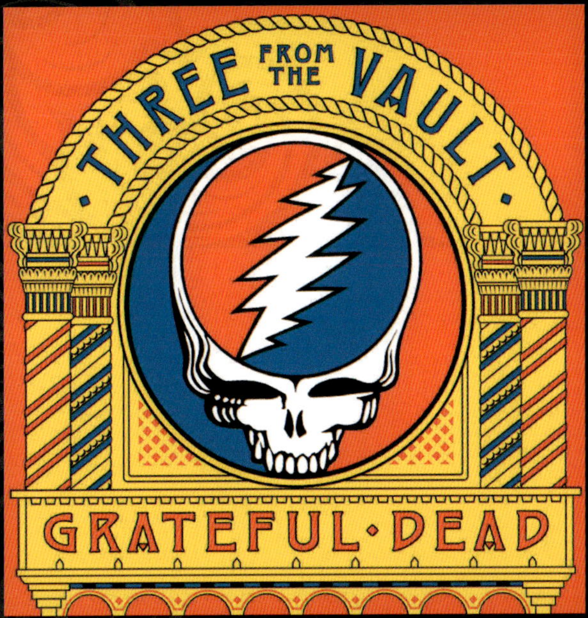

Three from the Vault, the Grateful Dead's 2007 album, was released fifteen years after *Two from the Vault*.

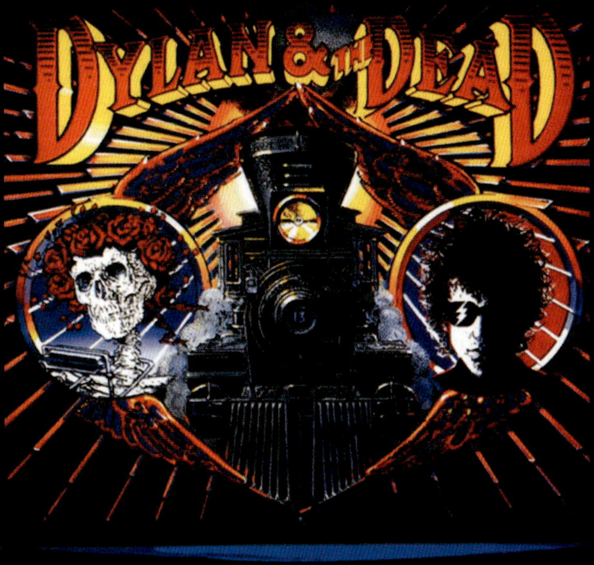

Album cover from 1989's *Dylan and the Dead*, a collaboration between Bob Dylan and the Grateful Dead.

Perhaps the best-known of the Grateful Dead's icons is the red, white, and blue skull logo with a lightning bolt that appears on 1976's *Steal Your Face*. It was designed by the Dead's soundman, Owsley "Bear" Stanley and artist Bob Thomas. It has been reused on other albums since then, including 2007's *Three from the Vault*.

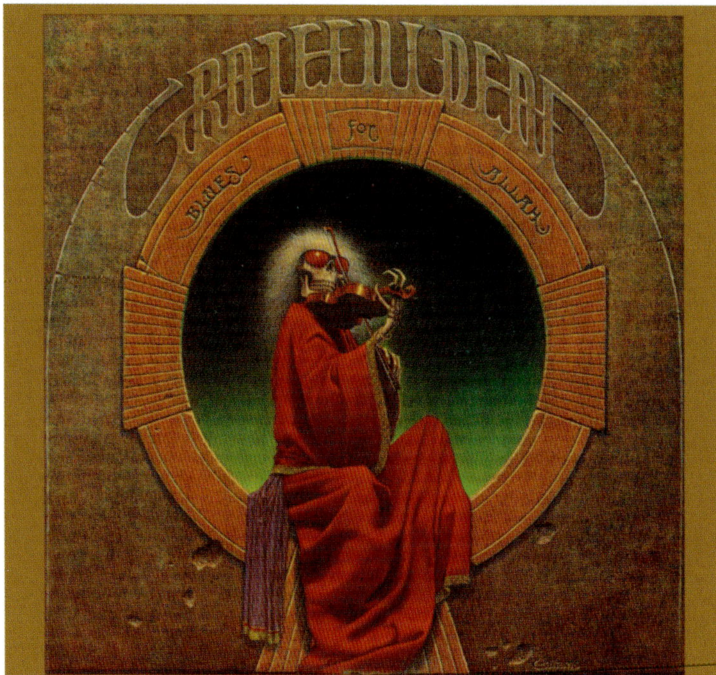

Album cover from 1975's *Blues for Allah*, from the Grateful Dead.

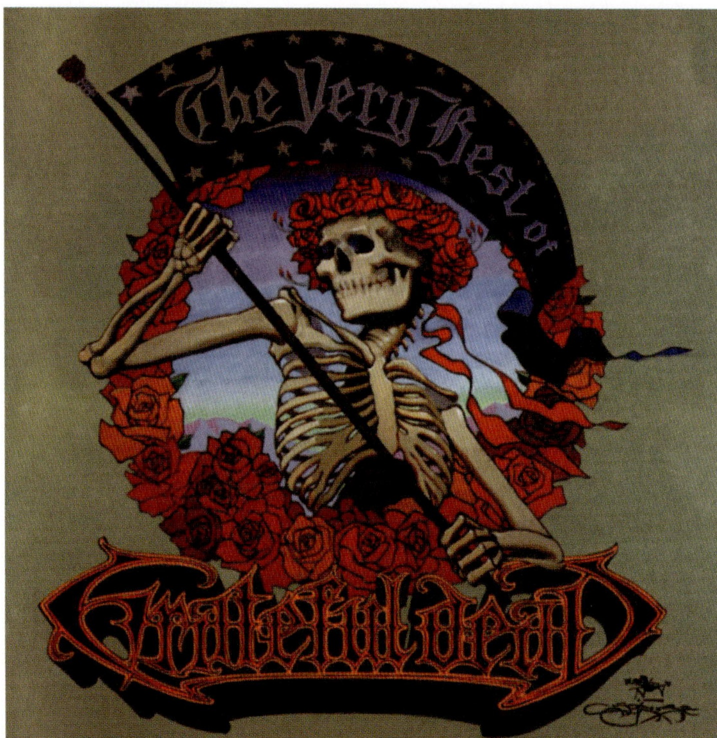

The cover of the 2003 compilation album, *The Very Best of the Grateful Dead*.

These two album covers from the Grateful Dead are for 2012's *Spring 1990* and 2014's *Spring 1990 (The Other One)*.

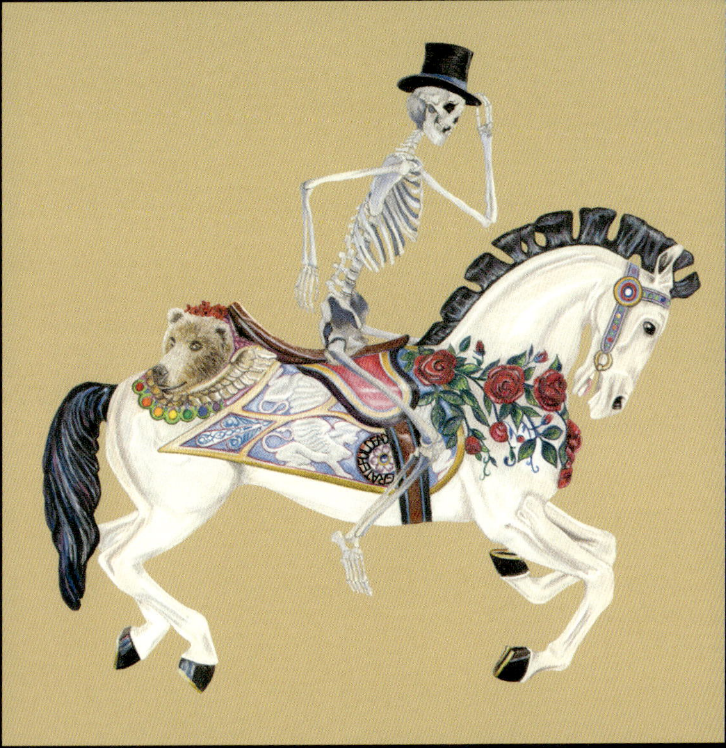

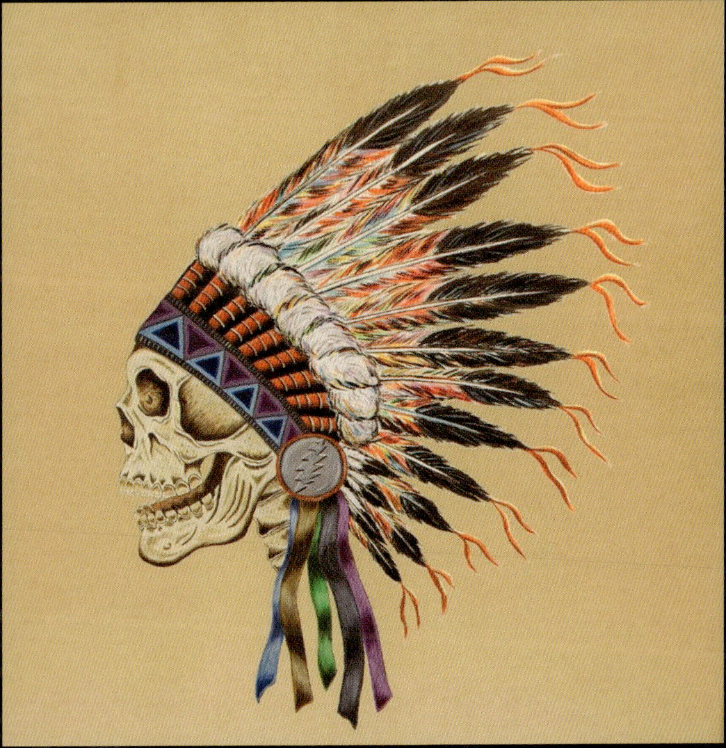

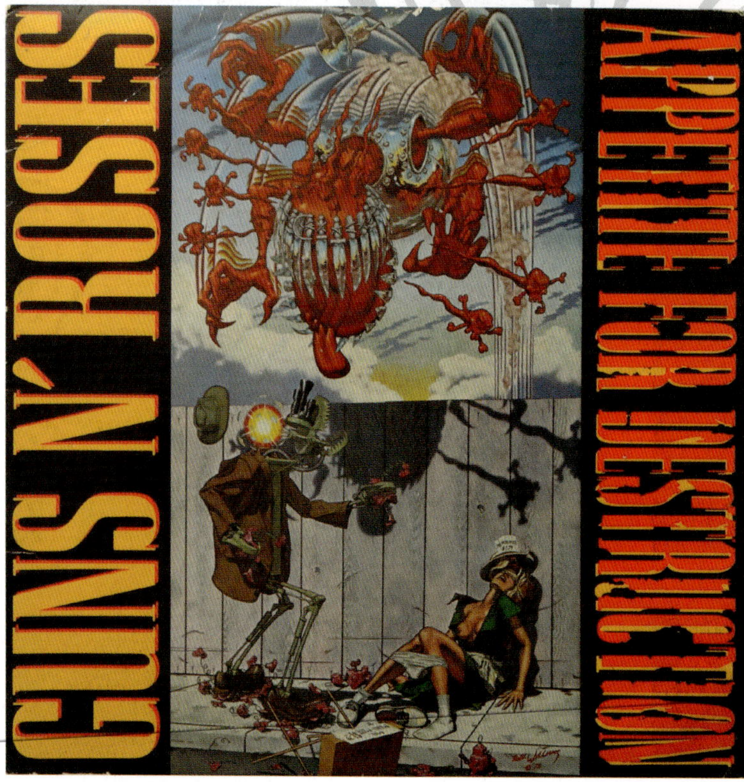

The controversial original cover for Guns N' Roses' 1987 album, *Appetite for Destruction*. It was rejected by the retailers, forcing the band to use a different cover; the original artwork was instead used on the inner sleeve.

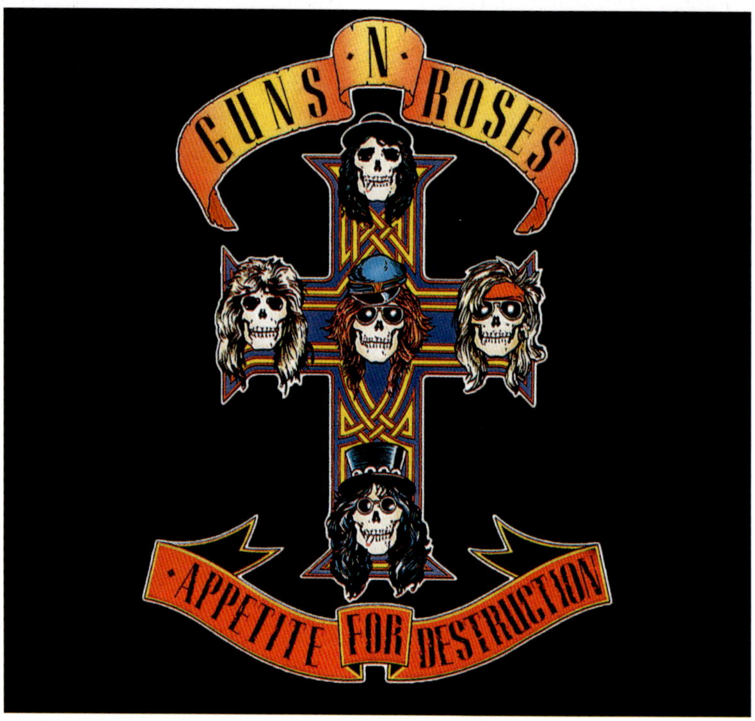

The revised cover for Guns N' Roses' 1987 album, *Appetite for Destruction*. This radically altered the original design and presented the five members of the band as cartoon-like skulls.

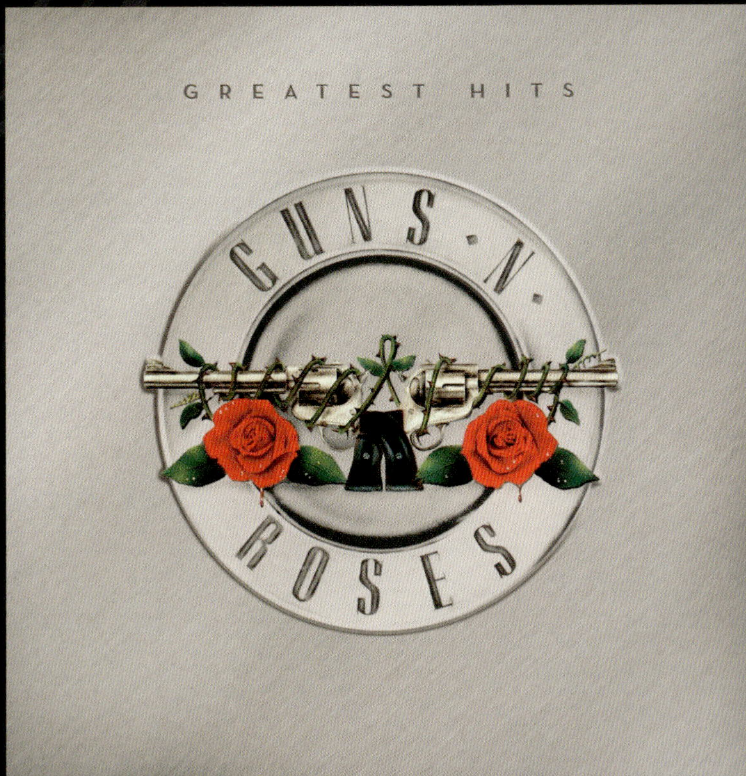

The cover from 2004's Guns N' Roses' compilation album, *Greatest Hits,* features a hidden skull formed by the two pistols.

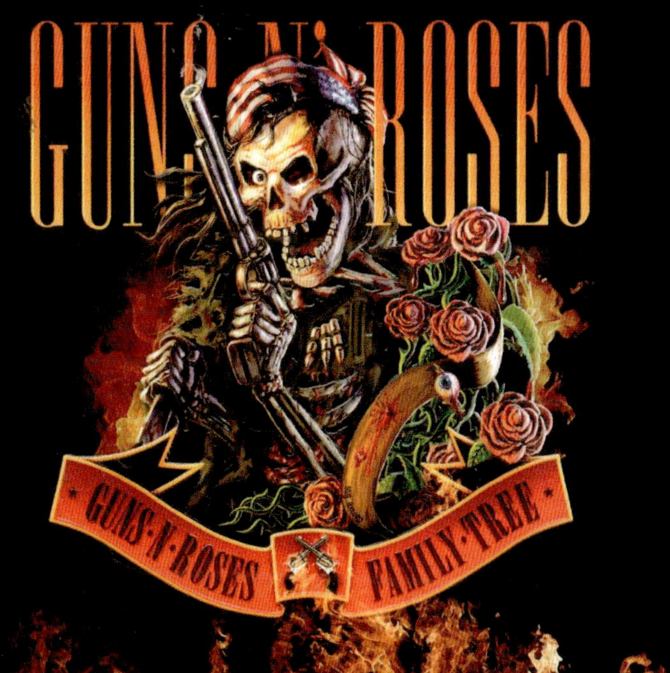

Family Tree is a compilation album released by Guns N' Roses in 2010.

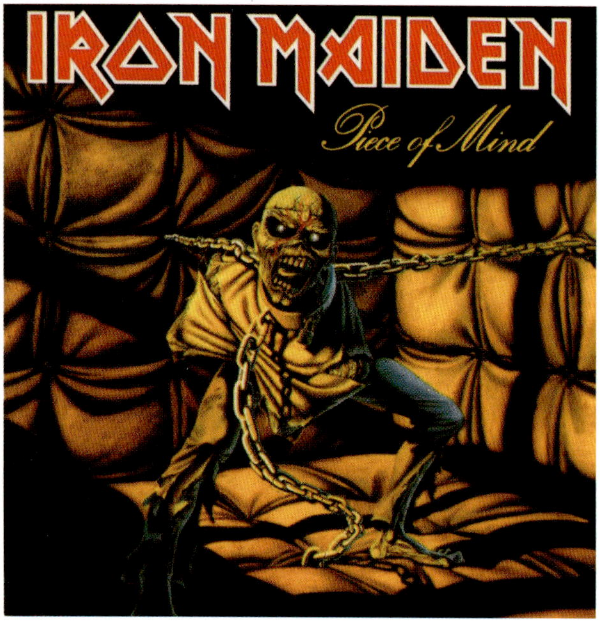

The album cover from Iron Maiden's 1983 album, *Piece of Mind*.

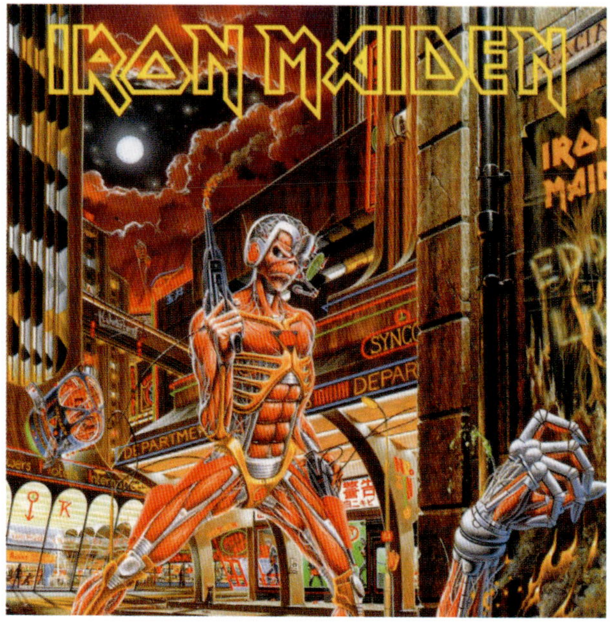

The album cover from Iron Maiden's *Somewhere in Time*, released in 1986.

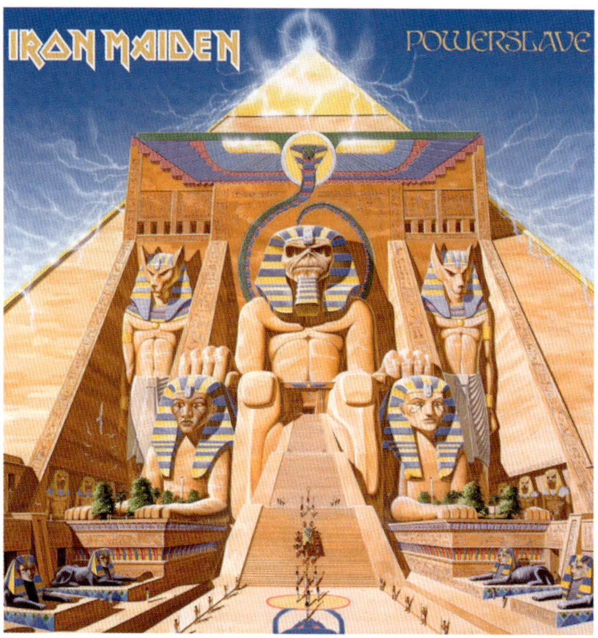

The artwork from 1984's *Powerslave,* from Iron Maiden.

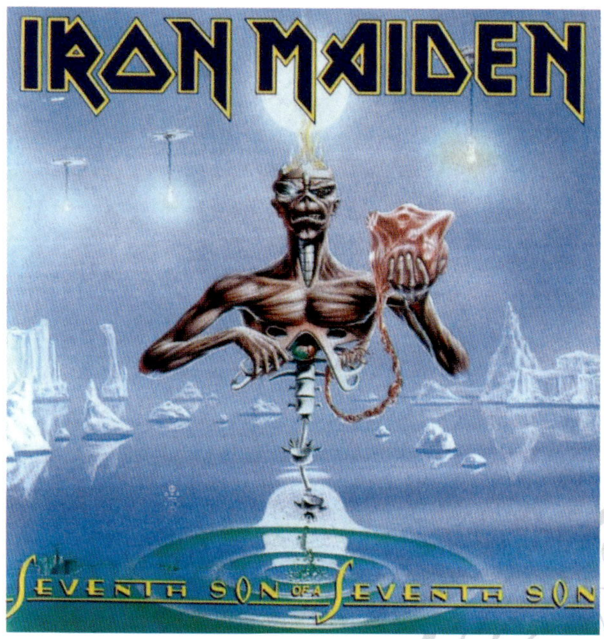

Album cover from 1988's *Seventh Son of a Seventh Son*, from Iron Maiden.

Eddie the Head, the skull mascot devised by Iron Maiden, features on their album cover in all sorts of guises: as a pharaoh (1984's *Powerslave*), a space explorer (2010's *The Final Frontier*), and more. Eddie was first drawn by Derek Riggs for the band's debut album cover.

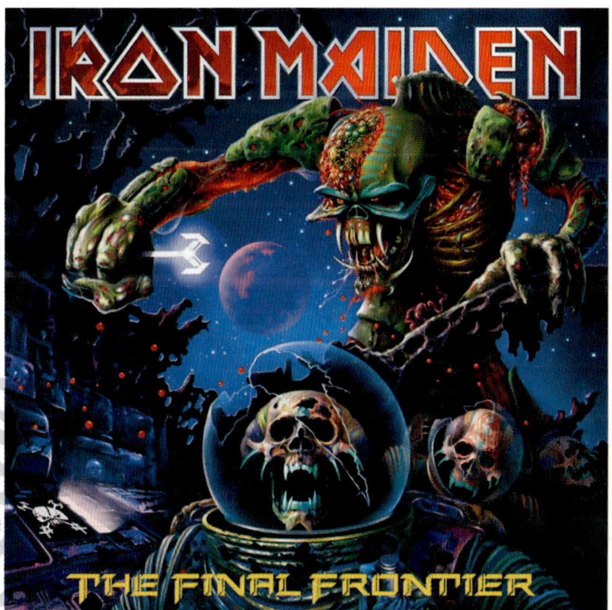

The album cover from Iron Maiden's 2010 album, *The Final Frontier*.

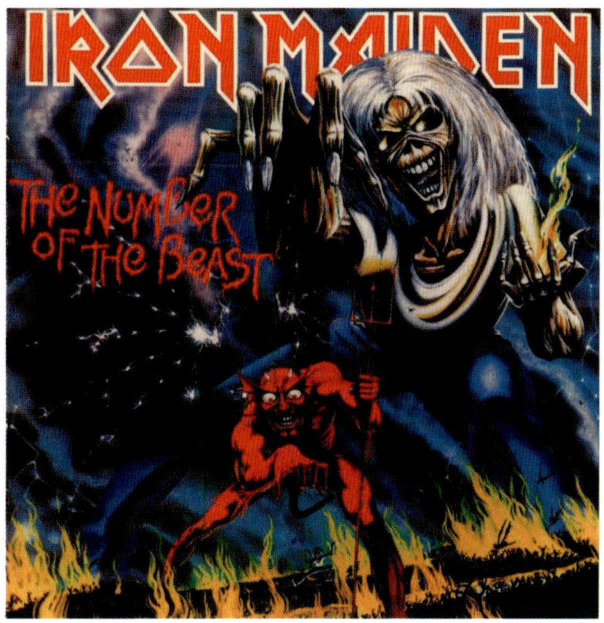

The album cover from *The Number of the Beast*, released by Iron Maiden in 1982.

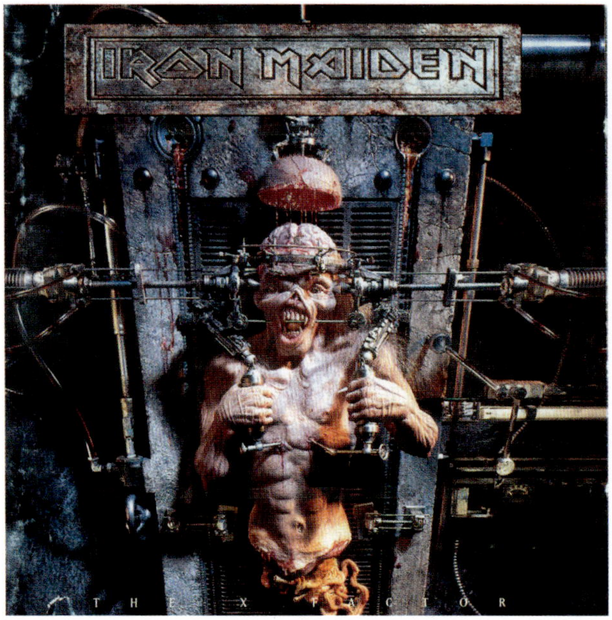

The album cover from Iron Maiden's 1995 album, *The X Factor*.

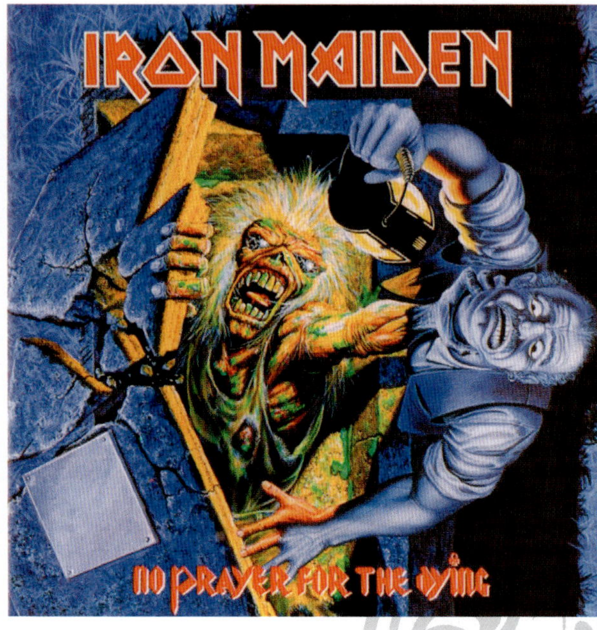

The album cover from Iron Maiden's 1990 album, *No Prayer for the Dying*

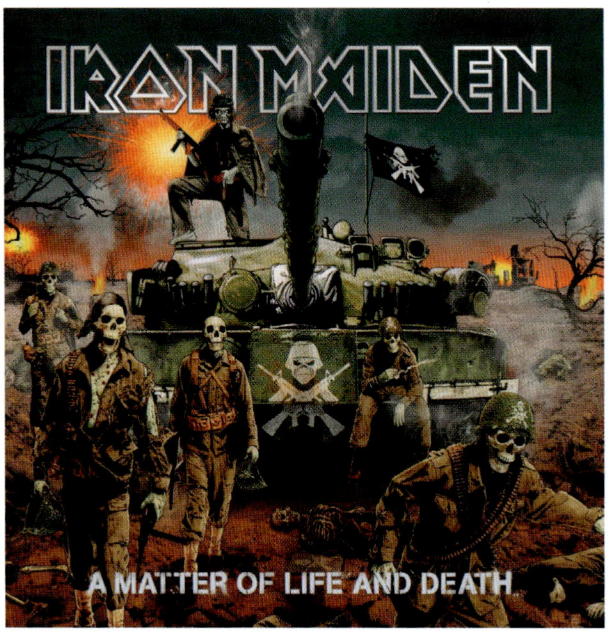

The album cover from Iron Maiden's 2006 album, *A Matter of Life and Death*.

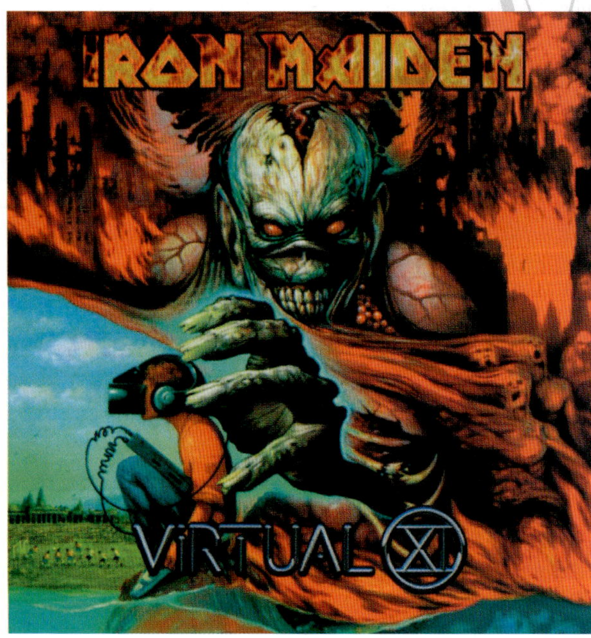

The album cover from 1998's *Virtual XI*, by Iron Maiden.

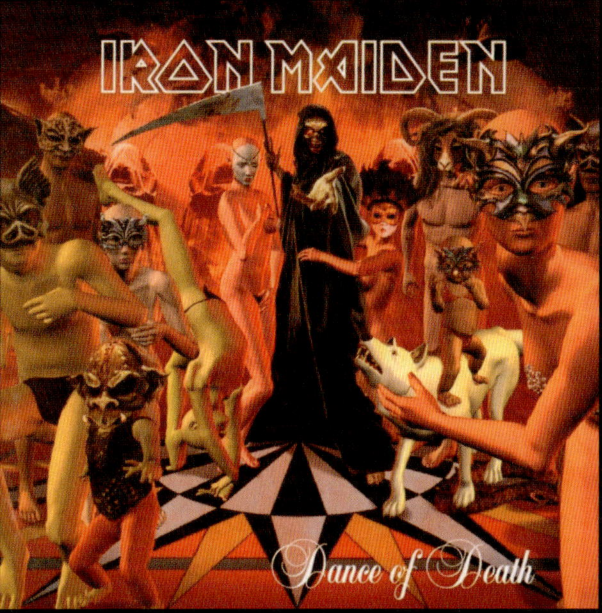

ron Maiden's album cover from 2003's *Dance of Death*.

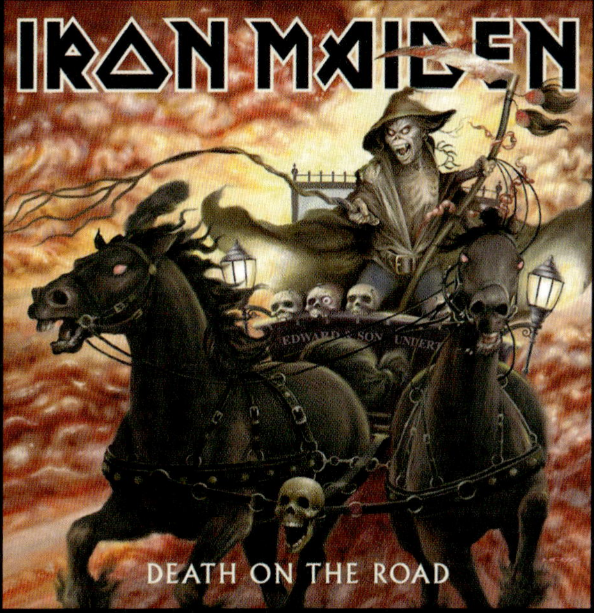

The album cover from Iron Maiden's *Death on the Road*, released in 2005.

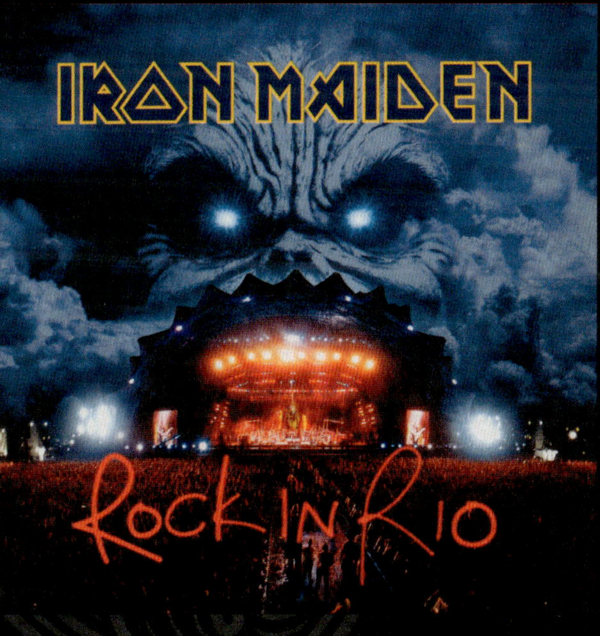

The album cover from 2002's *Rock In Rio*, by Iron Maiden.

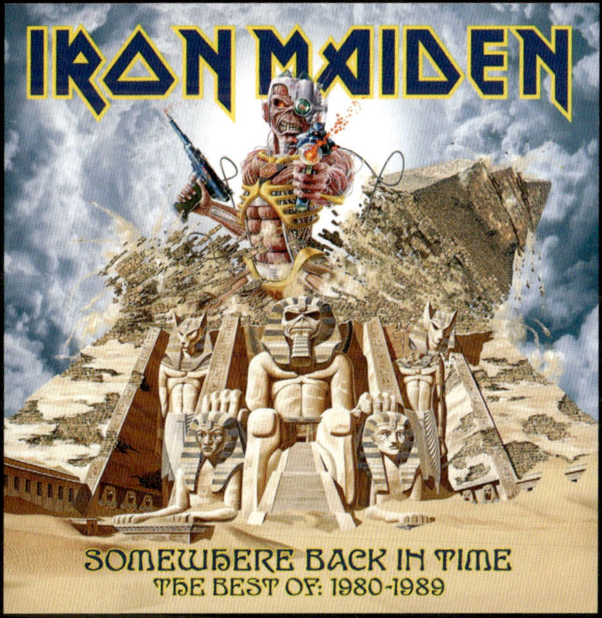

The album cover from Iron Maiden's 2008 album, *Somewhere Back in Time*.

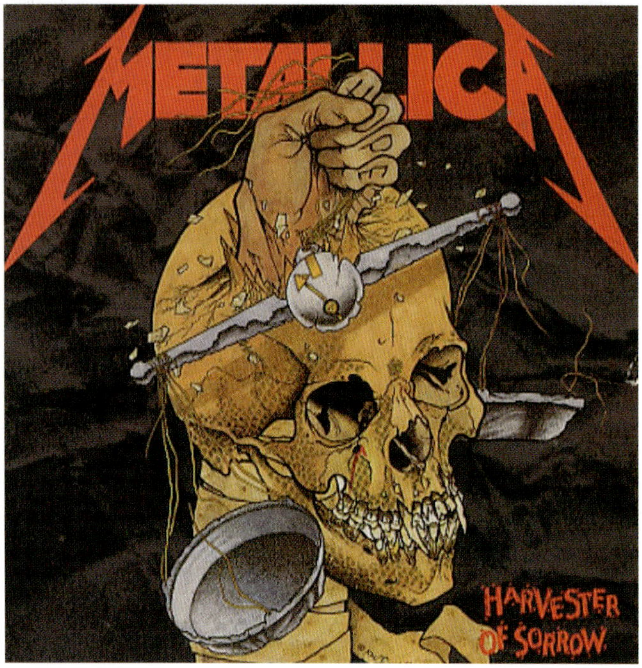

"Harvester of Sorrow" is a song by American heavy metal group Metallica that was released in 1988.

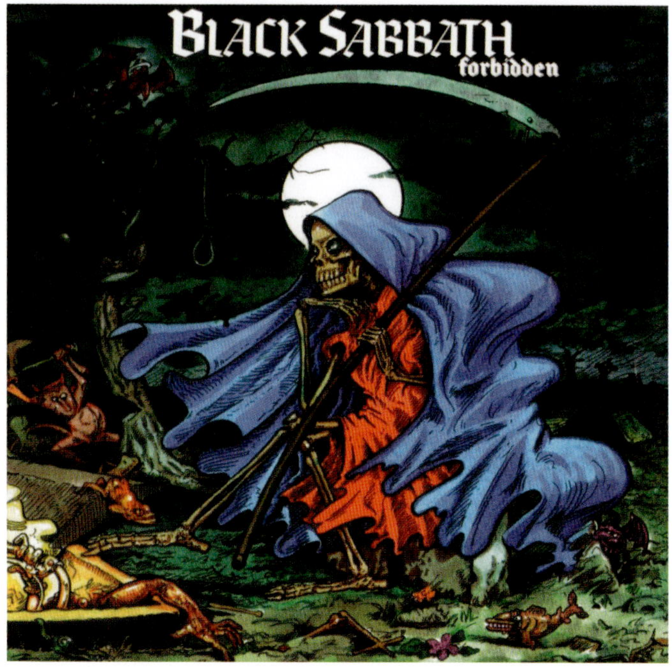

English heavy metal band Black Sabbath released *Forbidden* in 1995 with this album art.

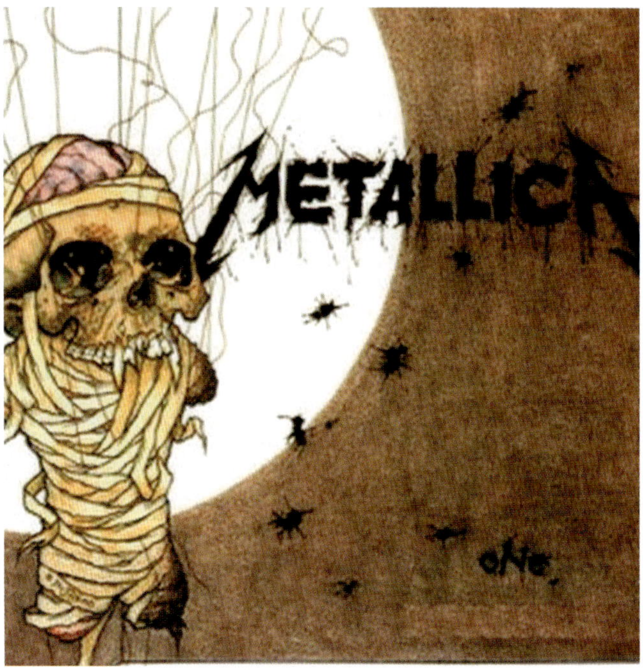

"One" was a song released by Metallica in 1989.

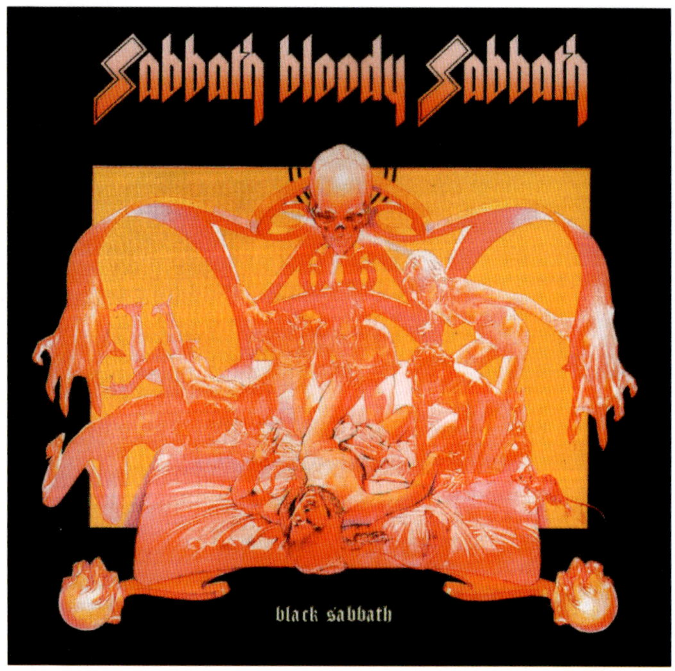

Black Sabbath released *Sabbath Bloody Sabbath* in 1973.

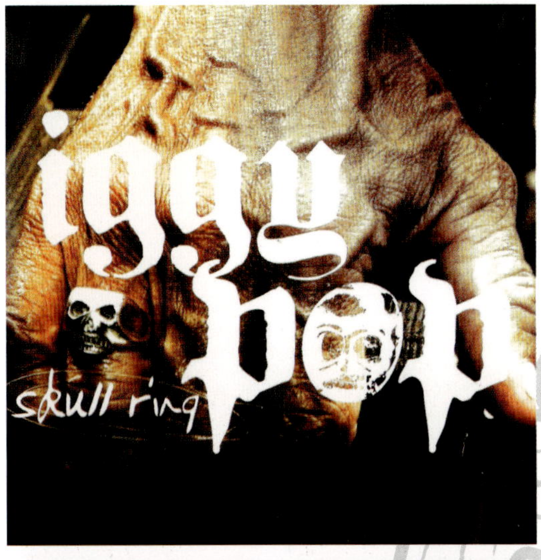

Skull Ring was released by American rock singer Iggy Pop in 2003.

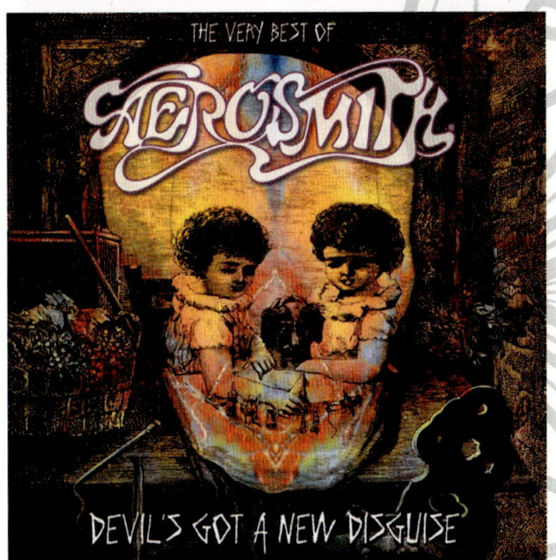

The Devil's Got a New Disguise was released by American hard rock band Aerosmith in 2006.

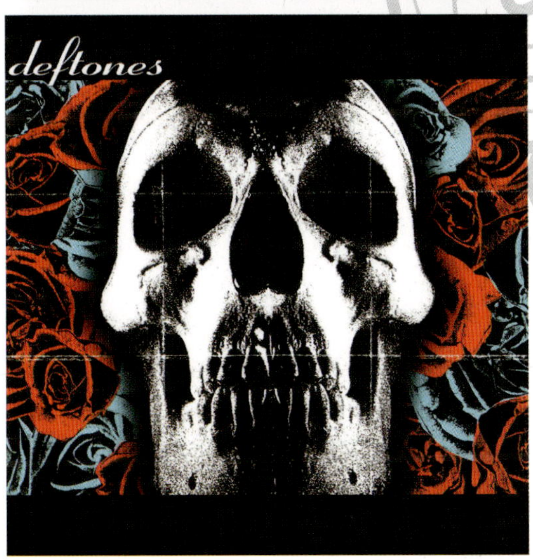

The Deftones eponymous 2003 release, *Deftones*, was the band's fourth studio album.

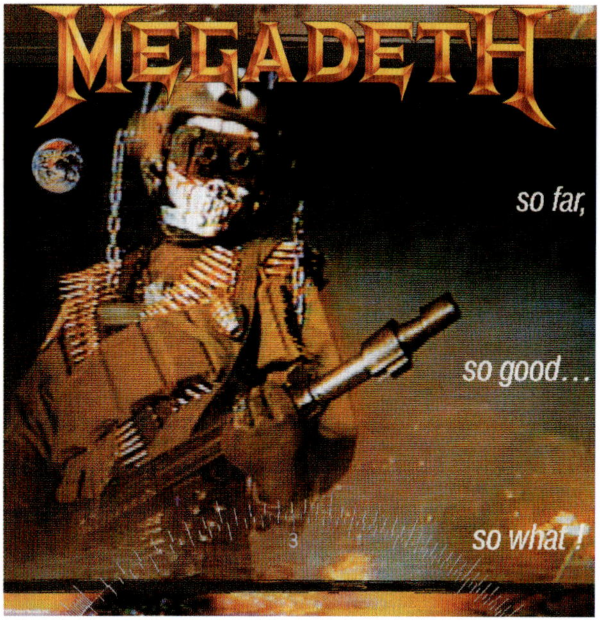

The album cover from Megadeth's 1988 release, *So Far, So Good . . . So What?*

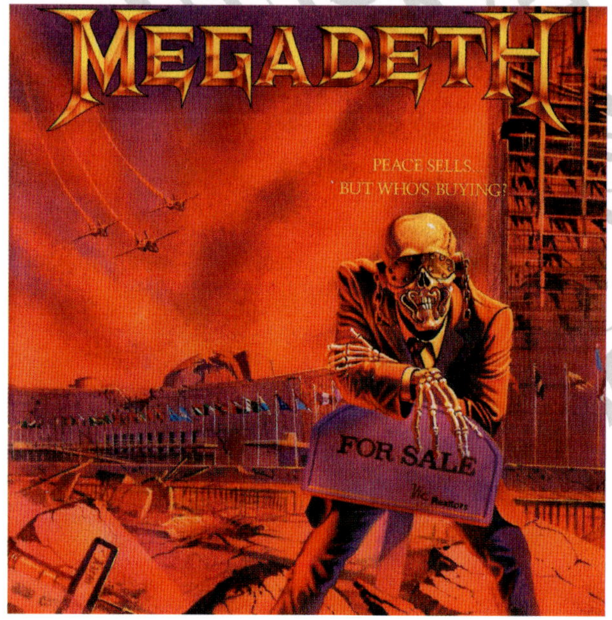

In 1986, Megadeth released *Peace Sells . . . But Who's Buying?*

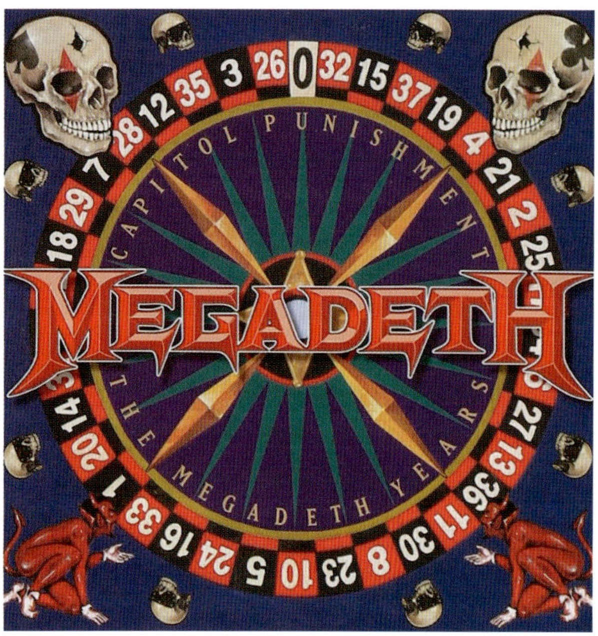

The artwork from 2000's *Capitol Punishment,* by Megadeth.

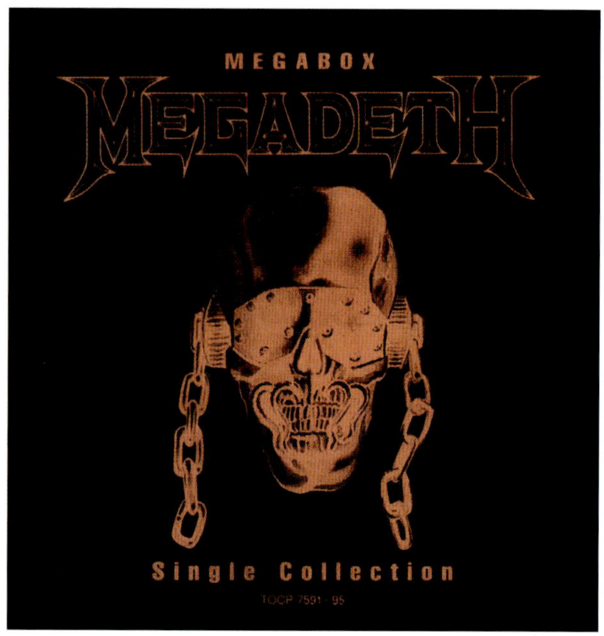

Megabox was released by Megadeth in 1993.

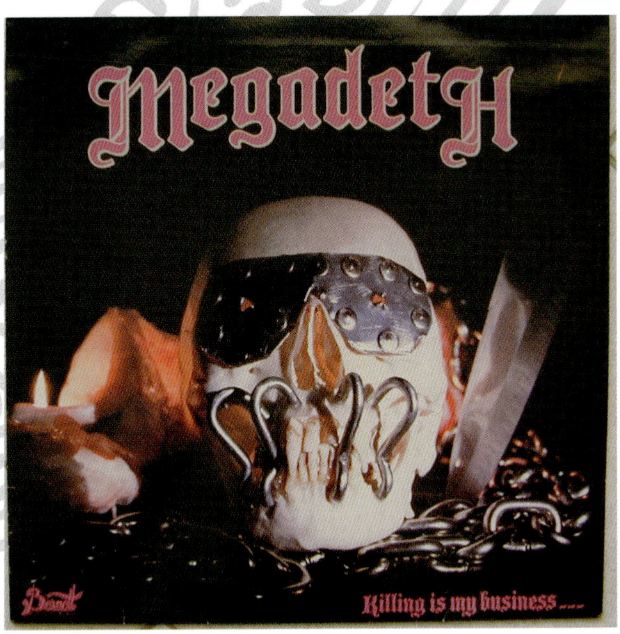

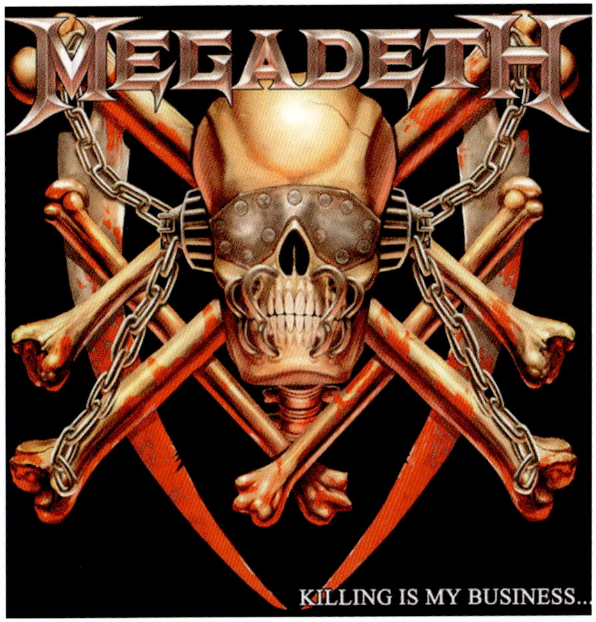

Alternate covers for Megadeth's first studio album, *Killing Is My Business . . . and Business is Good!*

Megadeth's mascot, Vic Rattlehead, is a skeletal figure that embodies the idea "See no evil, hear no evil, speak no evil." Band co-founder Dave Mustaine sketched the original drawing of Vic for the band's first album, *Killing Is My Business . . . and Business is Good!*, released in 1985.

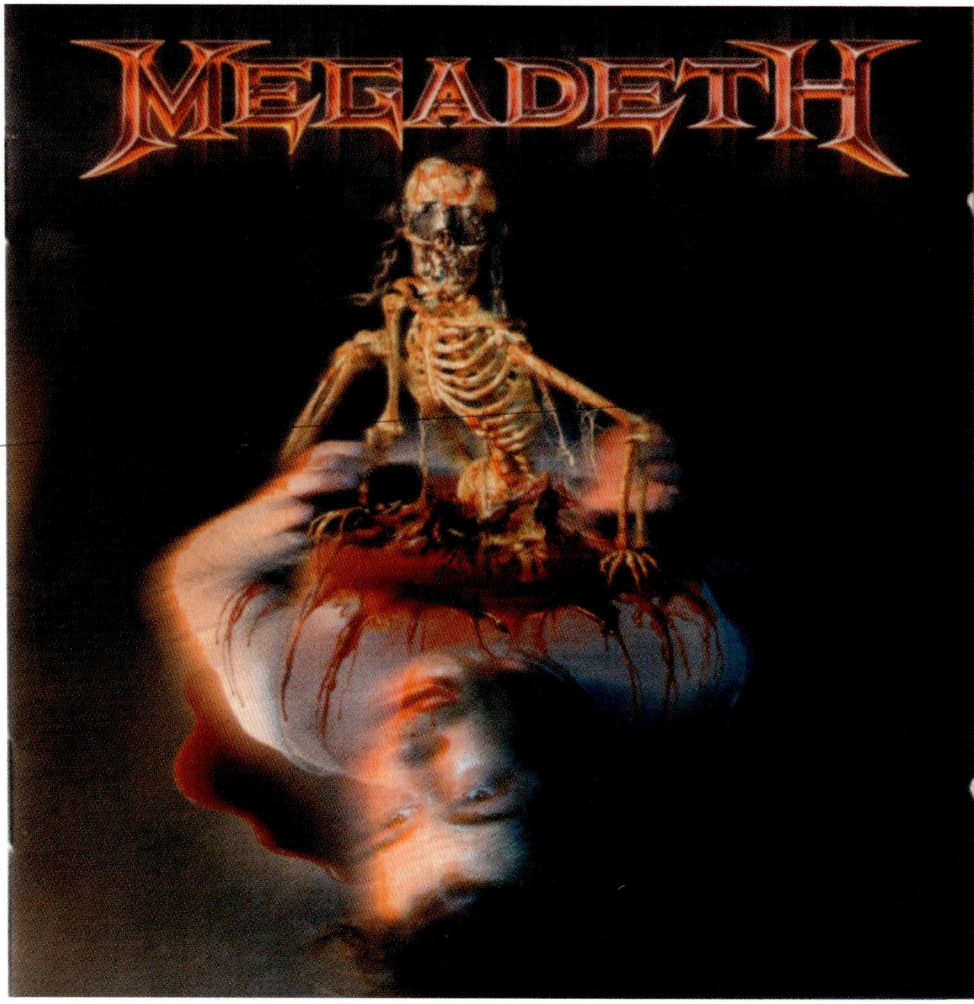

This album cover advertised Megadeth's 2001 release, *The World Needs a Hero.*

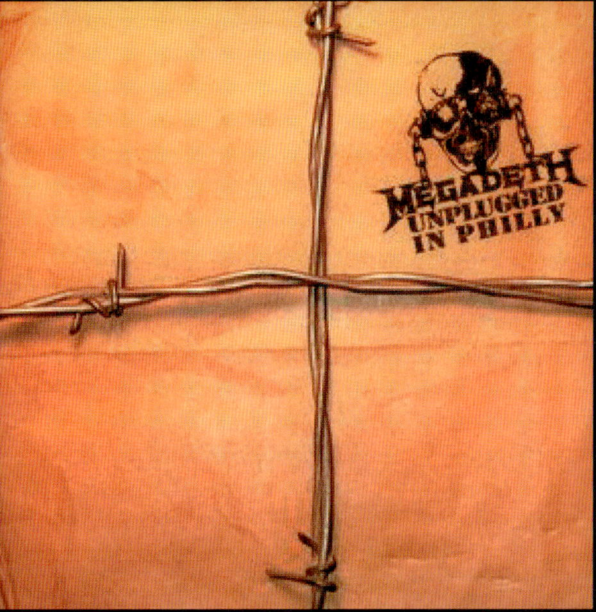

The artwork from 2005's *Unplugged in Philly,* by Megadeth.

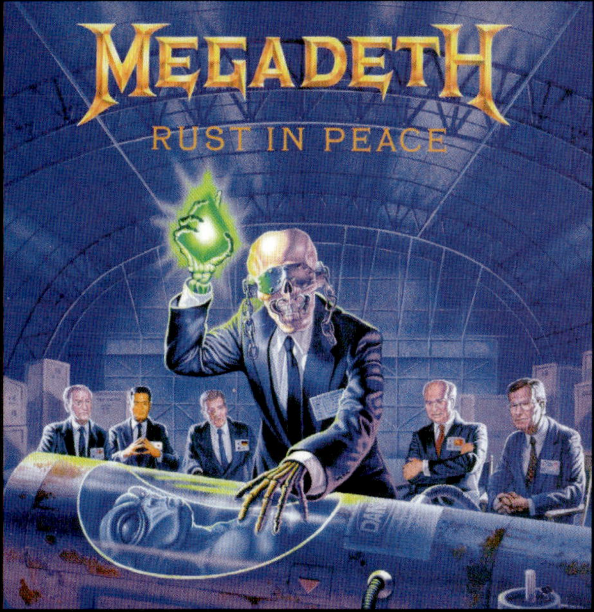

The album cover from Megadeth's 1990 release, *Rust in Peace*

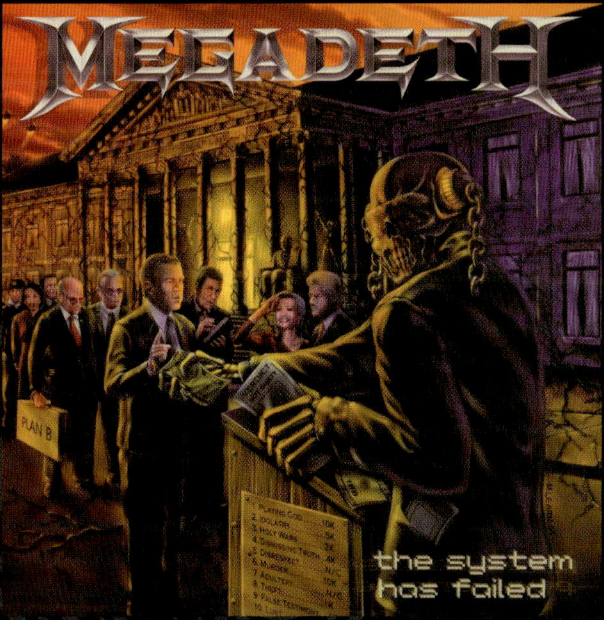

The System Has Failed was released by Megadeth n 2004.

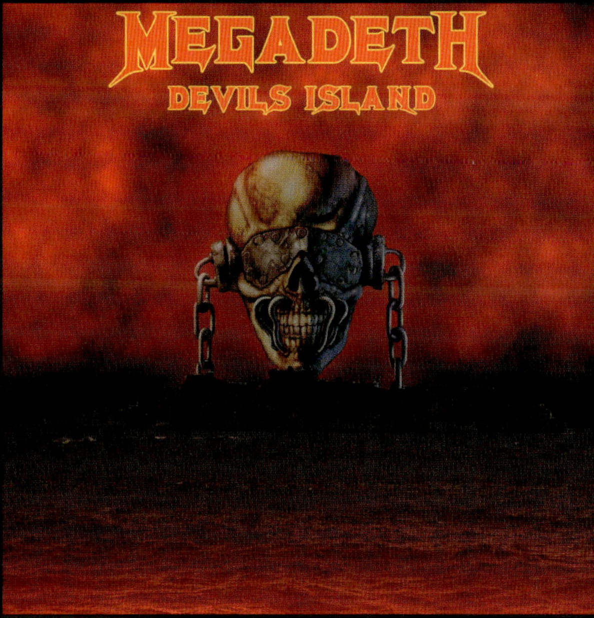

"Devils Island" is a song released by Megadeth in 1986.

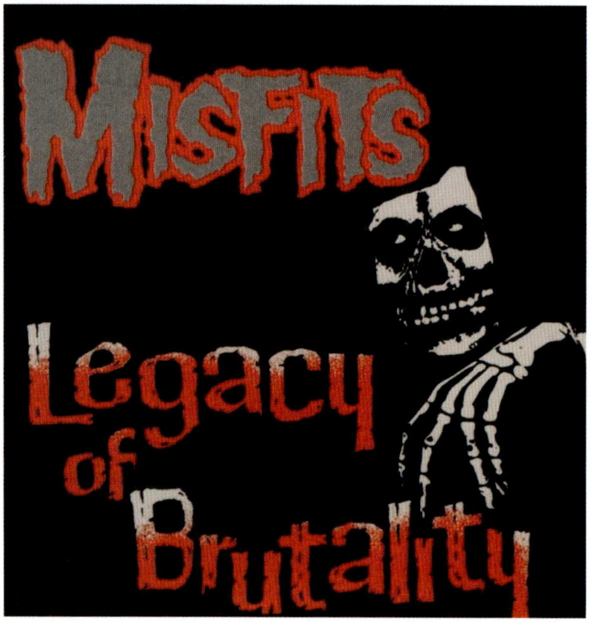

Misfits' album cover from 1985's *Legacy of Brutality.*

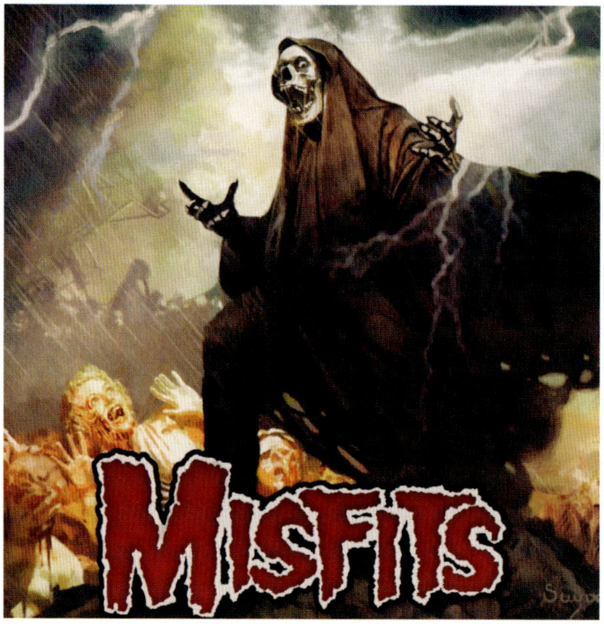

The album cover from the Misfits 2011 release, *The Devil's Rain*.

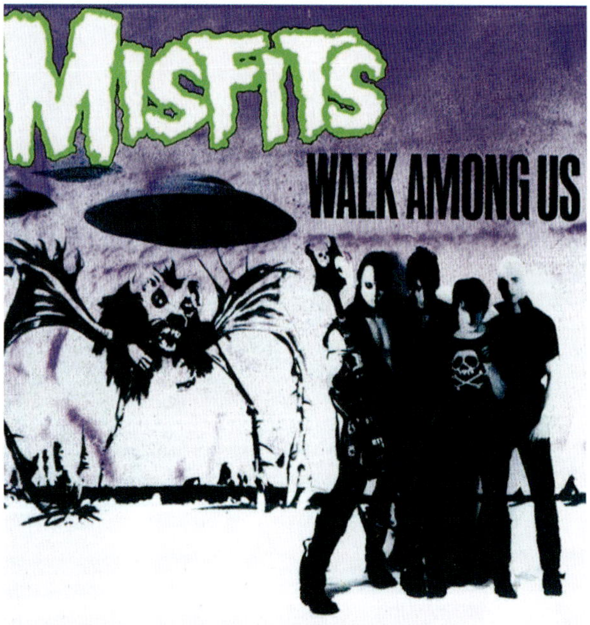

The artwork from 1982's *Walk Among Us*, by the Misfits.

The Misfits' logo is a stylized skull that was inspired by a poster for the 1946 film *The Crimson Ghost*.

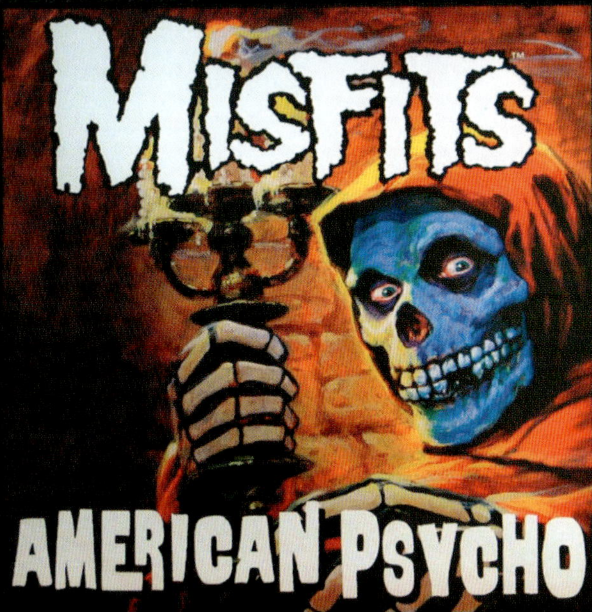

The cover for the Misfits' 1997 album, *American Psycho.*

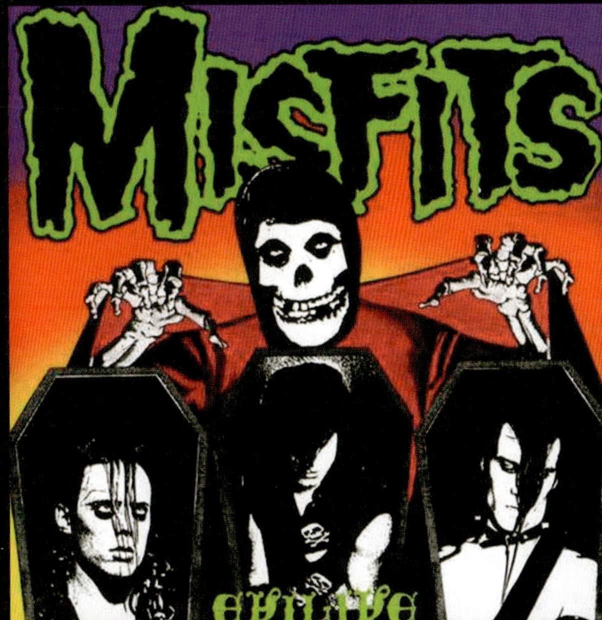

The original cover for the Misfits' 1982 album, *Evilive* (original).

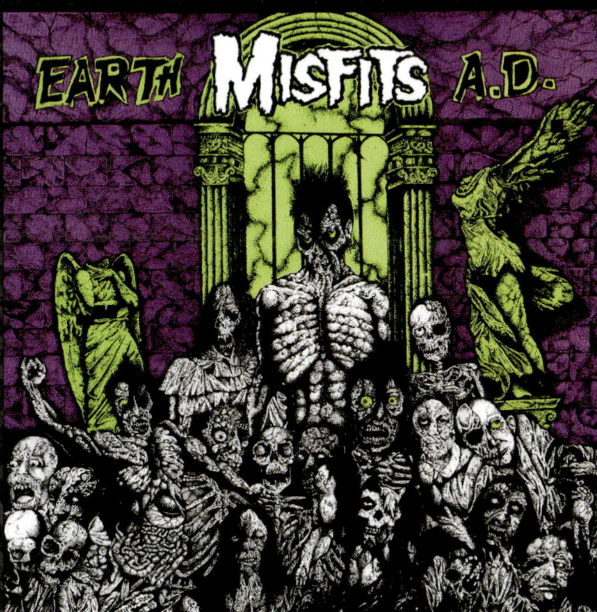

Earth A.D. was released by the Misfits in 1983.

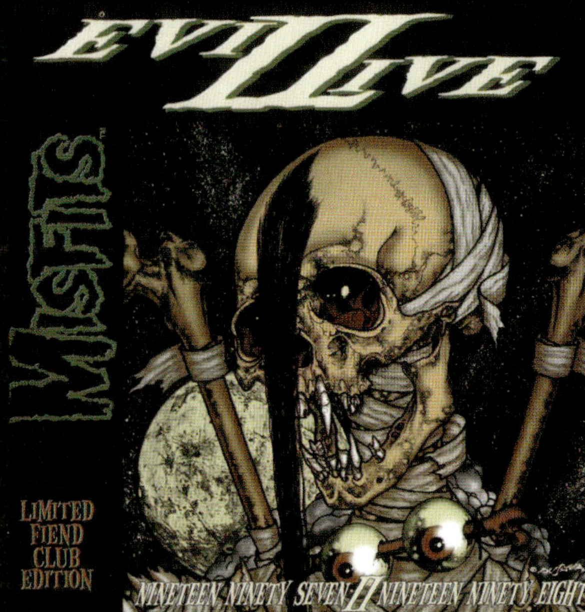

The limited edition club edition cover for 1982's *Evilive*, by the Misfits.

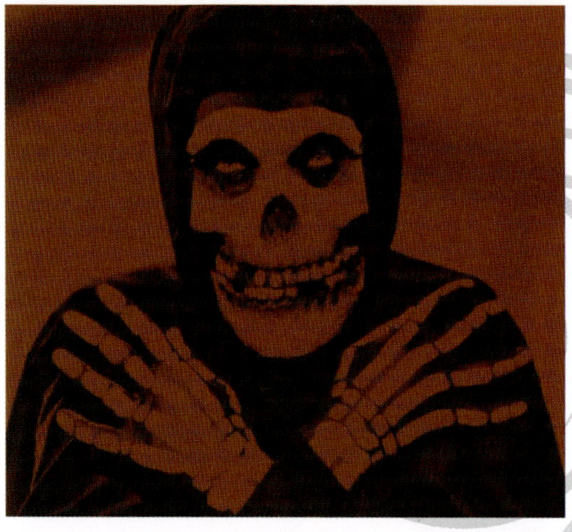

Collection II was released by the Misfits in 1995.

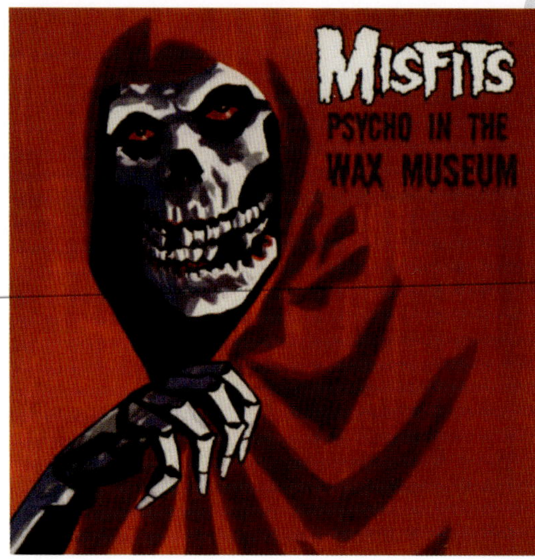

Psycho in the Wax Museum was released by the Misfits in 2006.

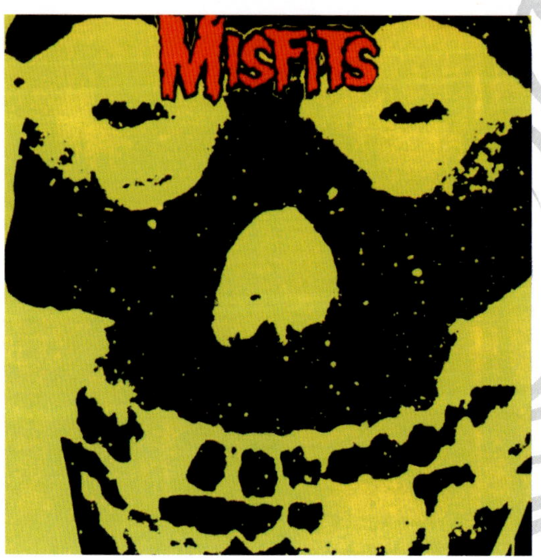

The album cover for 1986's eponymous *Misfits*.

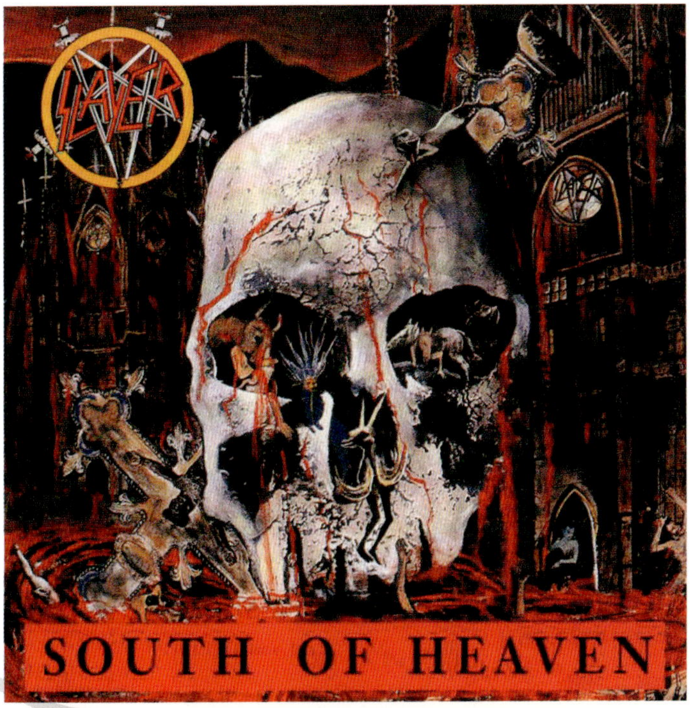

South of Heaven was released by Slayer in 1988.

American thrash metal band Slayer has several album covers with a consistent skull motif. *South of Heaven*, *Christ Illusion*, and *Seasons in the Abyss* were all designed by artist Larry Carroll.

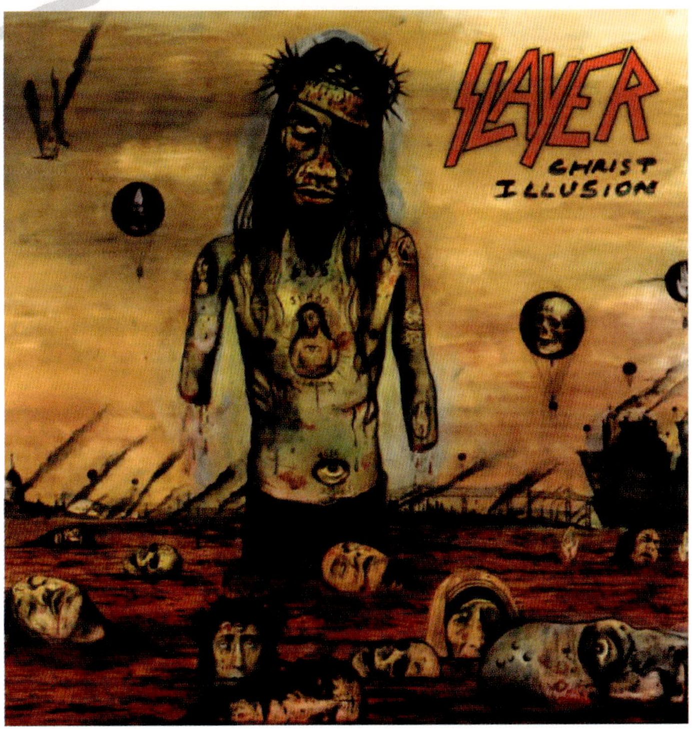

The album cover from Slayer's 2006 release, *Christ Illusion*.

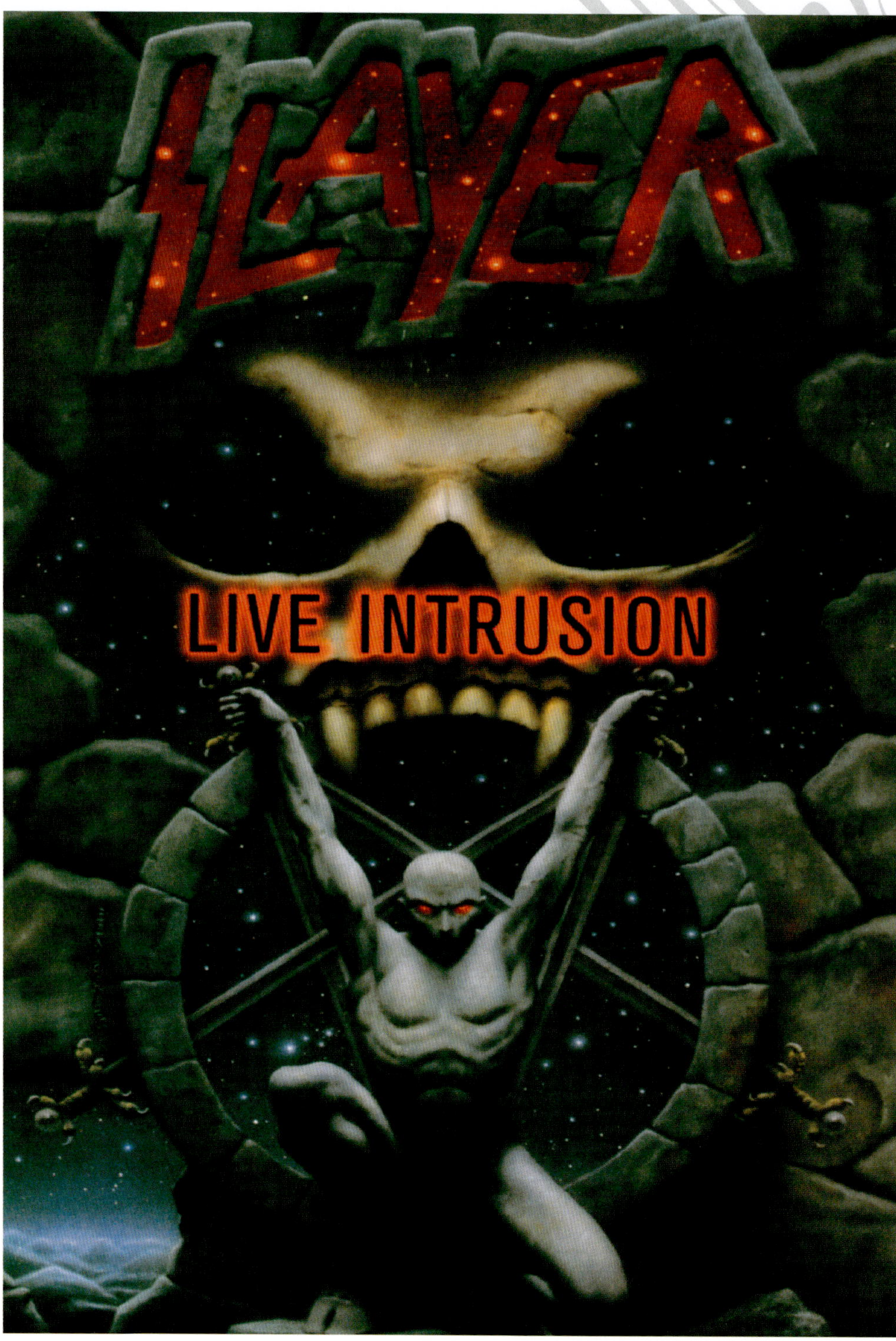

The album artwork for Slayer's 1995 release, *Live Intrusion*.

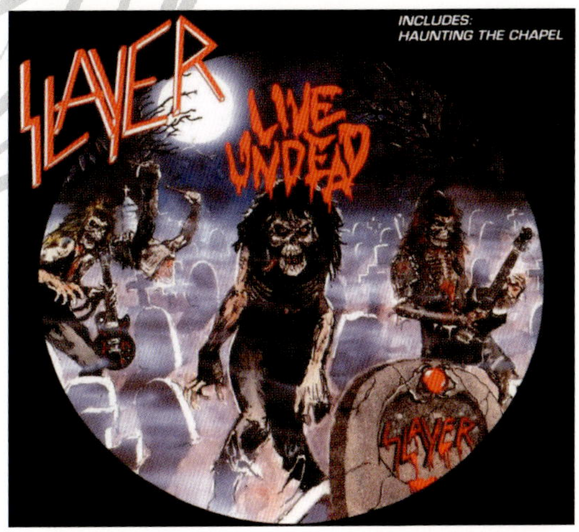

The cover for 1984's *Live Undead*, by Slayer.

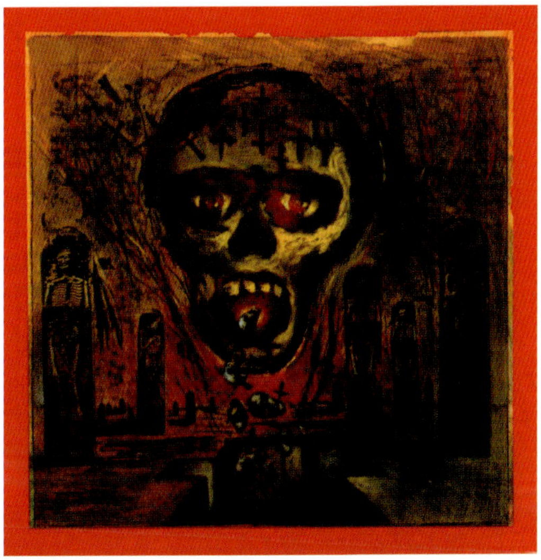

Slayer's album cover from 1990's *Seasons in the Abyss.*

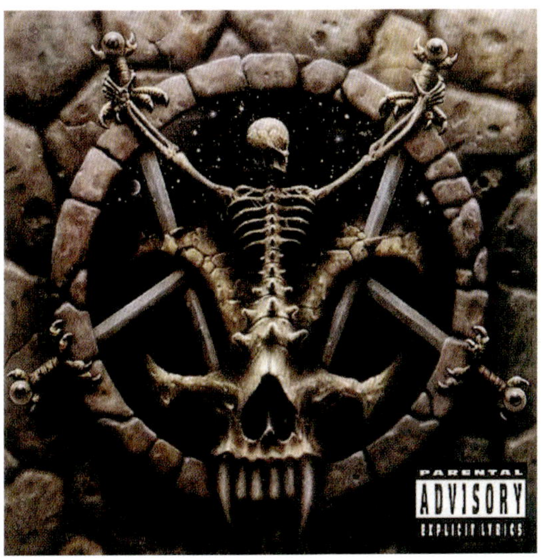

The artwork from Slayer's 1994 album, *Divine Intervention*.

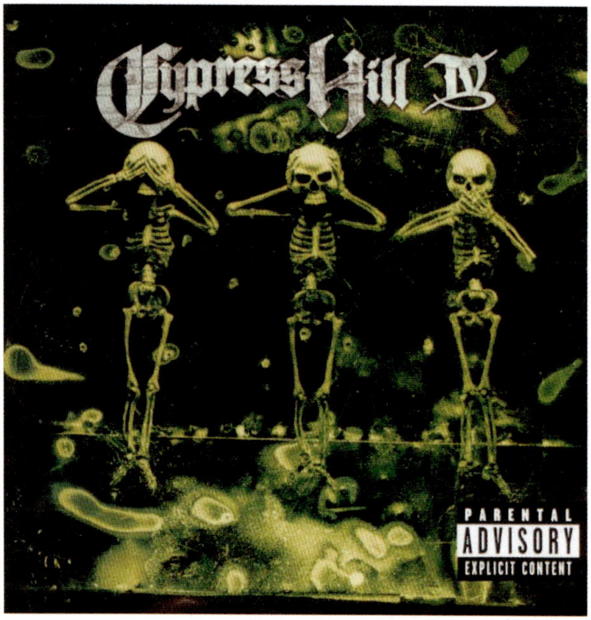

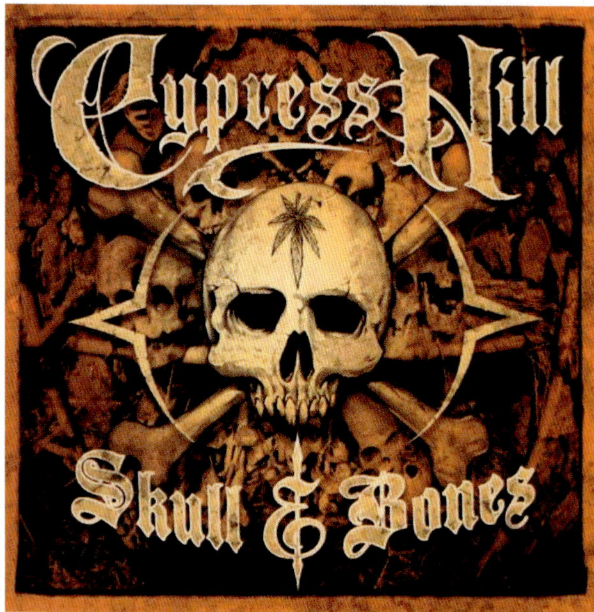

Black Sunday, by Cypress Hill, was released in 1993. The skull motif is consistent across these three albums by the rap group, giving it an easily recognizable style.

Skull & Bones was released by Cypress Hill in 2000.

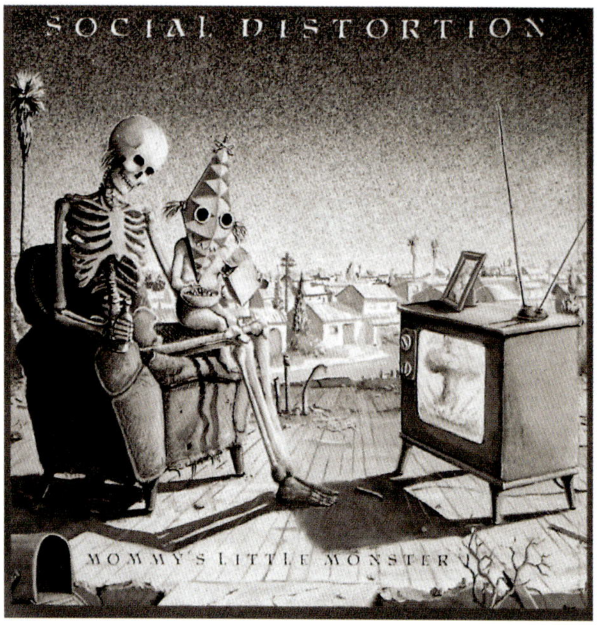

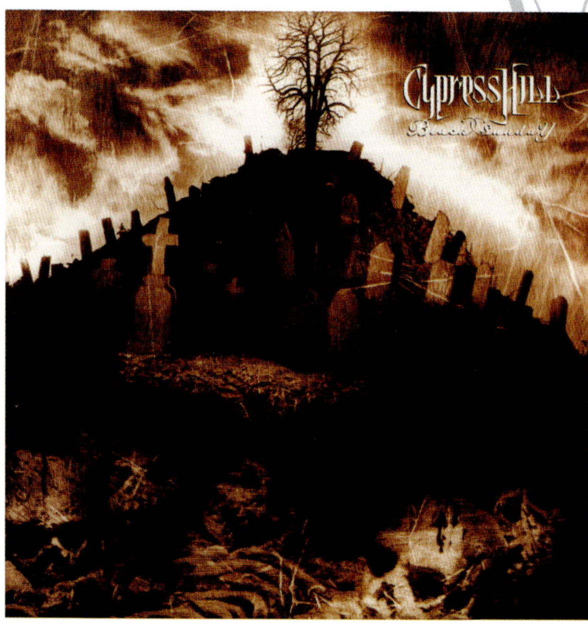

Social Distortion released their debut album, *Mommy's Little Monster*, in 1983.

Rap group Cypress Hill's album cover from 1998's *IV.*

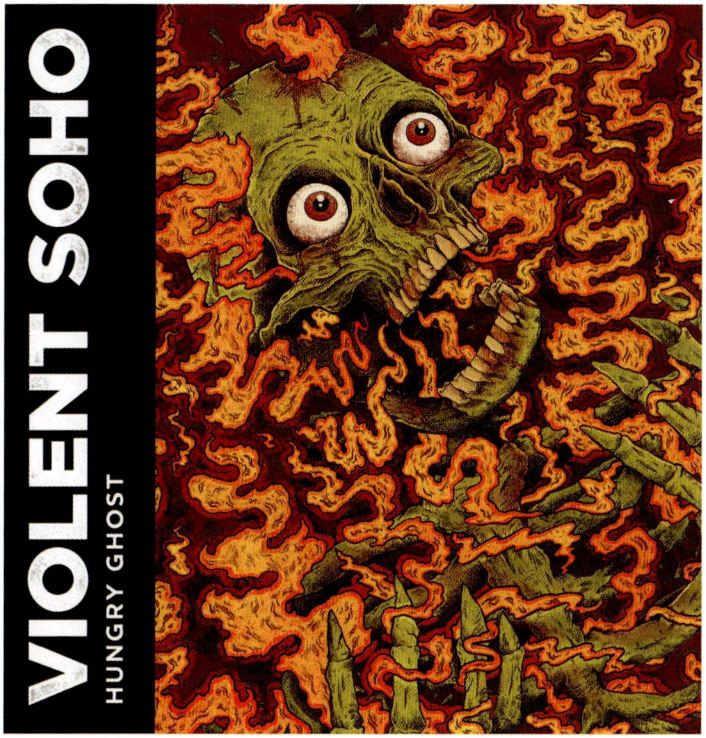

The album artwork from Violet Soho's 2013 album, *Hungry Ghost,* designed by Sam Dunn.

The logo of American hardcore punk band Kill Your Idols featured a stylized skull.

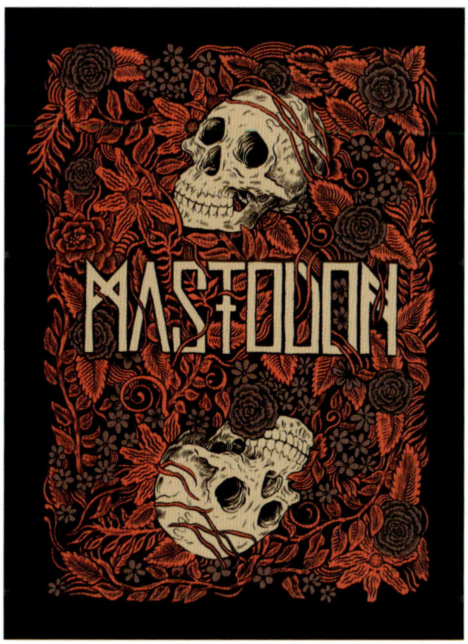

A promotional poster for American heavy metal band, Mastodon, designed by Sam Dunn.

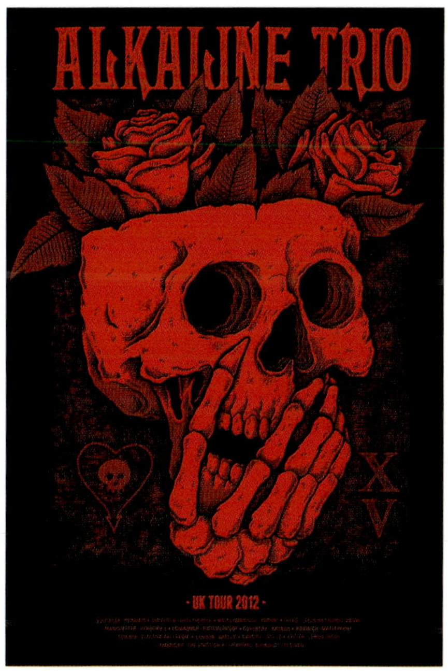

A 2012 UK tour poster from American punk rock band Alkaline Trio, designed by Sam Dunn.

SKULL TATTOOS

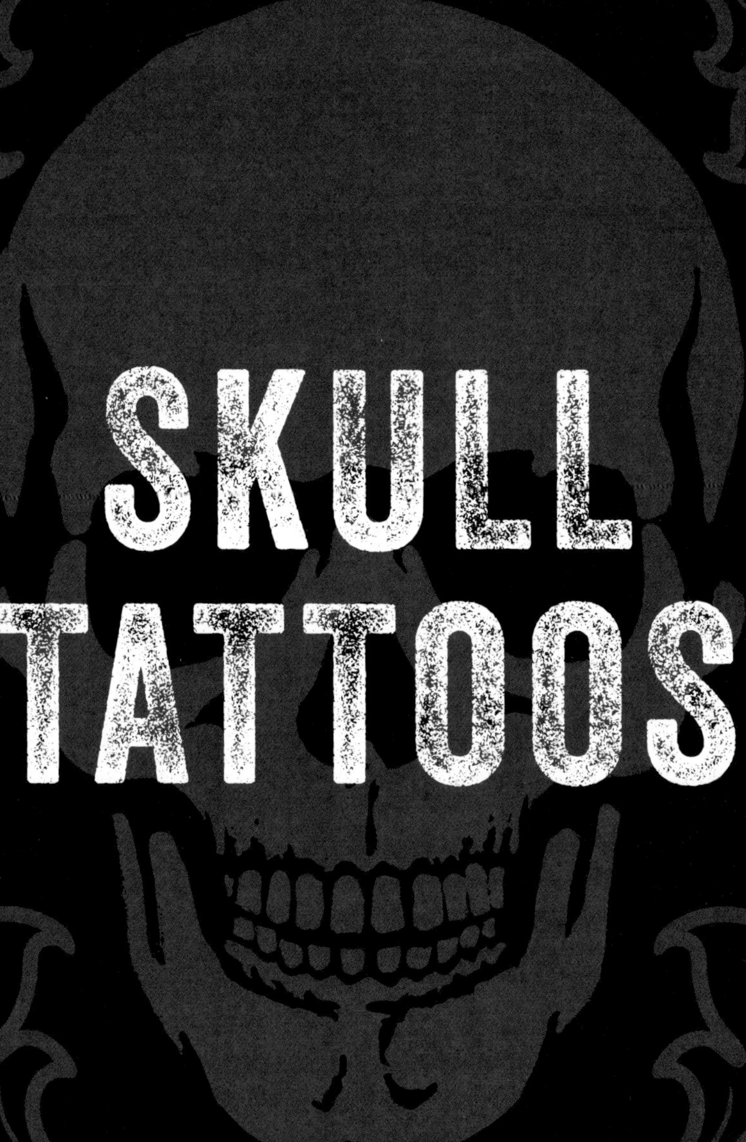

TODAY, THERE ARE THOUSANDS OF TATTOO DESIGNS FEATURING THE SKULL MOTIF. BUT, WHATEVER THE DESIGN, WHY DO WE HAVE TATTOOS IN THE FIRST PLACE? THE HISTORY OF TATTOOS, AND HOW CURRENT TATTOO DESIGNS DEVELOPED, IS FASCINATING.

The actual word "tattoo" was coined by the eighteenth century British explorer Captain Cook during his voyages in the South Pacific. It derives from the Polynesian "tatou," meaning "fitting," "balanced," or "appropriate." Having a tattoo was considered a serious matter, not to be lightly undertaken. Marking the skin was carried out on the Polynesian Pacific islands only by men specifically trained in the art, and each tattoo was accompanied by due ceremony and ritual.

We know that making permanent marks on the skin was practiced for various reasons all over the world, with occasional archaeological discoveries adding to our trove of knowledge. For example, a couple on vacation in the general area of the Austro-Italian Alps in 1991 couldn't possibly have expected to stumble upon a discovery which would make a major contribution to the world of tattoos. But that is exactly what happened. In the snow and ice of the mountains they came across mummified human remains that were so well preserved it was initially thought to be a relatively recent death; subsequent examinations revealed that the corpse was approximately 4,000 years old. Ötzi, named for the area in which he was discovered, was covered in some sixty-one tattoos. While there were no skulls amongst the images that decorated his body, Ötzi's skin was covered in lots of black lines (of various length and thickness) and dots, with bands around his wrists and crosses behind his knees. These skin markings are believed to have been made by scarifying or scratching the skin and then inserting soot or ashes into the scars. Ötzi may currently be the oldest example of tattooing, but he is not alone. Tattooed human remains have been discovered across the continents and from them we can learn something of the meaning and nature of skin markings. One of the more glamorous examples of tattoo art is an ice maiden from Siberia, who is displayed in a glass sarcophagus—like Sleeping Beauty—in a museum in central Russia. The "Ukok Princess" has glorious tattoos which look contemporary, especially since she lived in the fifth century B.C. The designs include bold, stylish hybrid animals and horses, which have been interpreted as a sign of her status as well as her spiritual connections.

In Ancient Greece and Rome, tattoos were considered degrading, and were reserved for the criminal classes. They were often used to brand the perpetrators and describe the crime and the name of the victim for all to see. In Asia, however, parts of the body were tattooed to cure illnesses and restore health.

A little more recently, records from Jesuit priests while exploring the Americas gives us another hint. In 1652 they wrote; ". . .Those who paint themselves permanently do so with extreme pain, using needles, sharp ails, or piercing thorns. . ." The native people would have used their tattoos in much the same way we do today, to define their tribe and to align with their spiritual animals and familiars. Also, the bloodletting during the process of getting the tattoo would be a libation of sorts, an honouring of the gods. The pain of the process itself was regarded as a rite of passage; the ability to grit the teeth and bear the pain would have been a mark of maturity. And of course, as anyone who has had a tattoo will know, the endorphins released during a painful tattoo can be extremely exhilarating; some even describe them as addictive.

We think nothing of the pigmented inks used in tattoos today. But again, we have to go back in time to find out how they were first created. It is difficult to point to a precise date, but it was in the fourteenth century B.C. in Japan that pigments were first used.

In the early 1800s, tattoos became a handy form of identification for sailors and other seafarers, especially in the United States. This led to the proliferation of shore-side tattooists. A tattoo could be used as identification, which meant that it was unnecessary for sailors to carry papers. The tattoo would also provide a means of identifying the sailor should he fall foul of the sea, and drown.

It wasn't until the hazy hippy days of the 1960s and 1970s that tattoos started to become as diverse as they are today, with designs limited only by the imagination of the bearer as well as the skills of the tattooist. Skull tattoo designs were influenced at that time by the use of the image by bands and musicians.

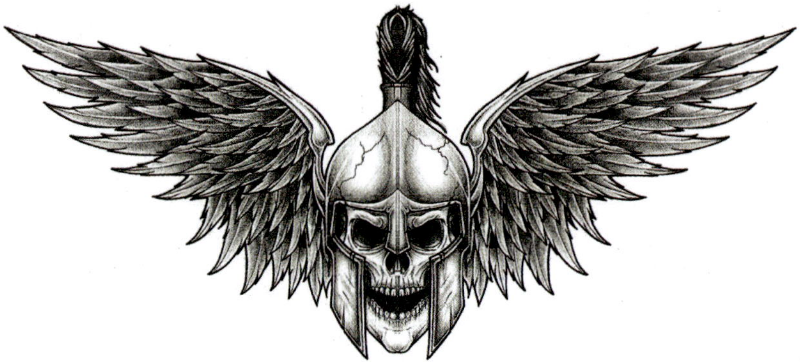

Spartan Warrior Skull with Wings by aquili-filius

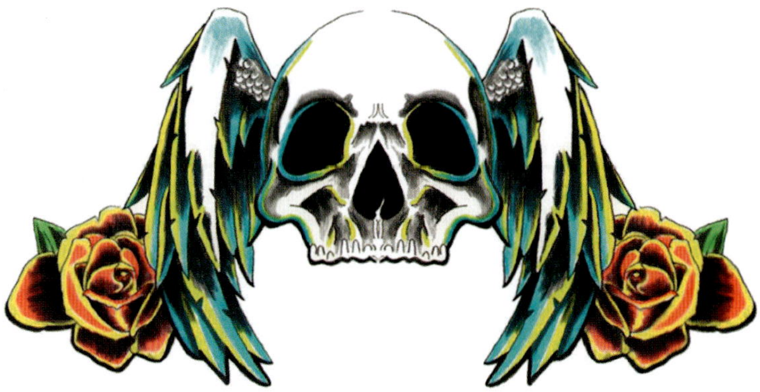

Skull with Wings by ladytat2

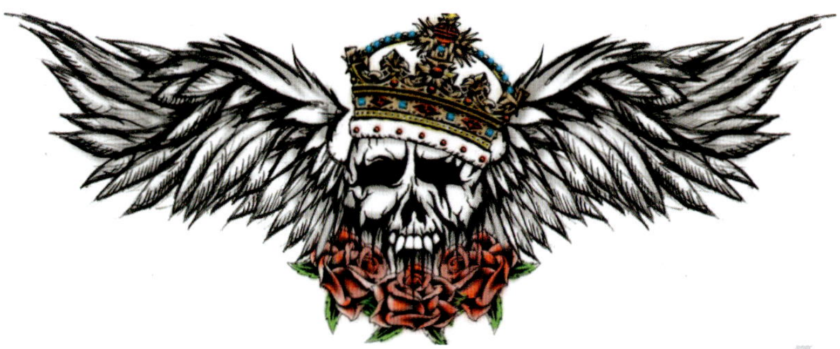

Spartan Warrior Skull with Wings by aquili-filius

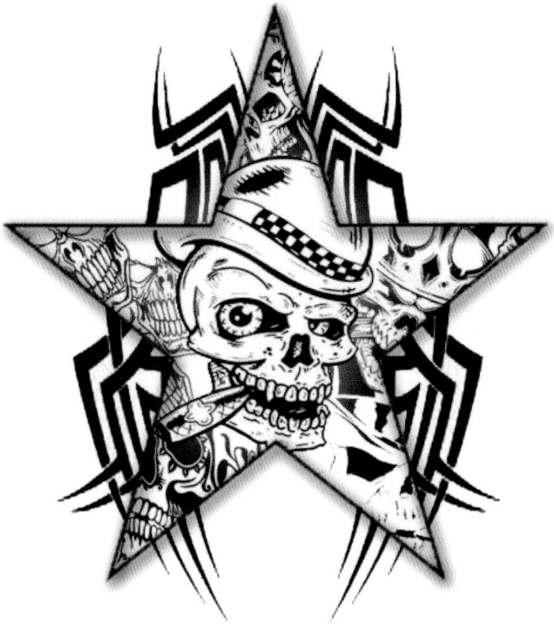

Rock Stars & Skulls by lild_customs

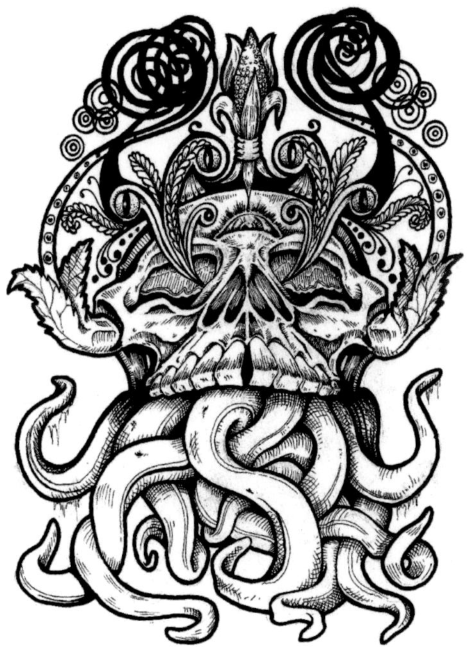

Octopus Skull by Justinwinter

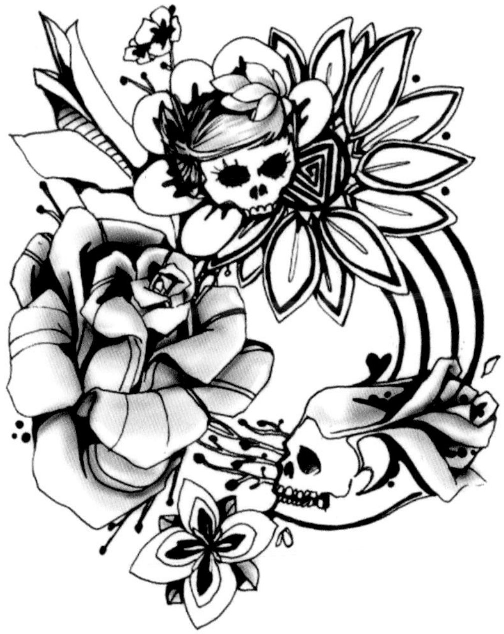

Skulls Roses Daisy by Mythology

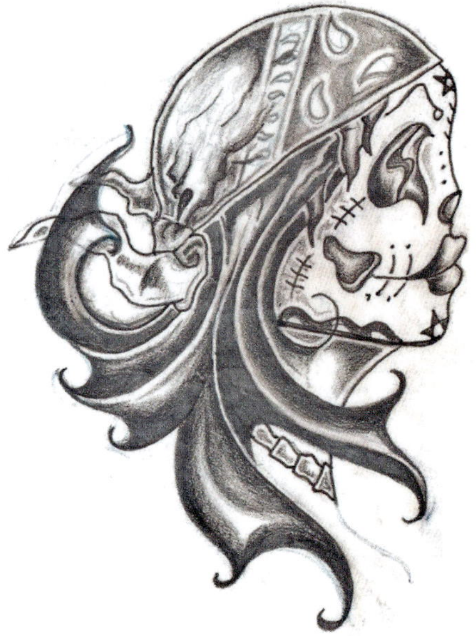

Sugar Skull Bandana by PecuniamArt

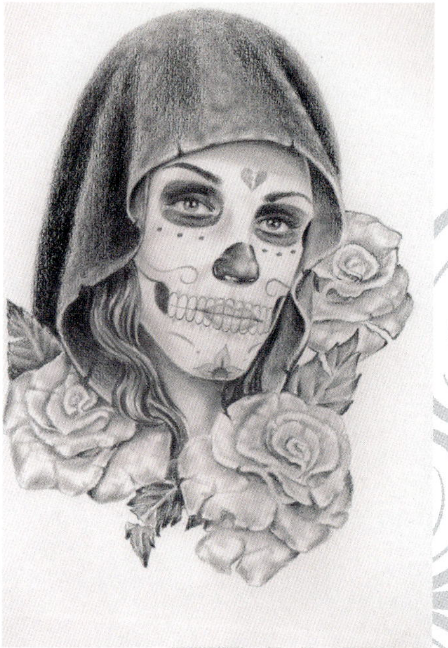

Day of the Dead by shady

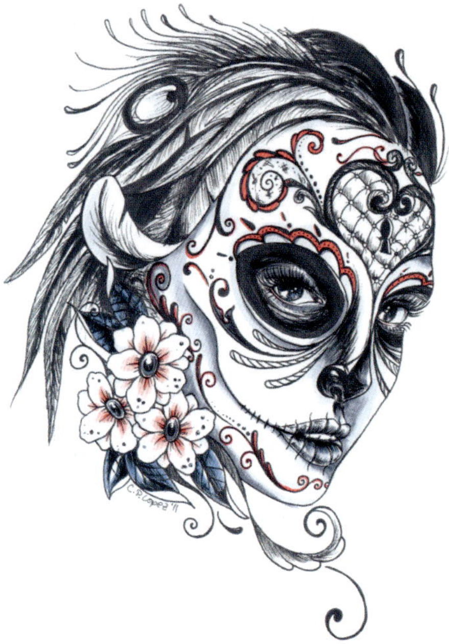

Sugar Skull Girl by kasini

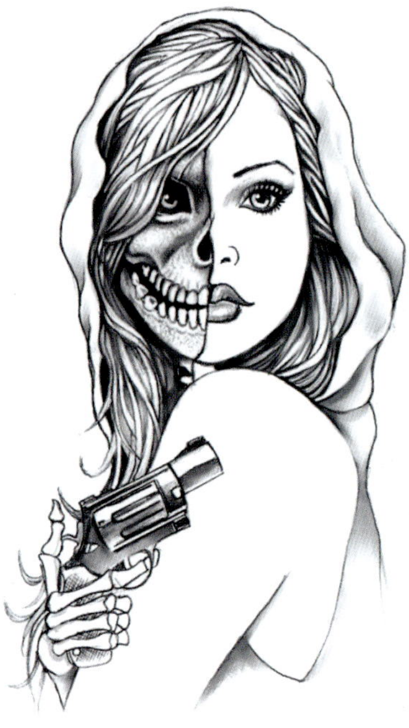

Sexy Reaper by chicktattoo

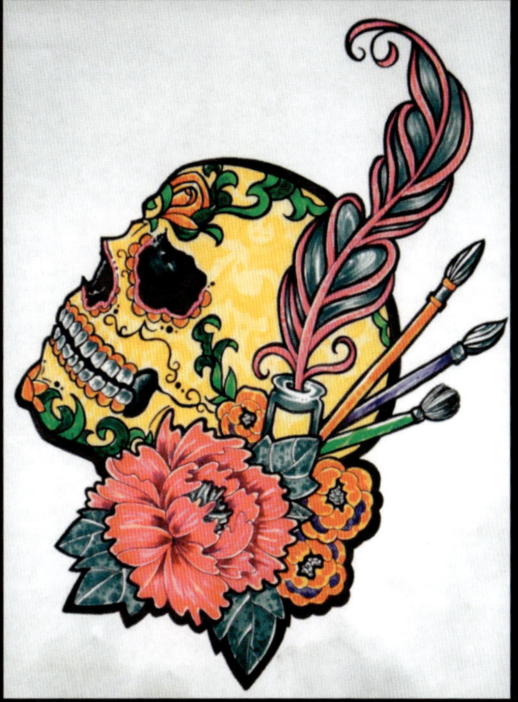

Artist for Life by ShaunaHope

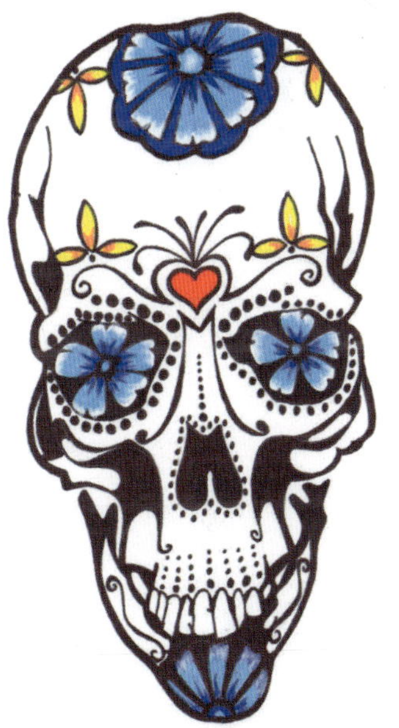

Sugar Skull by bmills

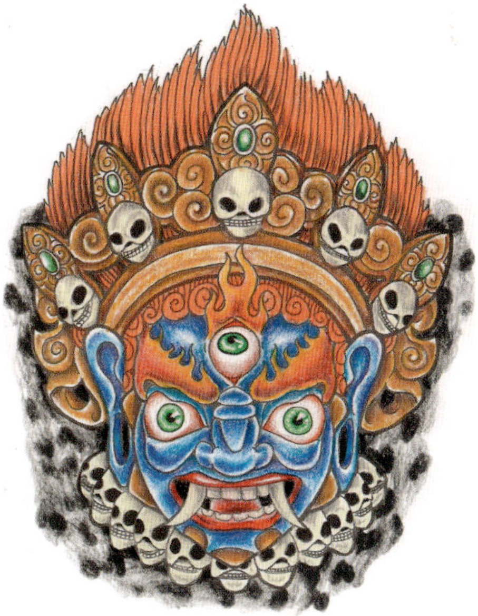

Mask by Inksage

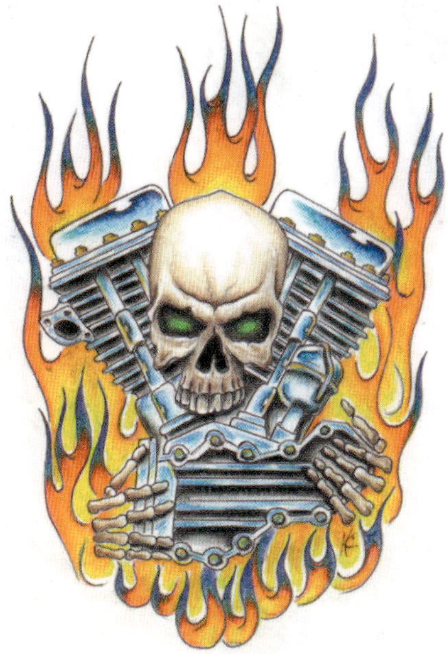

Pan Head & Skull Head by KCool

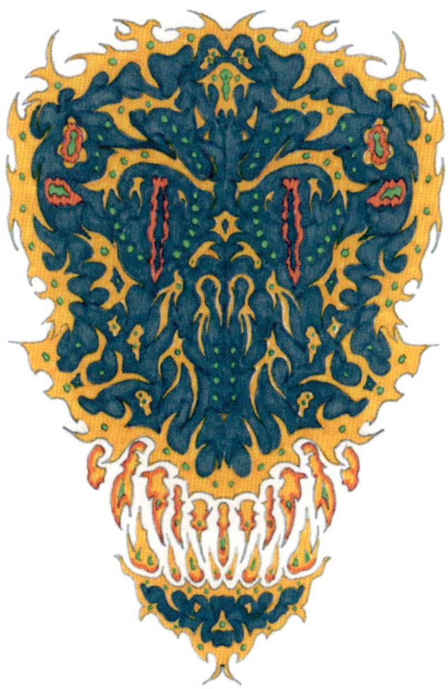

Ornate Monkey Demon Skull by kaneykreative

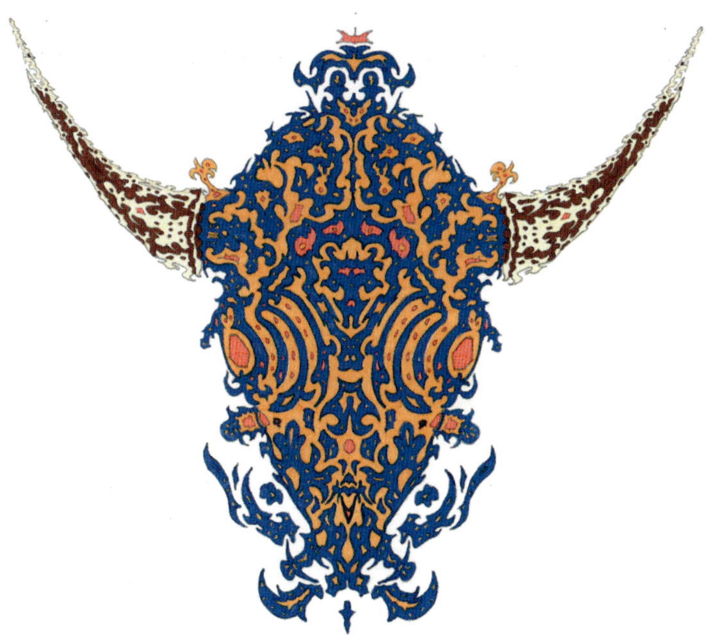

Ornate Western Cow Skull by kaneykreative

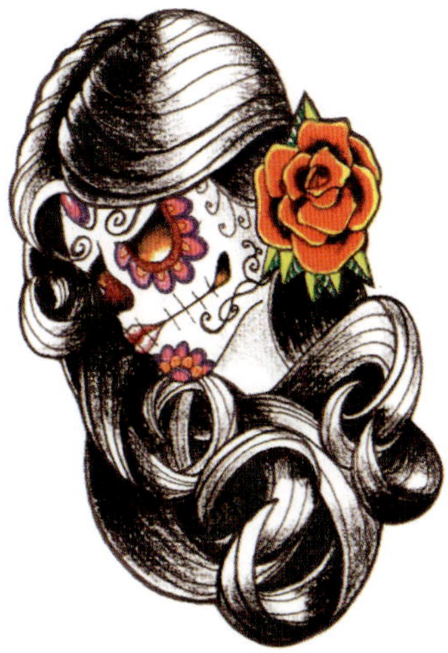

De Los Muertos #2 by Dim-Dim

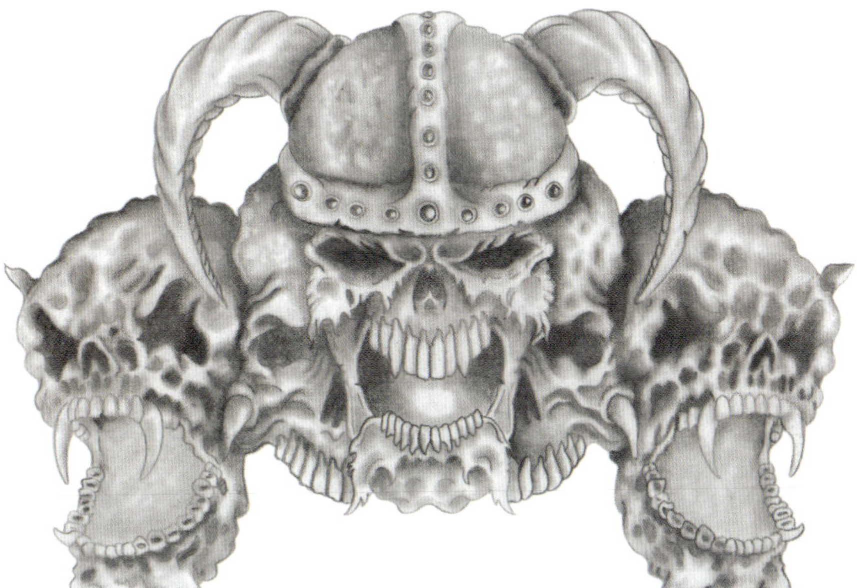

Three Skulls with Horn by mfrank

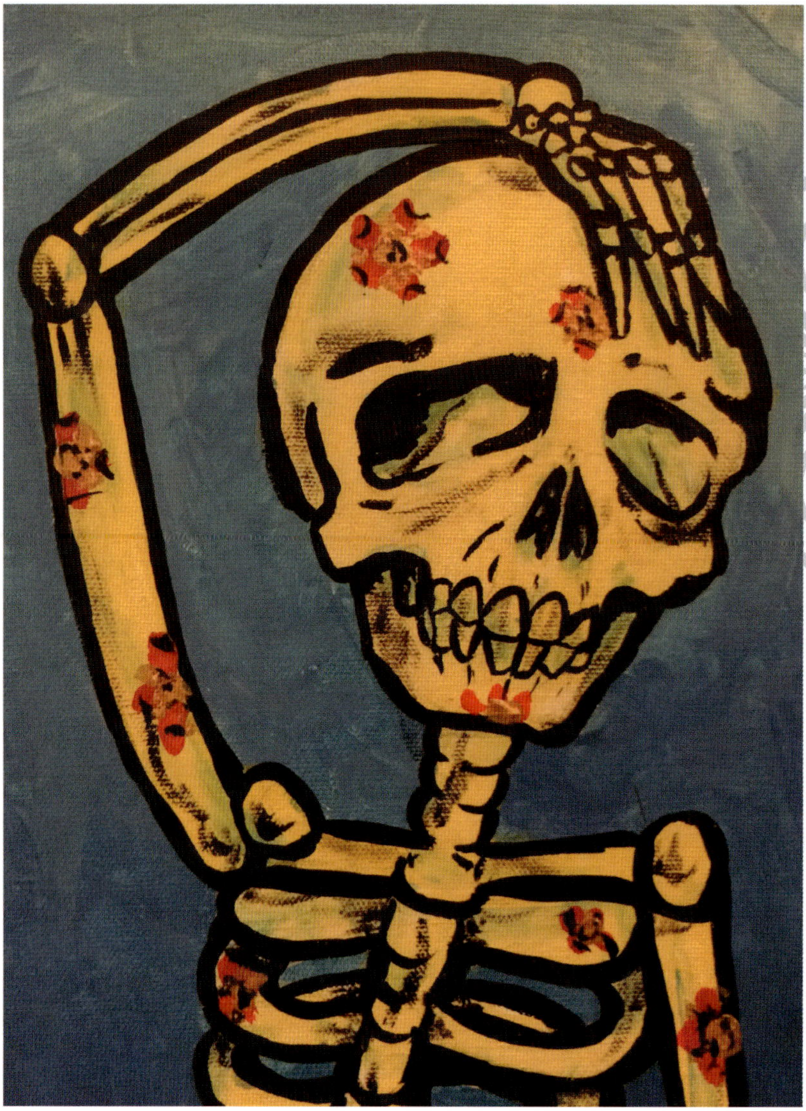

This spread: Barb Allen-Bolger from Inka Tattoos in Brighton, England, designed these skull tattoos.

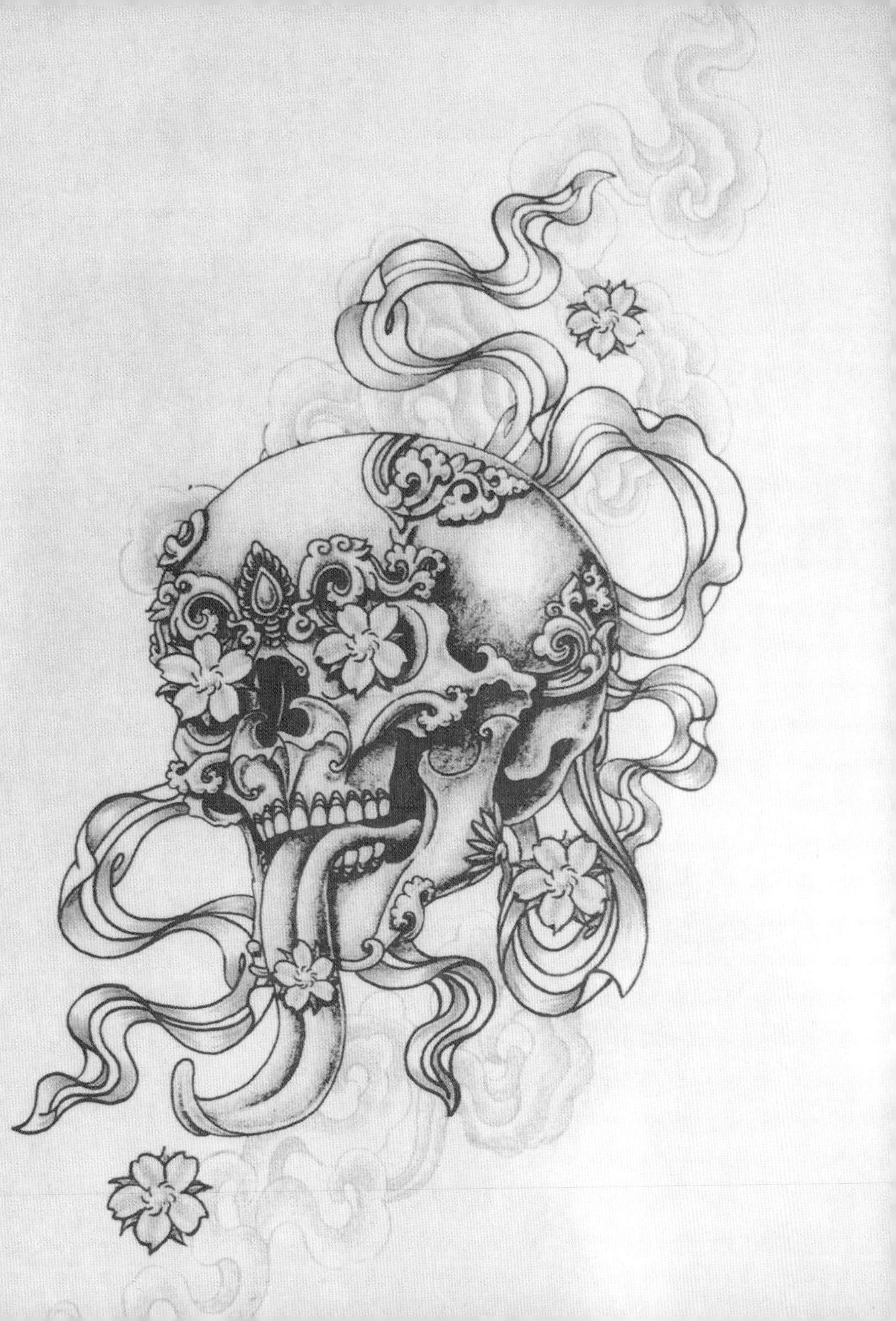

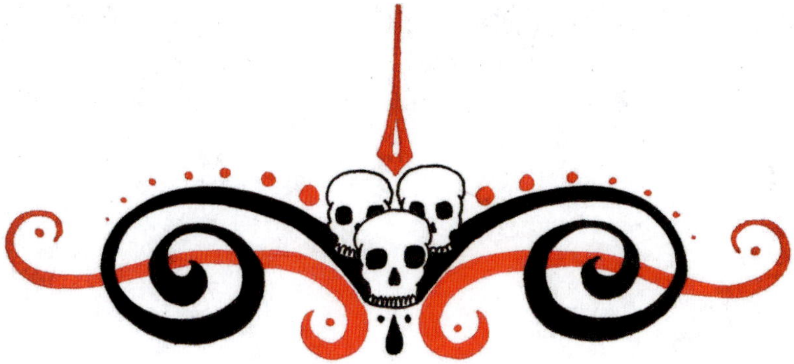

Triple Tribal Skull by MissRita

Tribal Skull by MacsTats

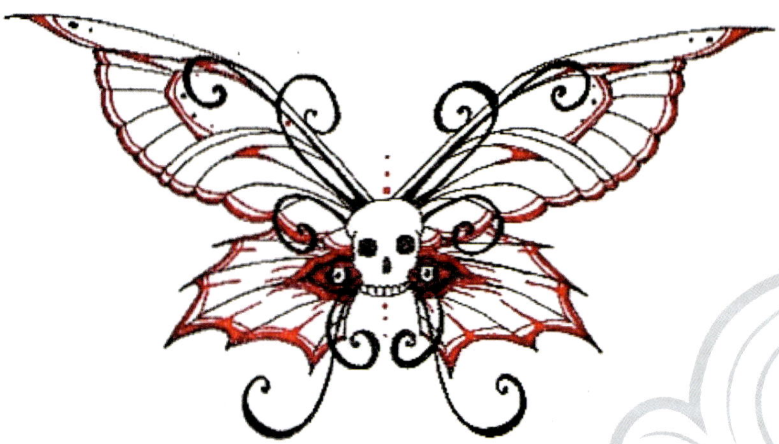

Fluttering Death by Miss Rita

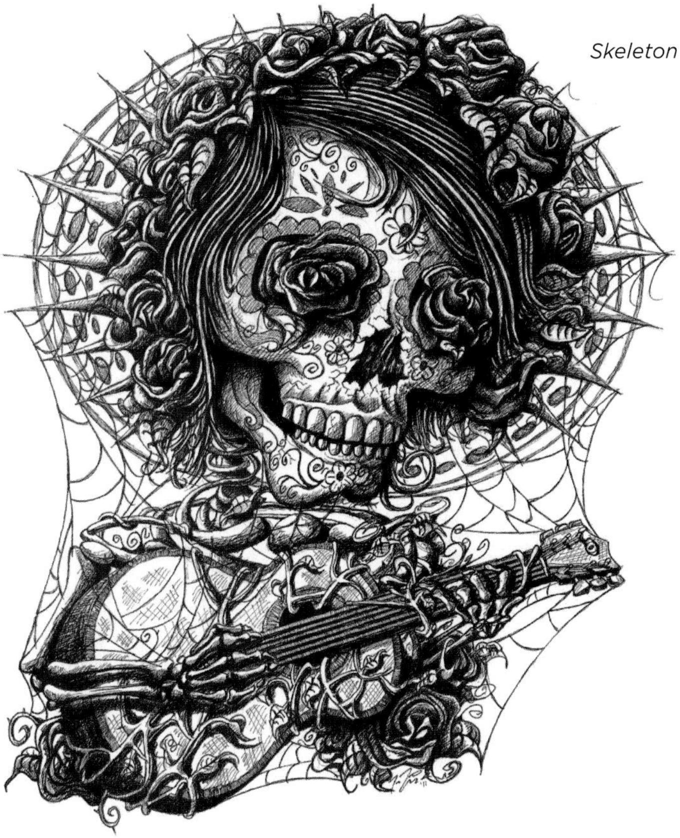

Skeleton Musician by Toddric76

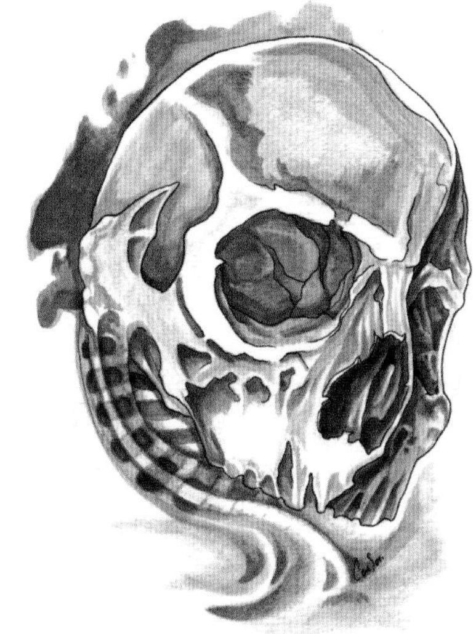

Skull Abstract Realistic by tat2Carson

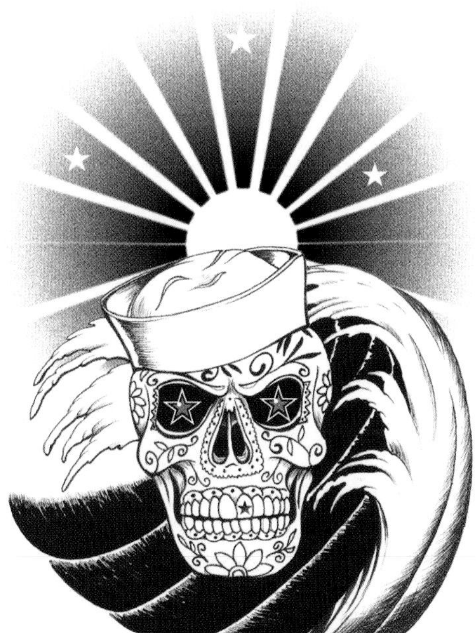

Navy Skull with Wave by chicktattoo

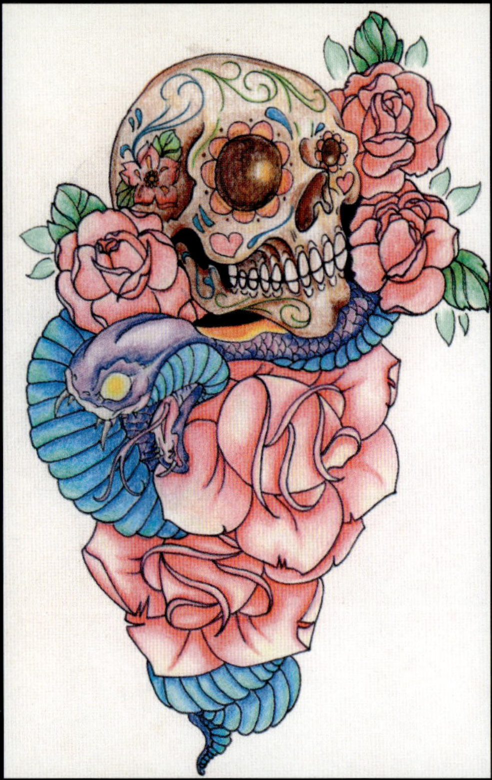

Day of the Dead by Ghost13

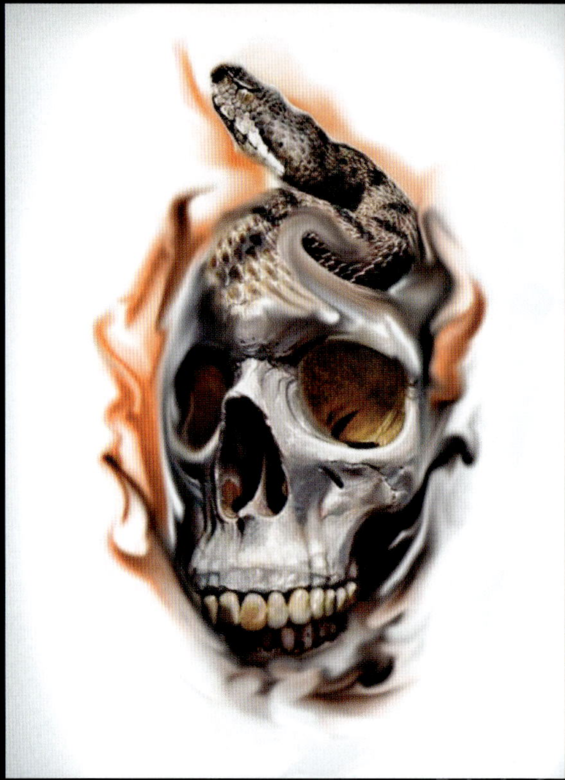

Matrix Draws Tattoo by Gianluca

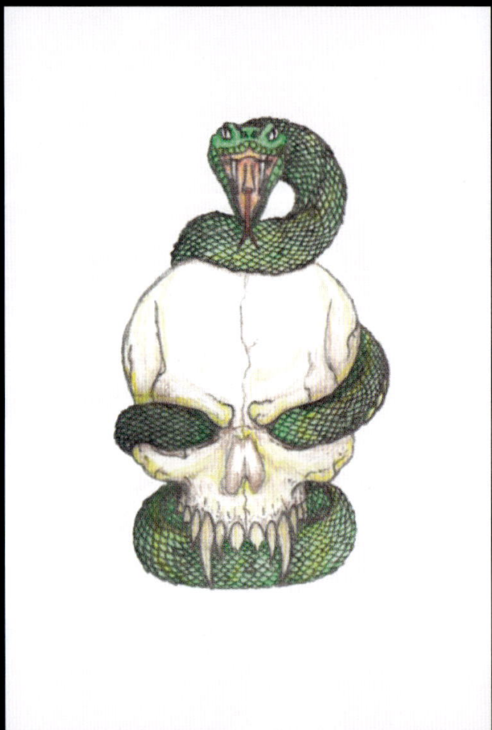

Skull with Snake by rport35

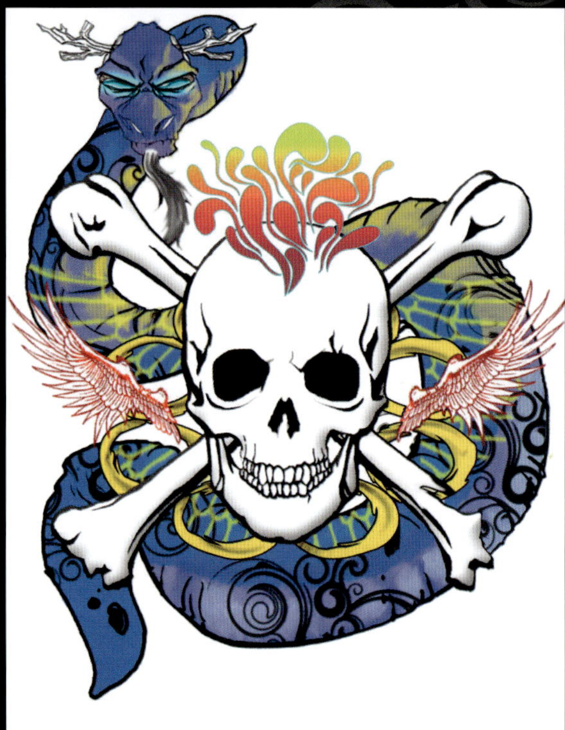

Dragon Skull by bcaso

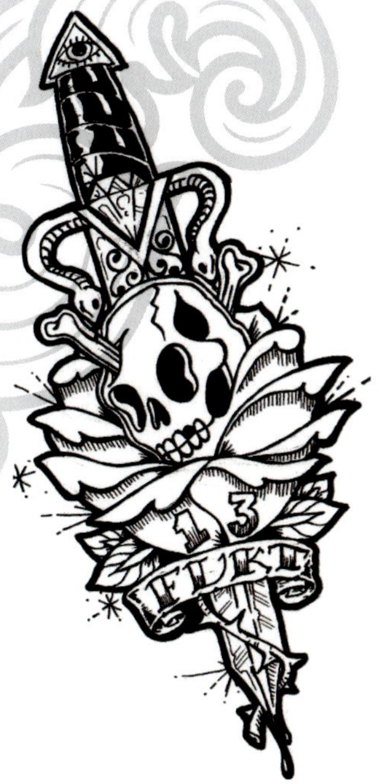

Fukt by Justinwinter

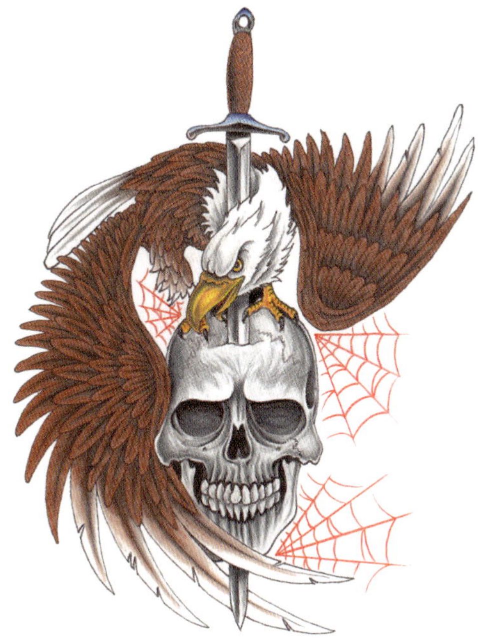

Eagle Skull by inkaholick

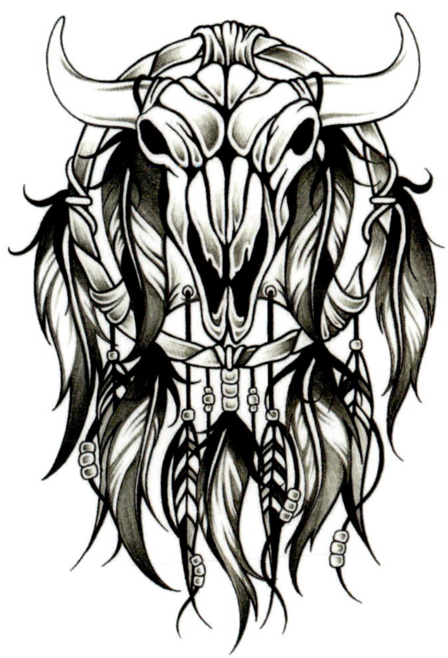

Bison Skull Dreamcatcher by achemical

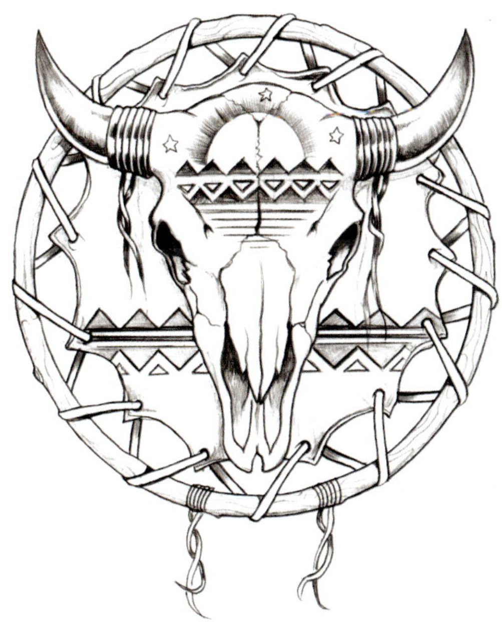

Bison Mandela by chicktattoo

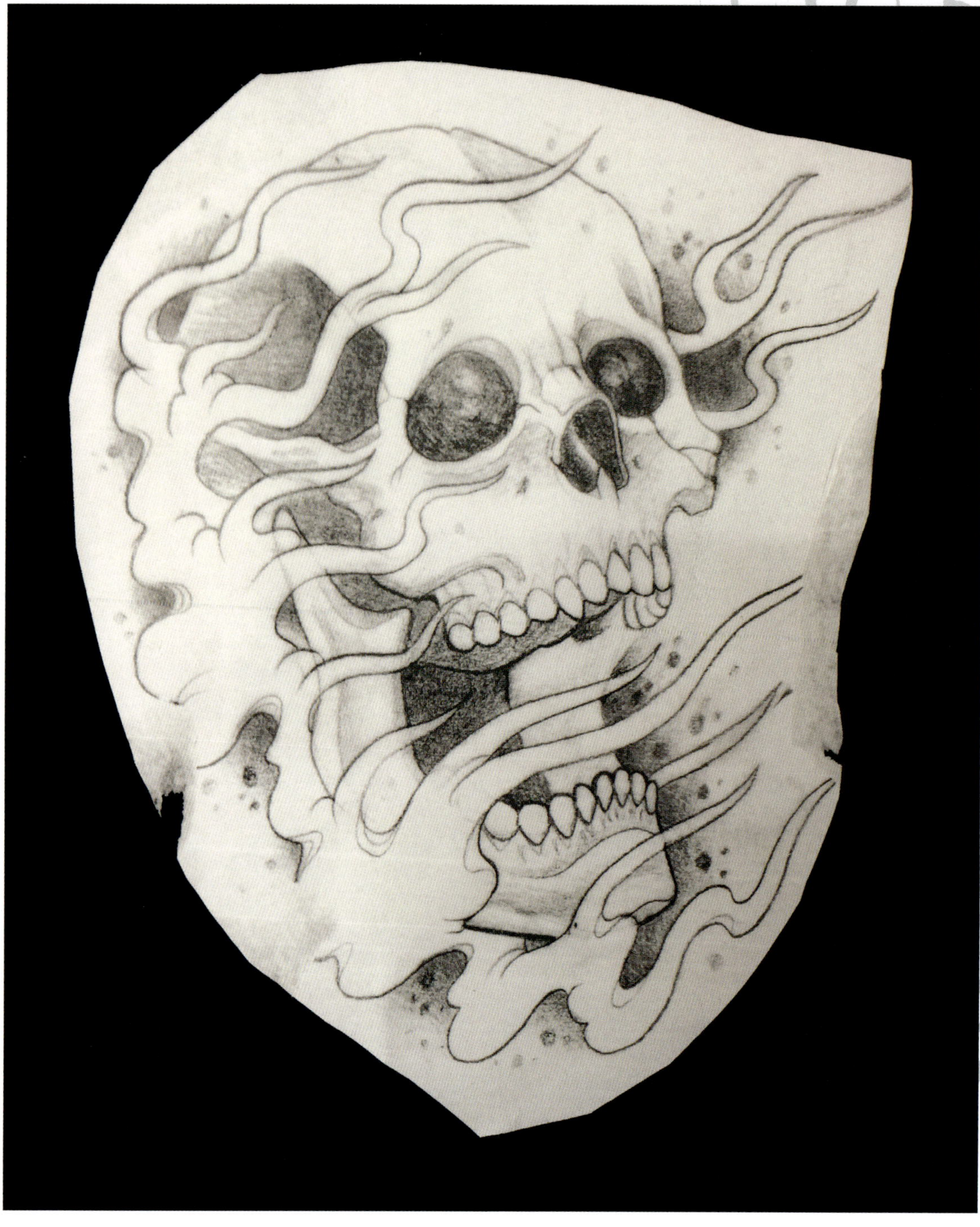

This tattoo was designed by Gustavo Castro, a Brazilian tattoo artist who travels around Europe.

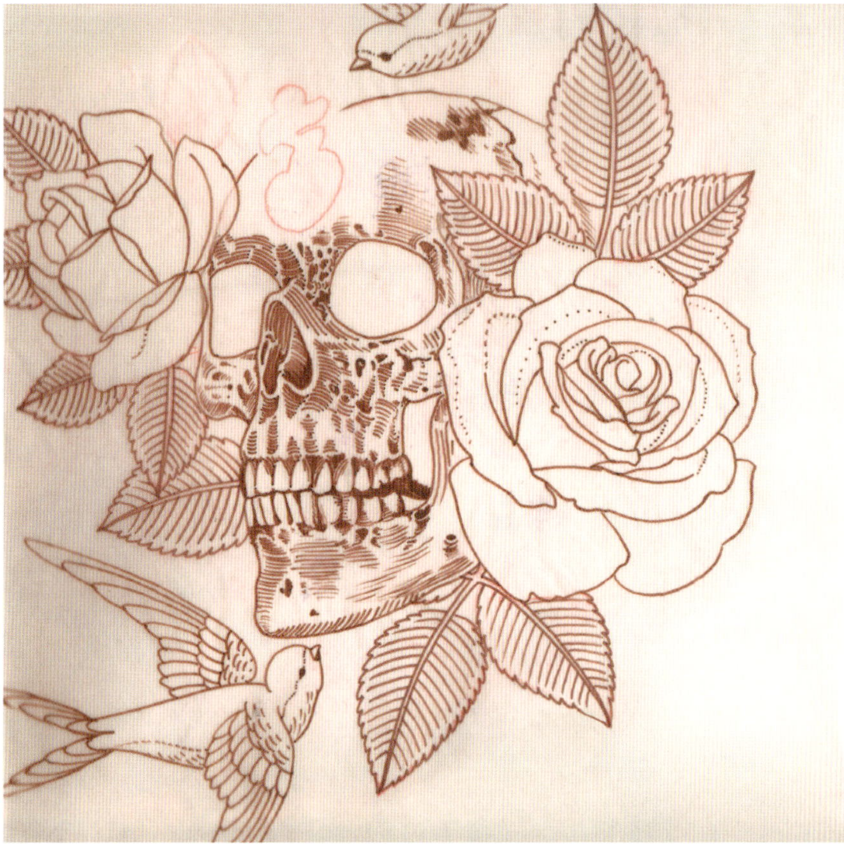

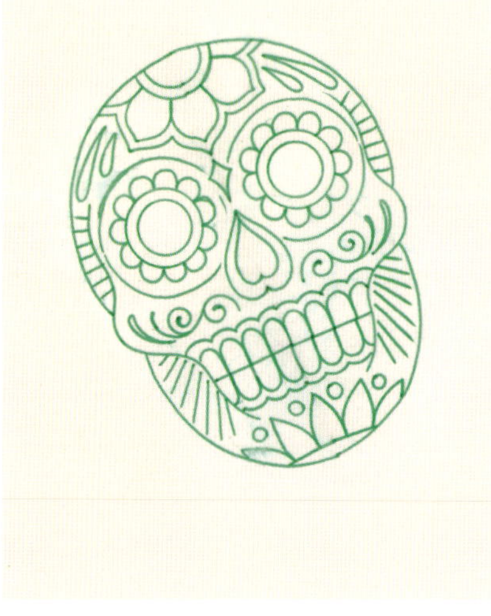

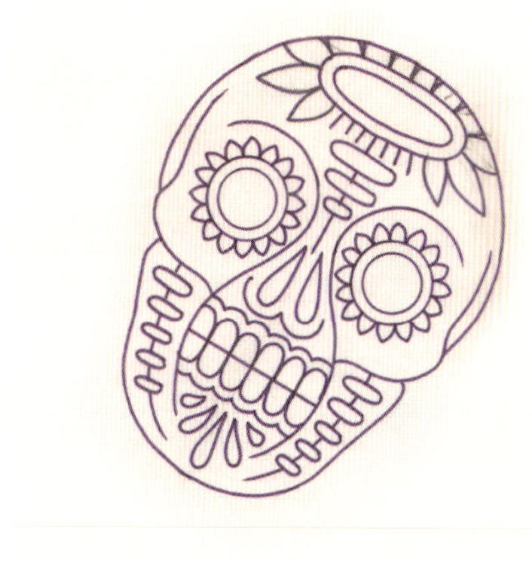

These tattoos were designed by Ben T. Fiedler of Royal Tiger Tattoo in Austin, Texas.

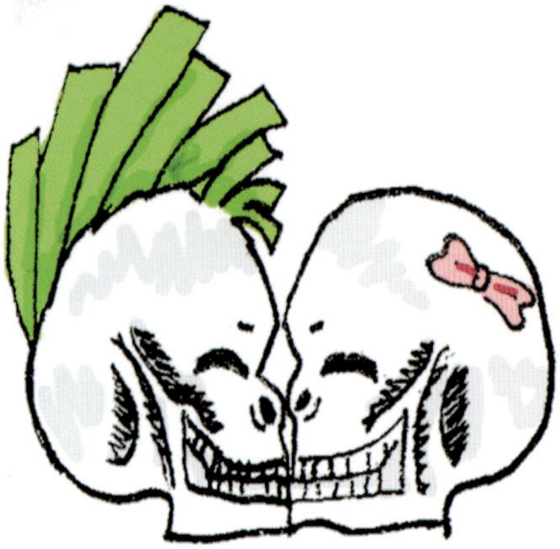

Skull Lovers by Patricia

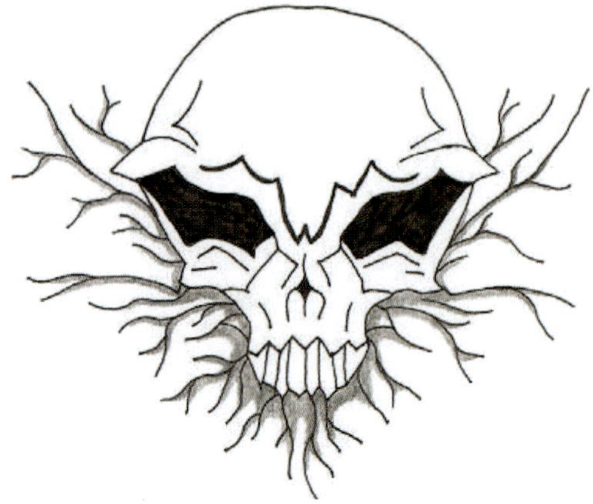

Skull with Cracks Around It by wildheart

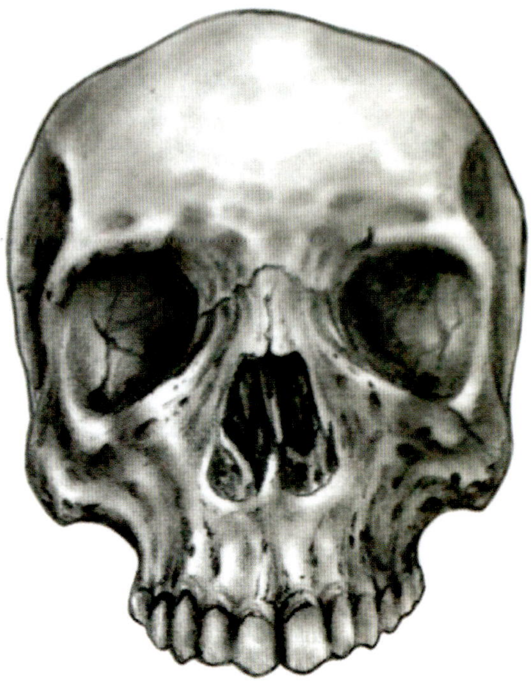

Skull by Dwiwilson

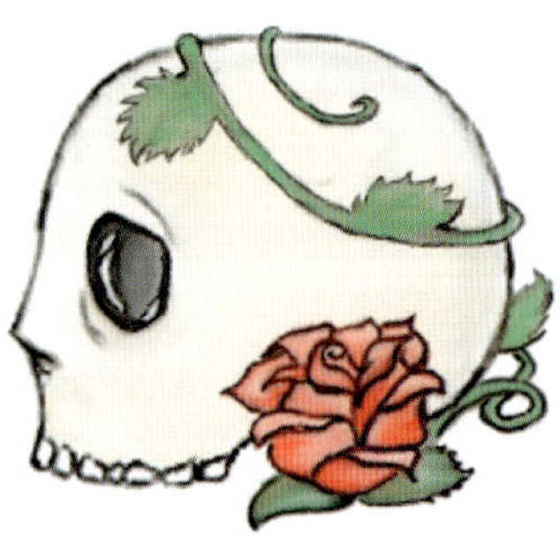

Skull & Rose by ShyPersephone

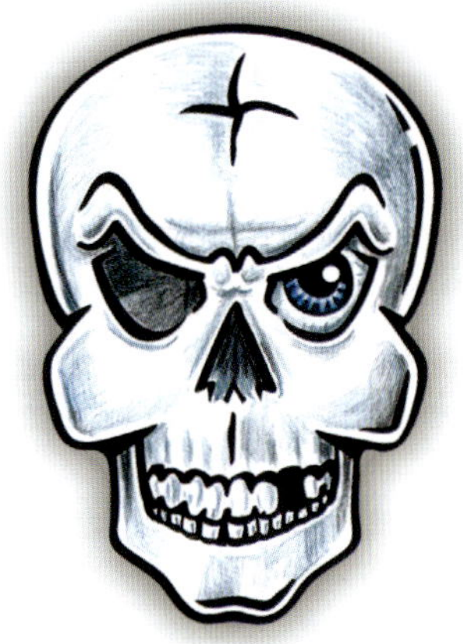

Skull Black and Grey by OMB

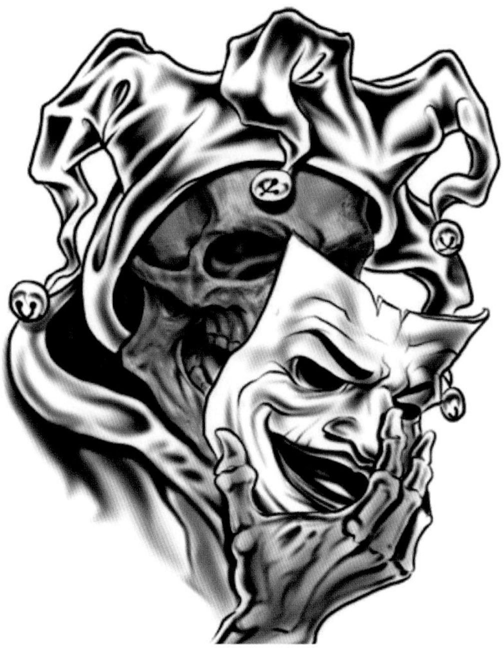

Skull Joker by MANCAVE

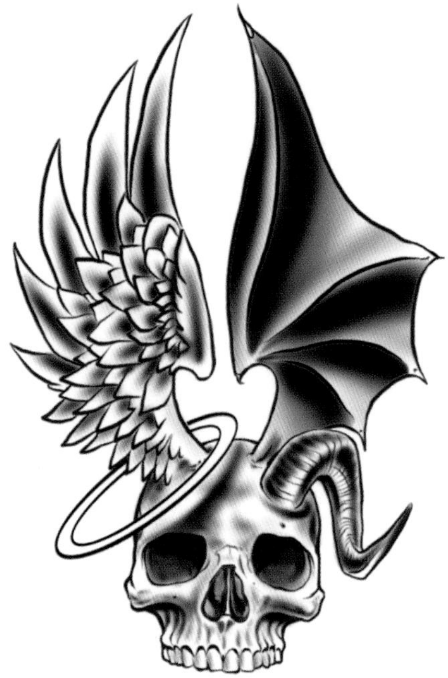

Skull by MANCAVE

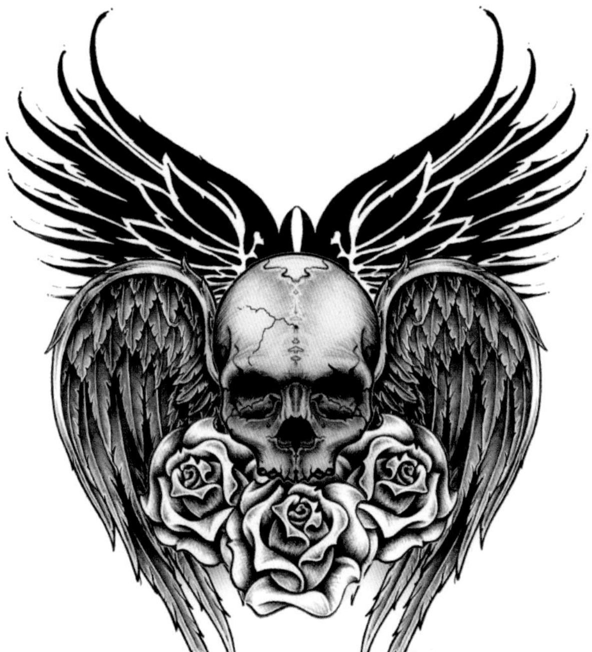

Skull with Wings and Roses by aquili-filius

Louder by hannonartink

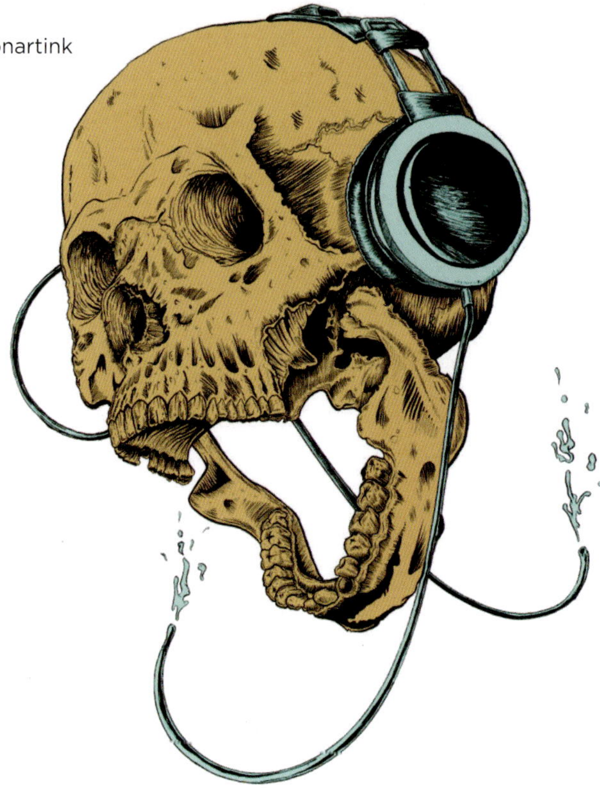

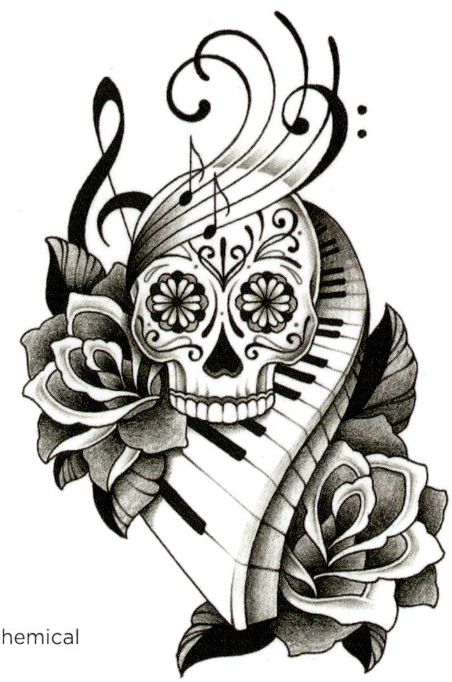

Piano Music Sugar Skull by achemical

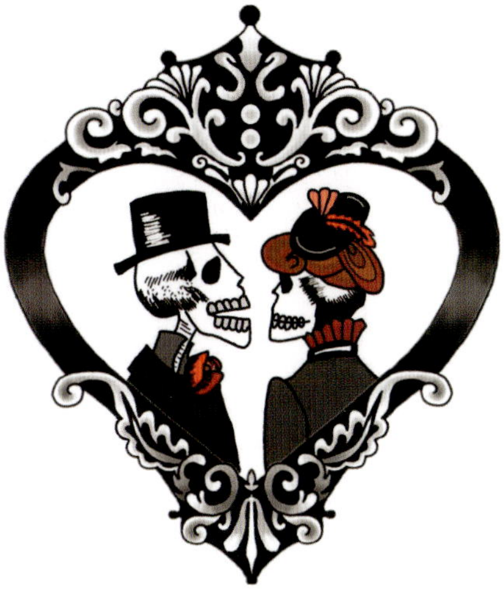

Vintage Skeleton Couple by Porcelaine

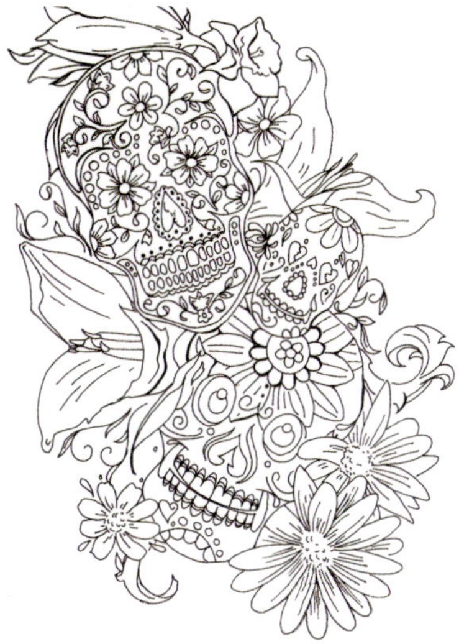

Sugar Skull Flower Side Piece by Mckayamber

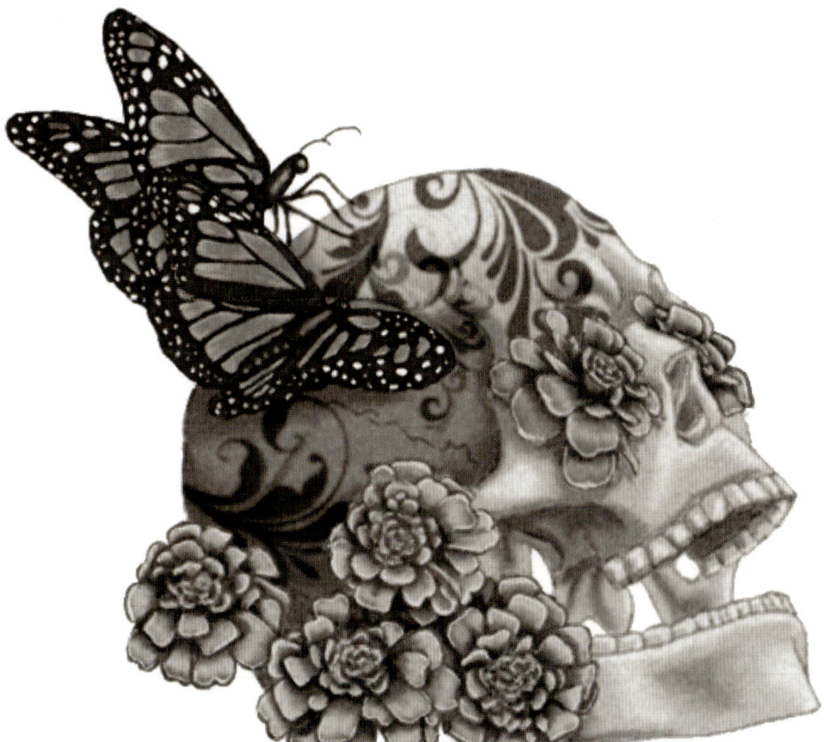

Sugar for the Dead by devonsw

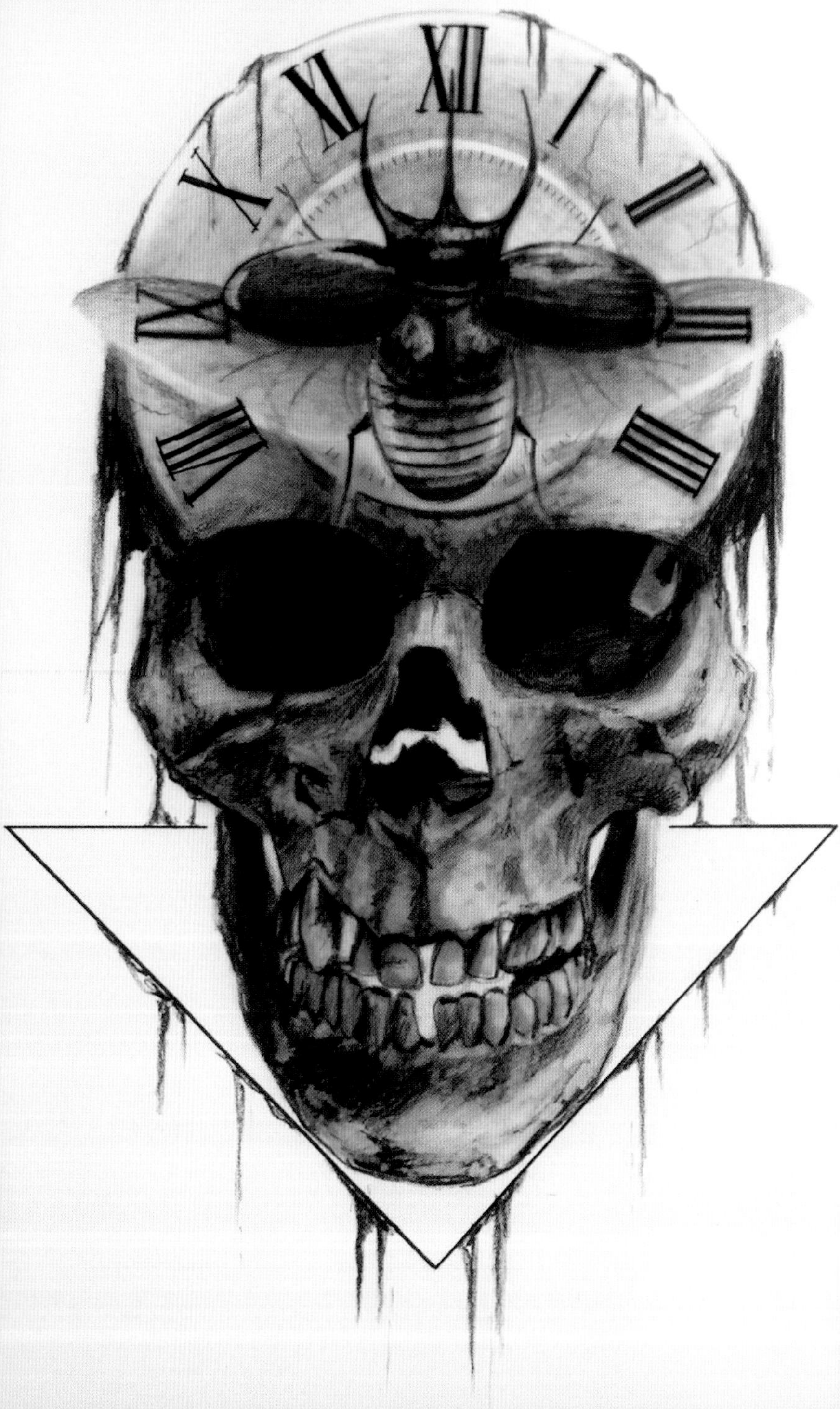

This spread: These skull tattoos were designed by Dino Tomic, a Norwegian tattoo artist.

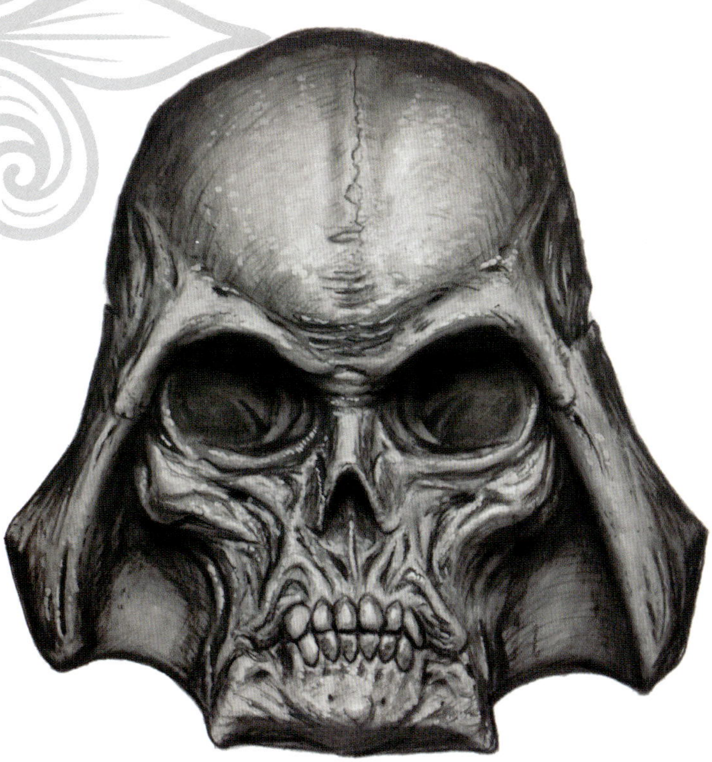

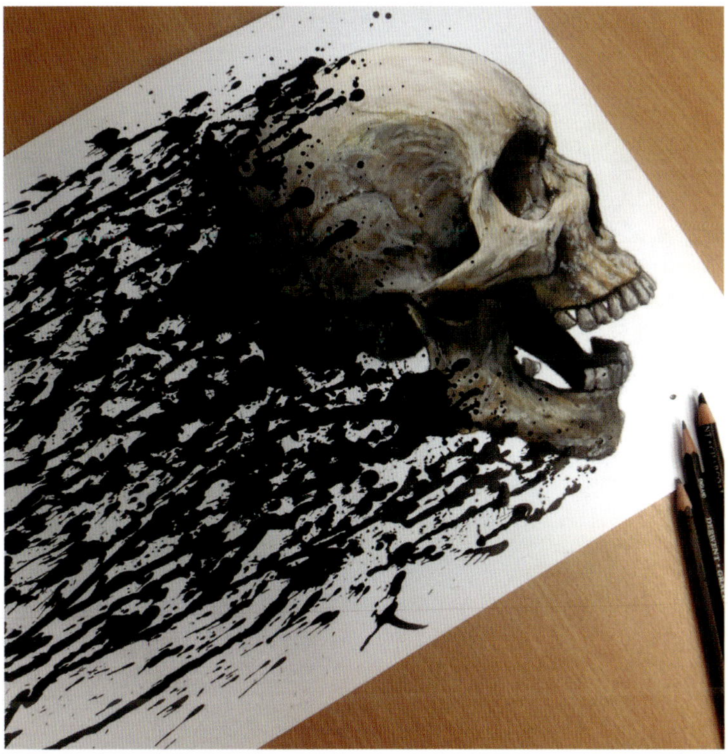

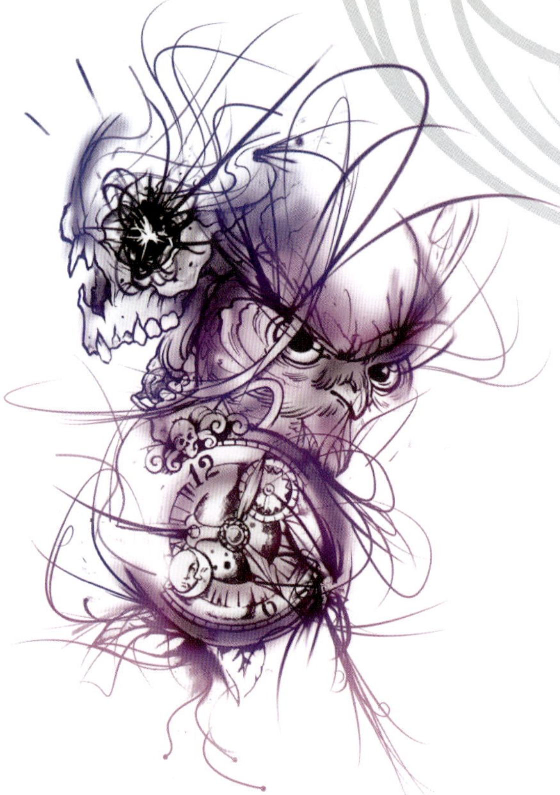

Abstract Watercolor Tattoo with Skull, Owl, and Clock by vmslv

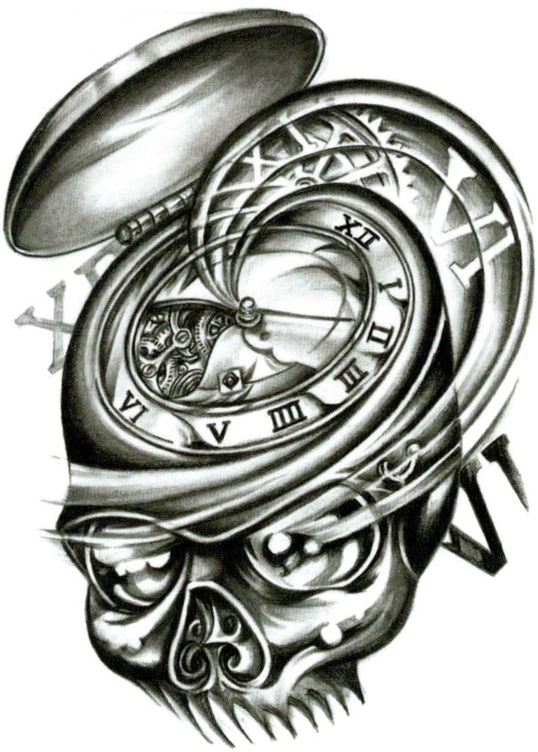

Skull Timepiece by gisty73

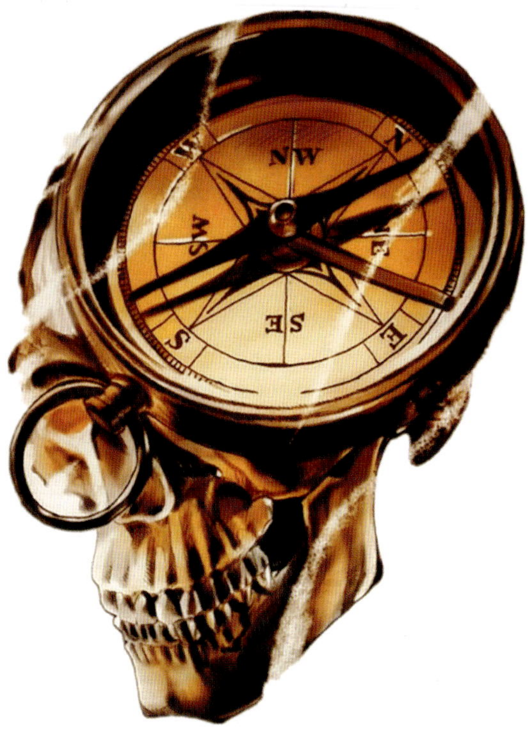

Skull and Compass by Porcelaine

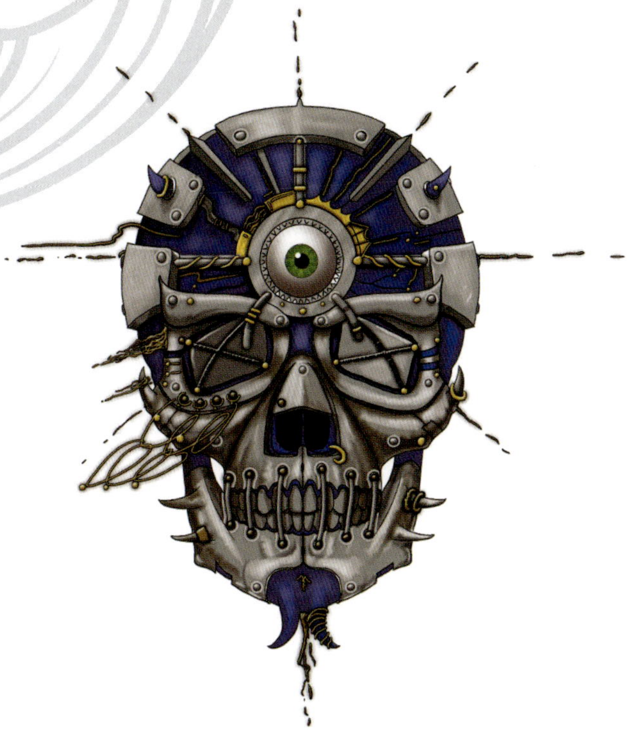

Voodoo Skull by SeanG

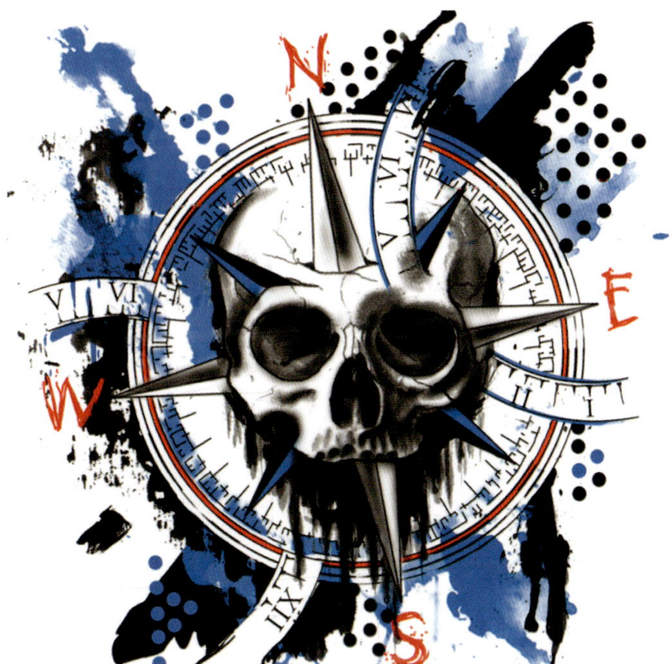

Compass by GrechkaArt

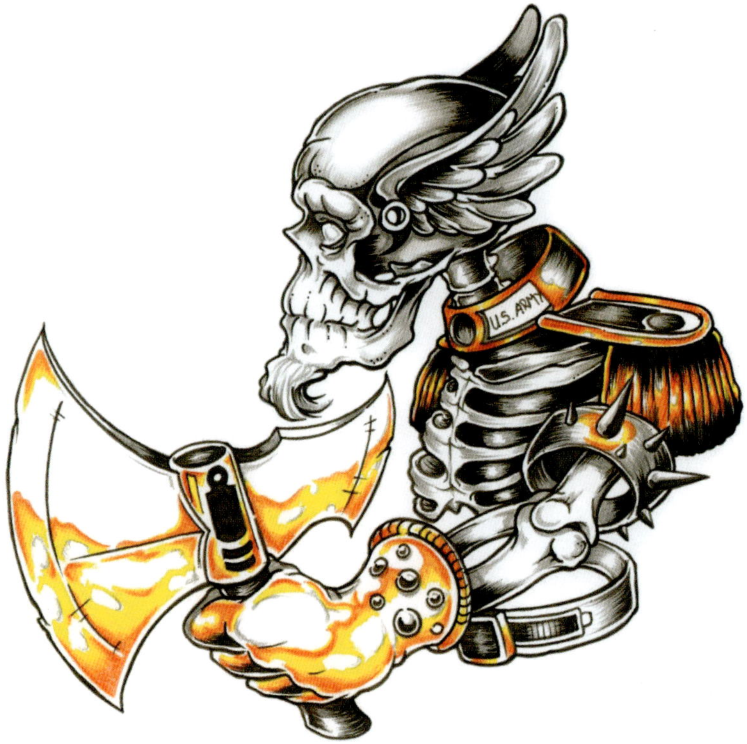

Skeleton Soldier by Tattoodesigne

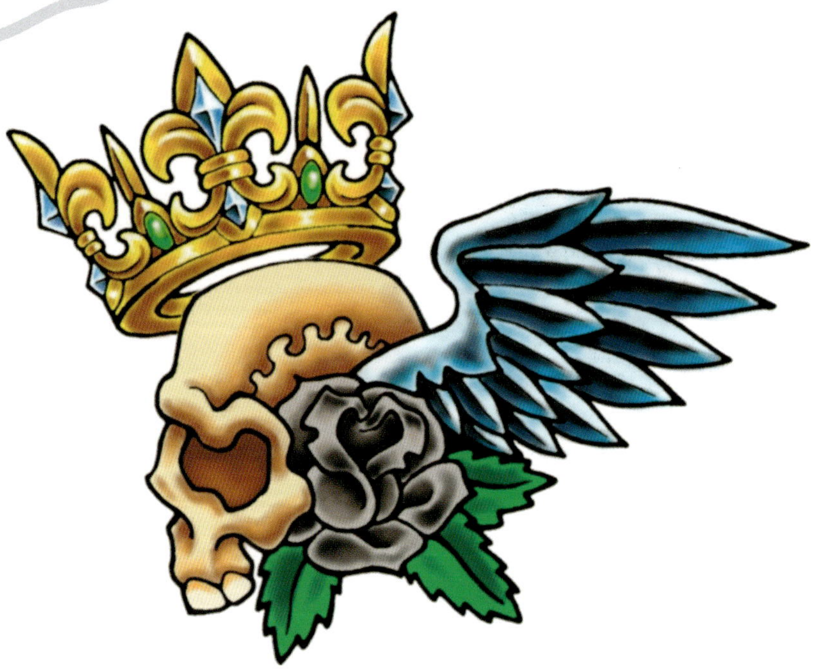

Skull & Rose by PatrickSellers

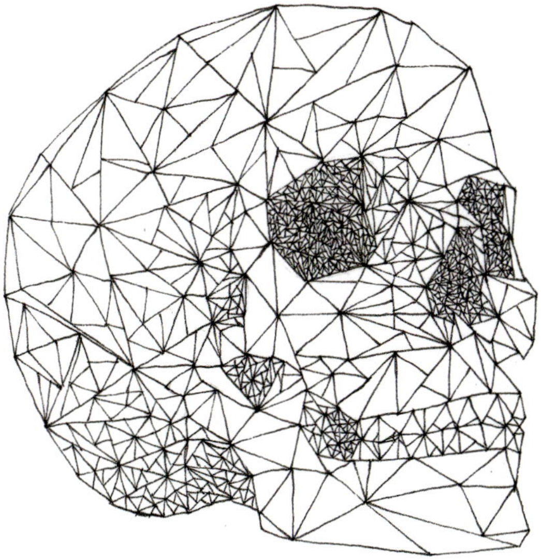

Triangle Skull by carotdraws

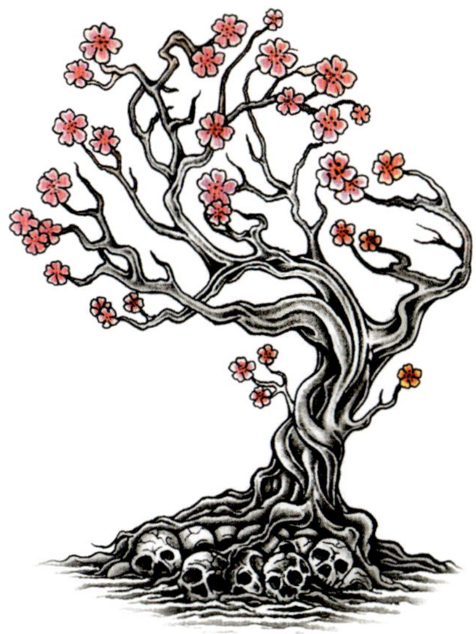

Japanese-Style Cherry Tree of Life and Death by aquili-filius

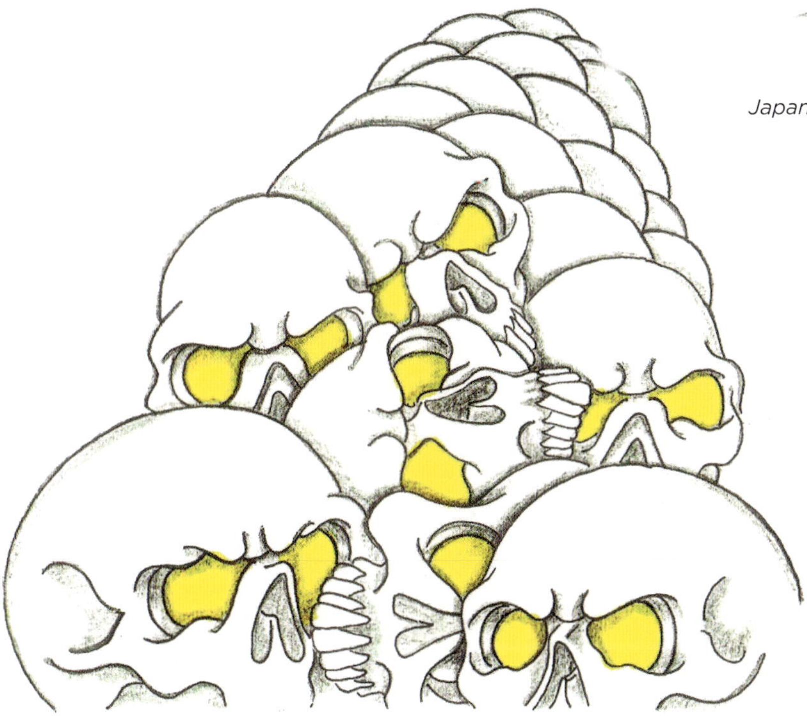

3-D Skull Road by mfrank

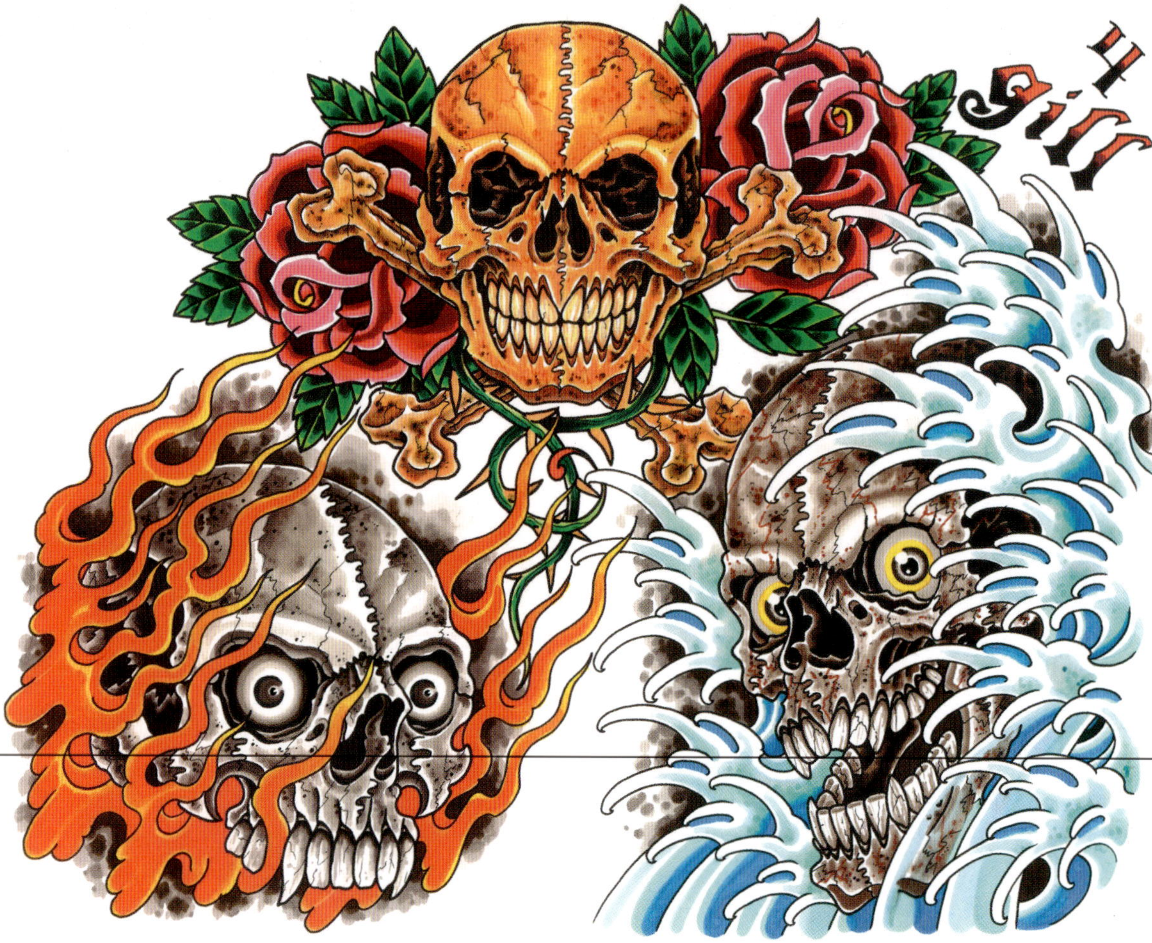

This spread: Henning Jørgensen, a tattoo artist and founder of Royal Tattoo in Helsingør, Denmark, designed these skull tattoos.

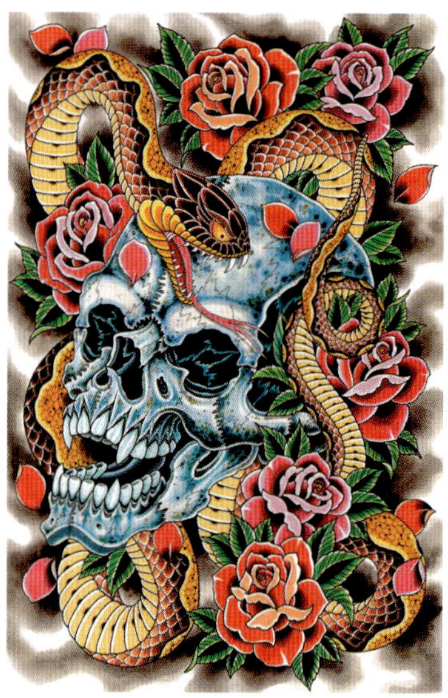

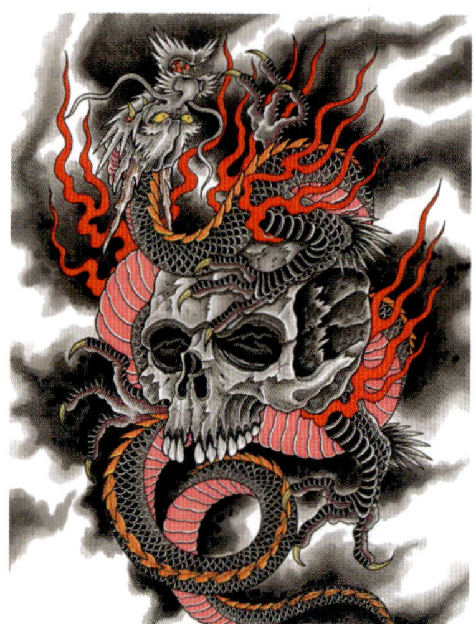

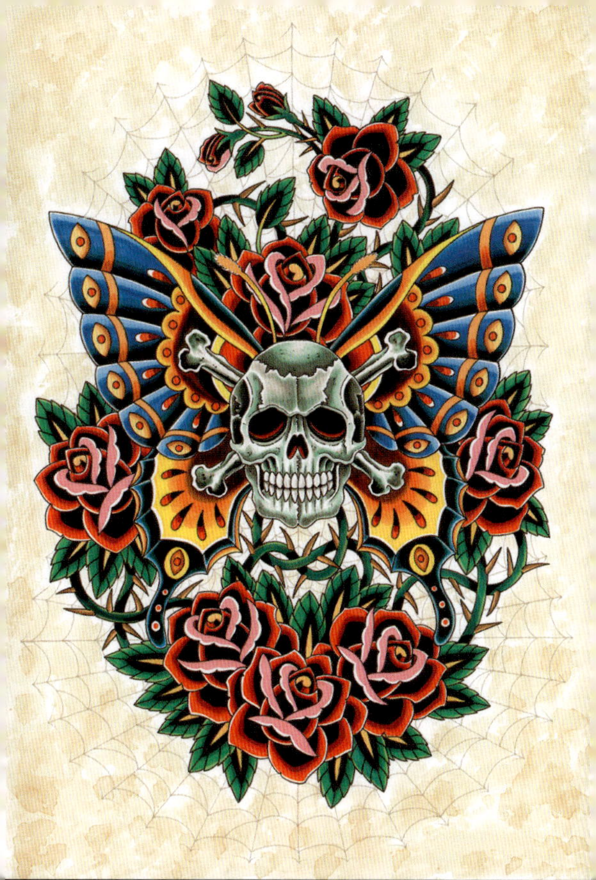

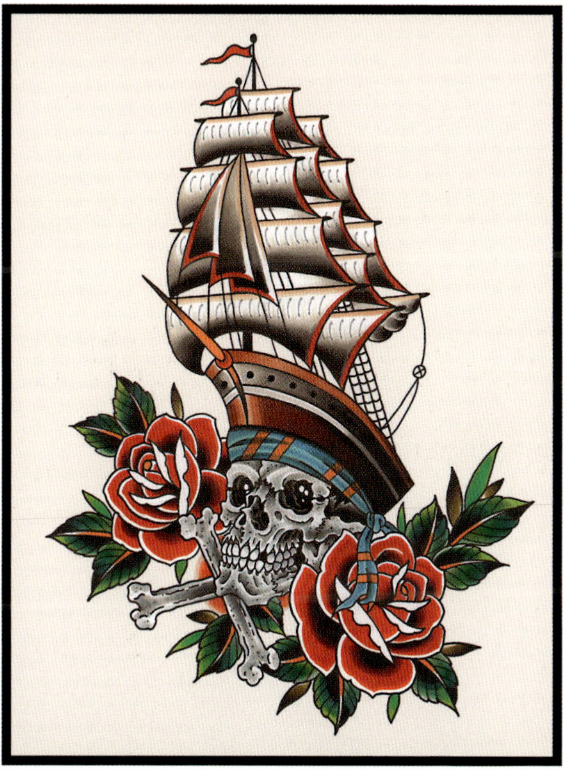

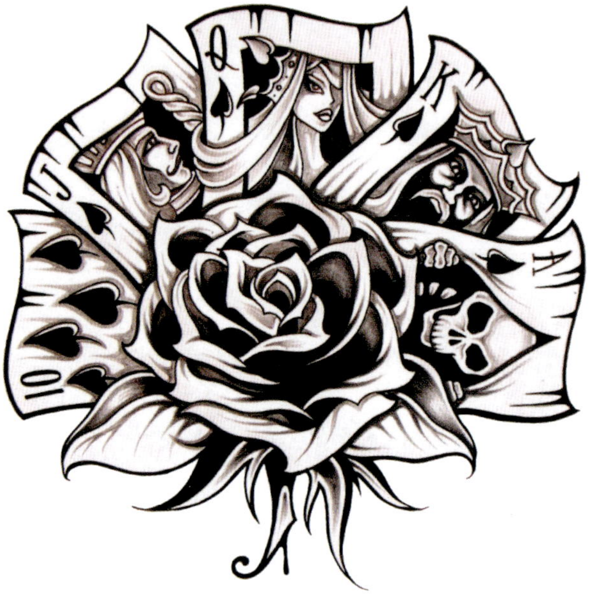

Royal Flush by achemical

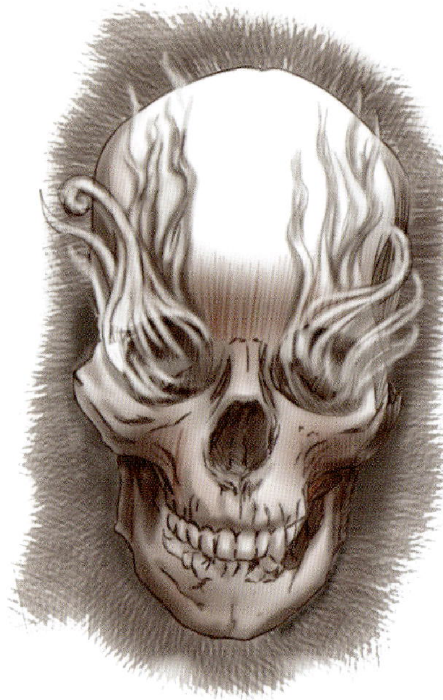

3D Skull by CHeRFe

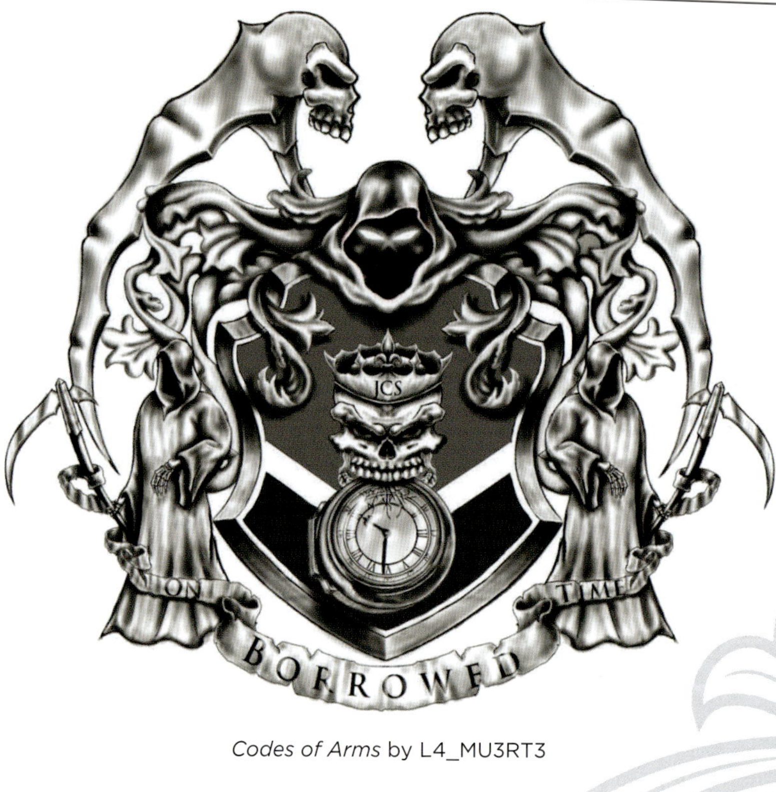

Codes of Arms by L4_MU3RT3

Grim Reaper by bmills

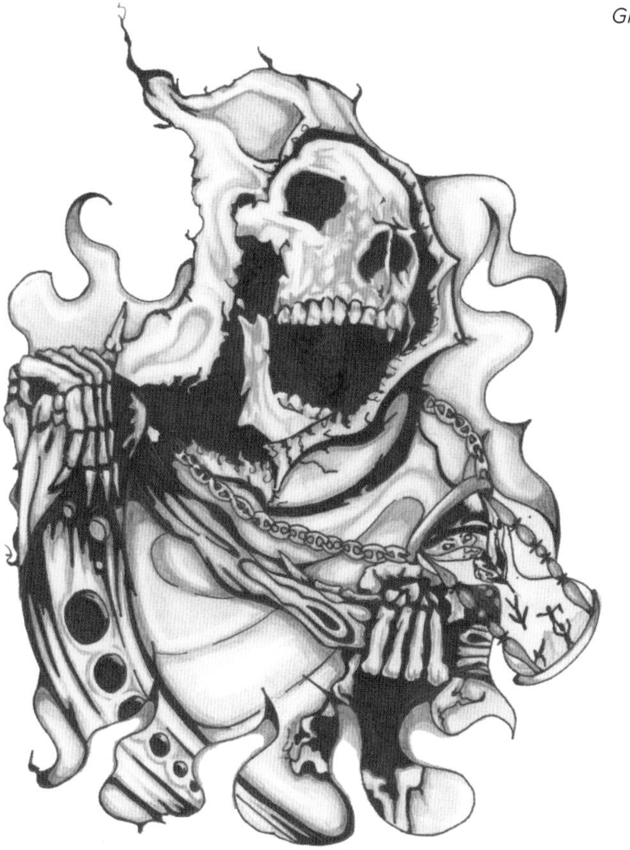

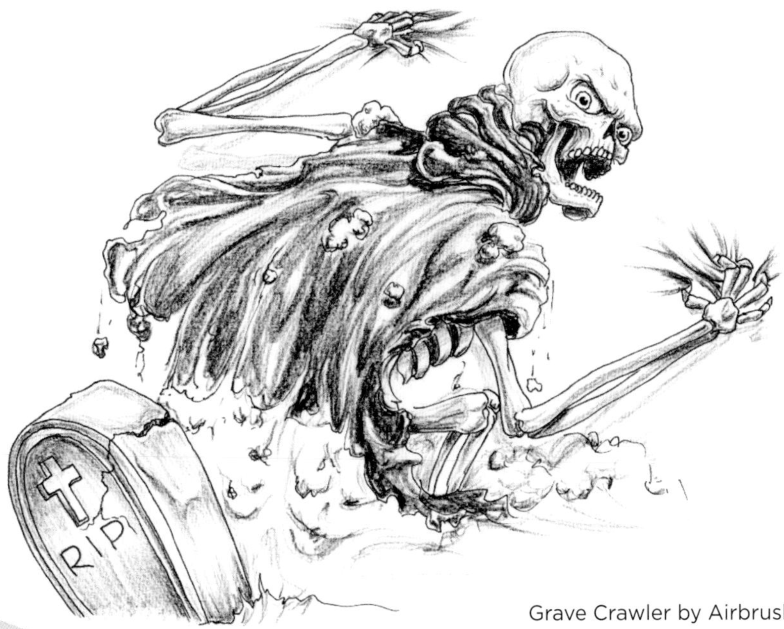

Grave Crawler by AirbrushMike

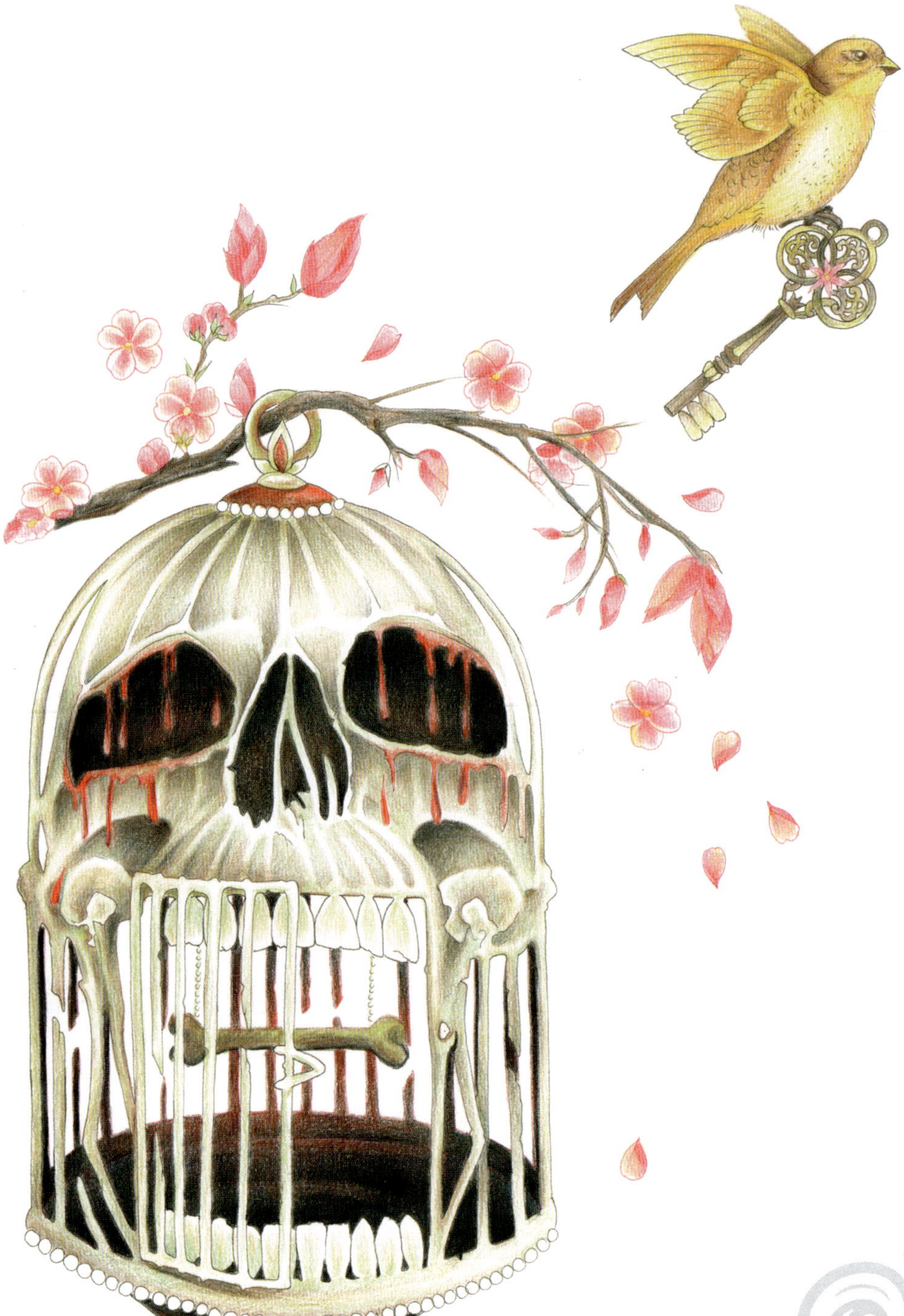

This spread: These tattoos were designed by James Robinson of Gilded Cage Tattoo Studio in Brighton, England.

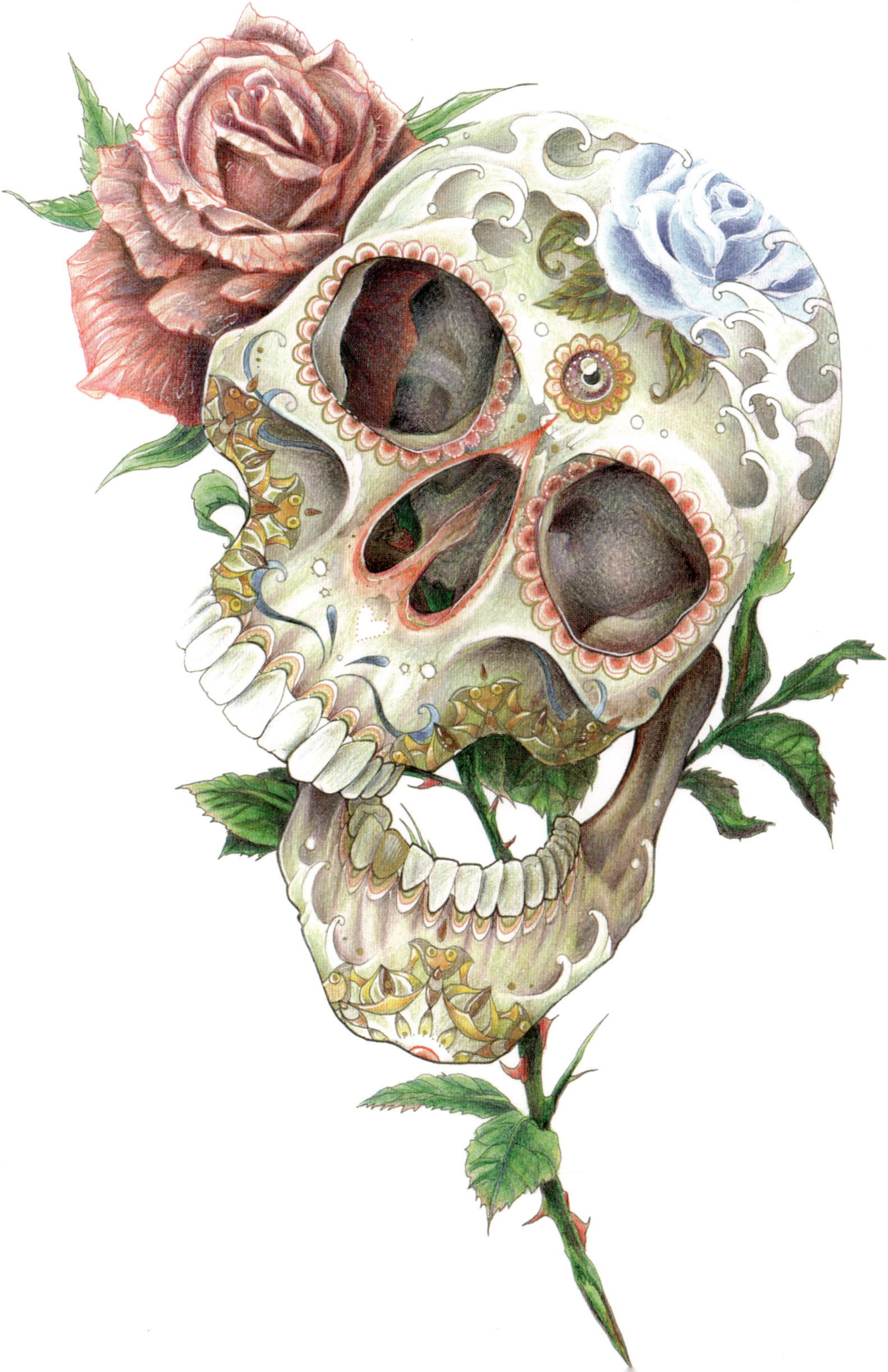

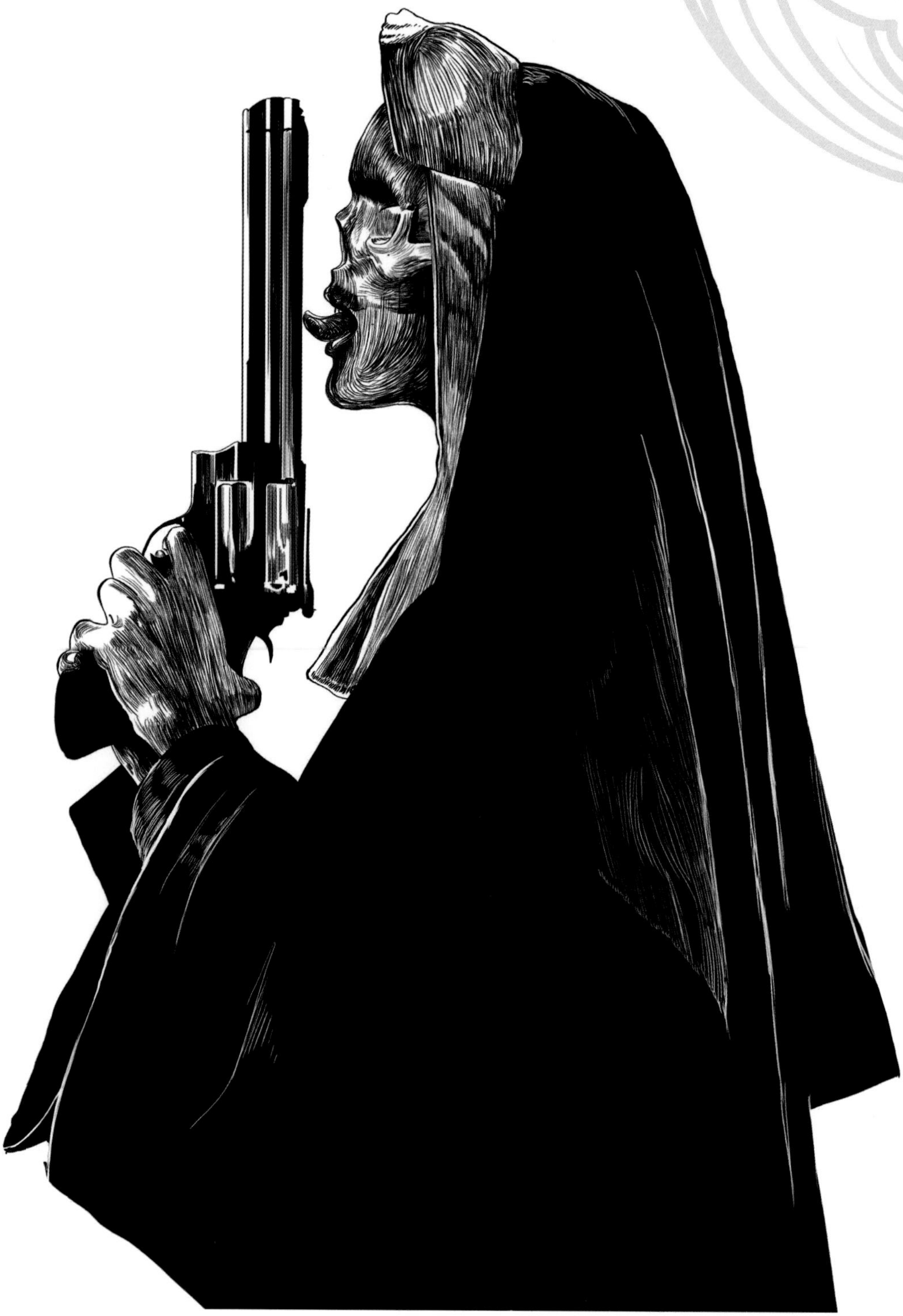

Paris-based artist Julien Lemoine designed this tattoo of a skull-faced figure licking a gun.

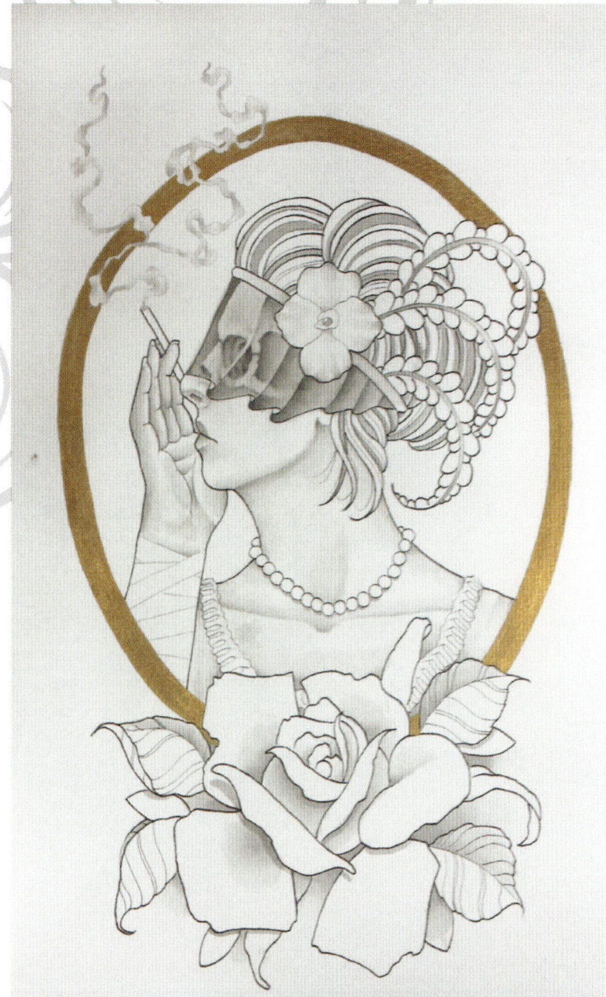

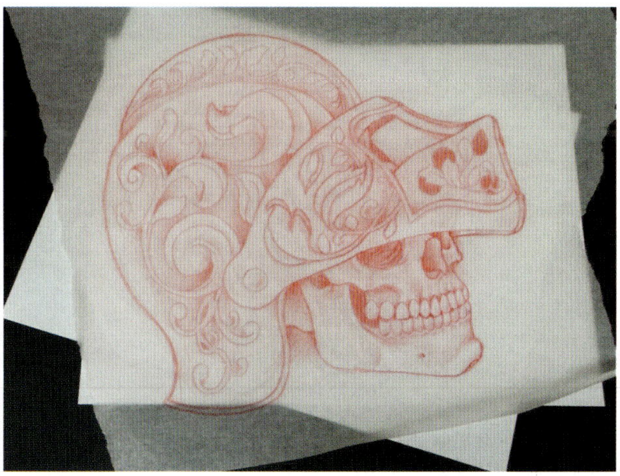

Doug Hand of Northern Liberty Tattoo in Philadelphia, Pennsylvania, designed these tattoos.

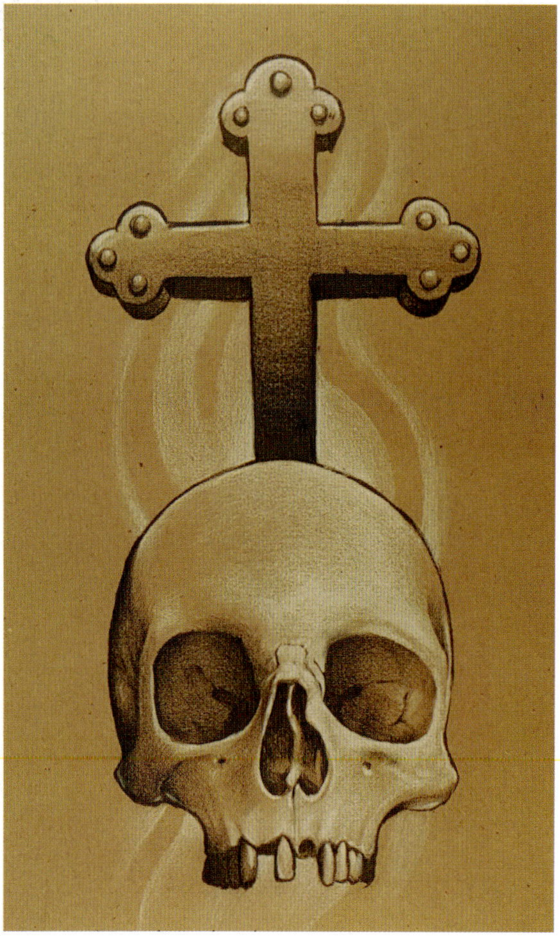

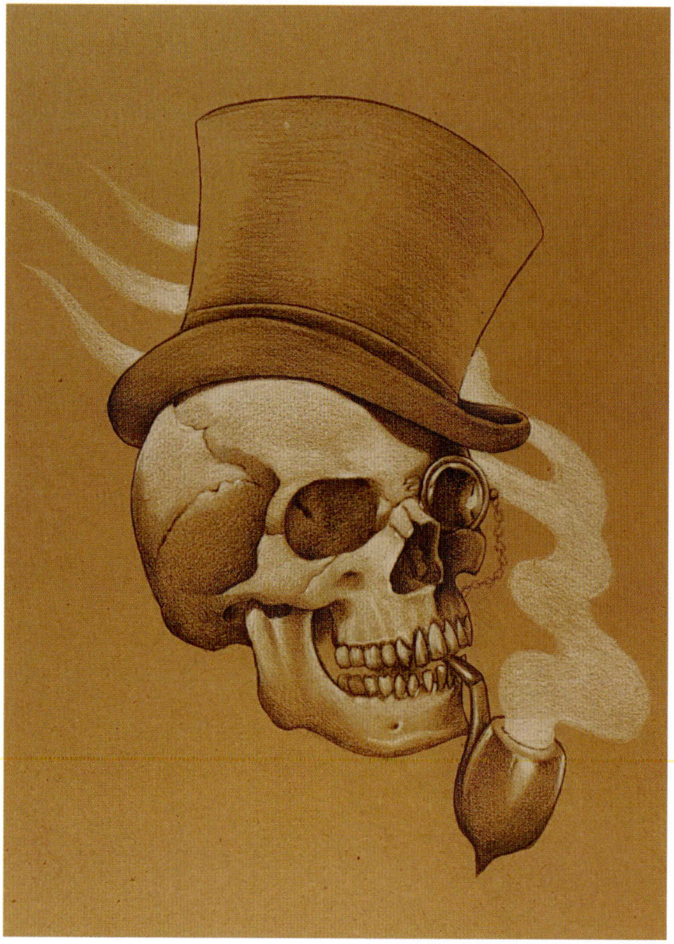

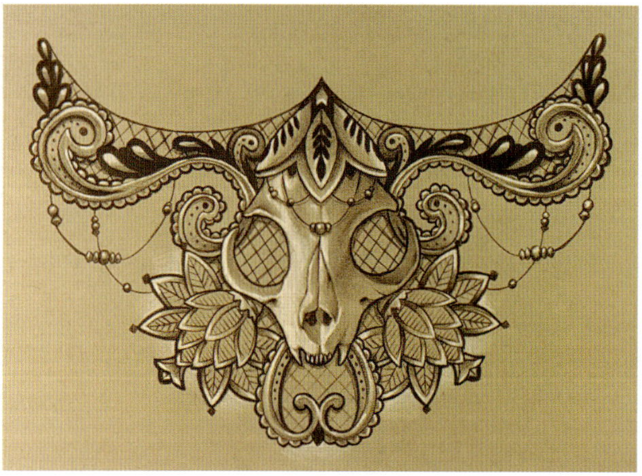

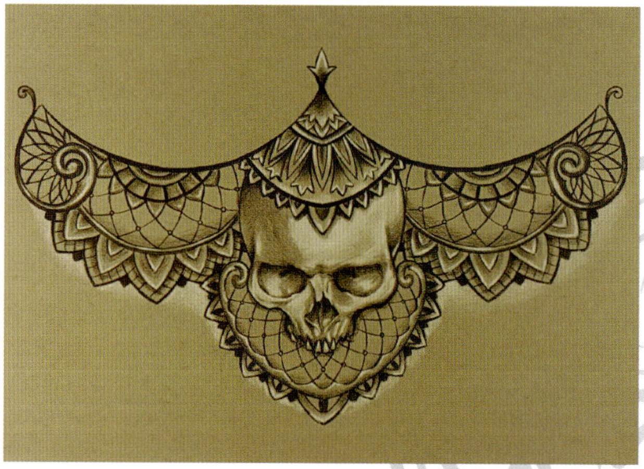

This spread: These skull tattoos were designed by Mario Delgado, of Petaluma, California.

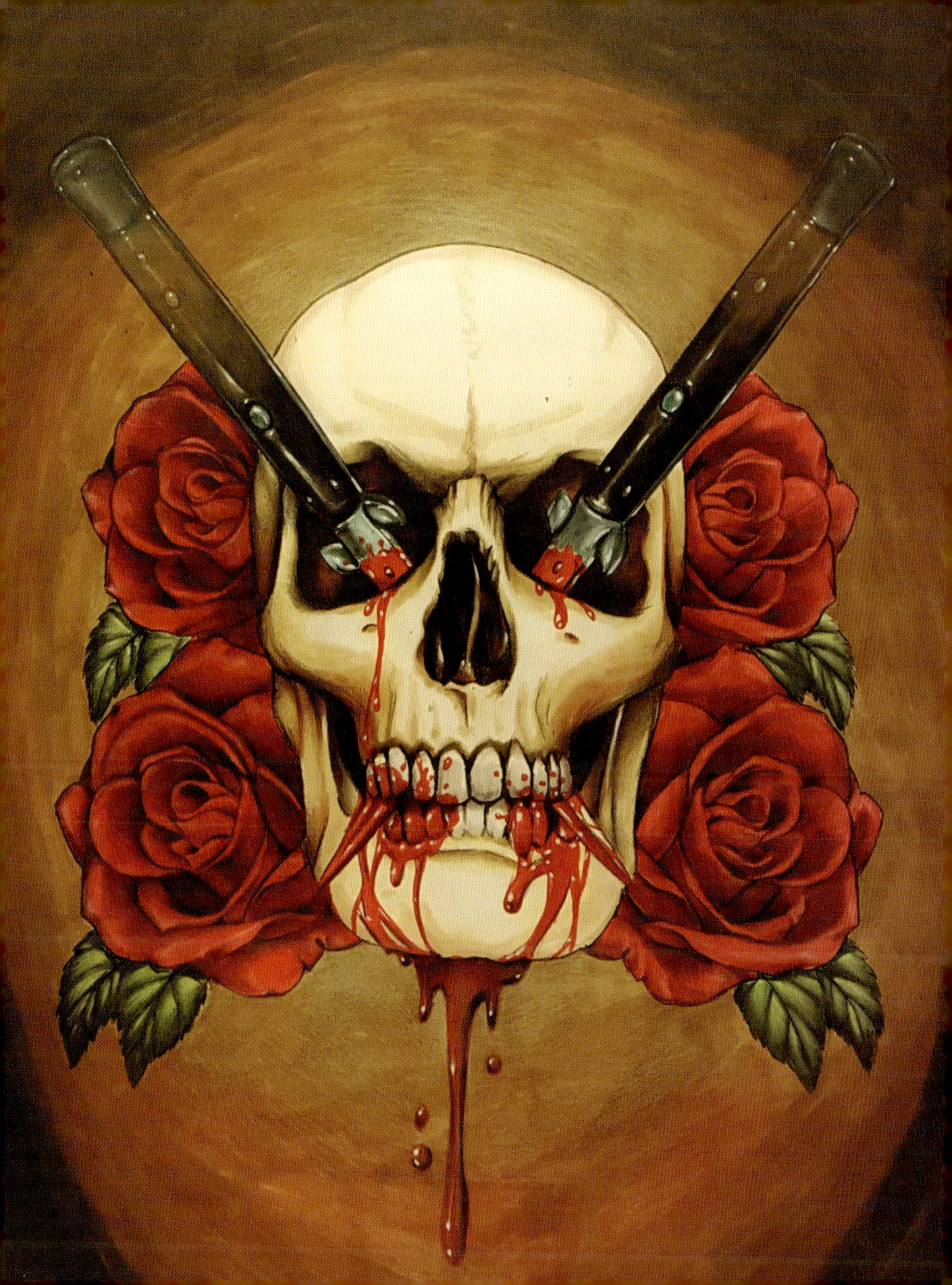

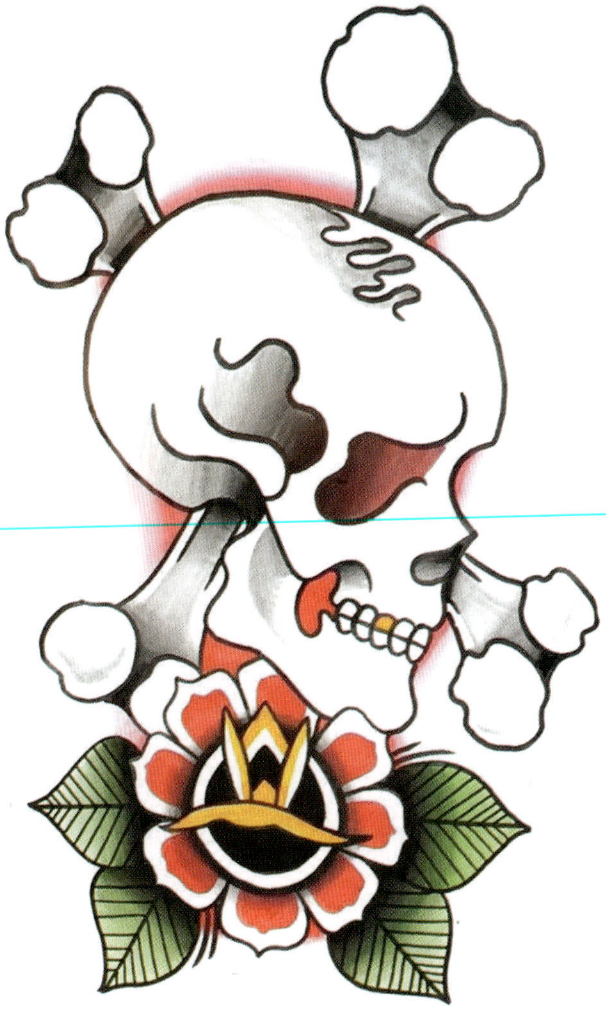

This tattoo was designed by Salvatore Ffiti of Northern Liberty Tattoo in Philadelphia, Pennsylvania.

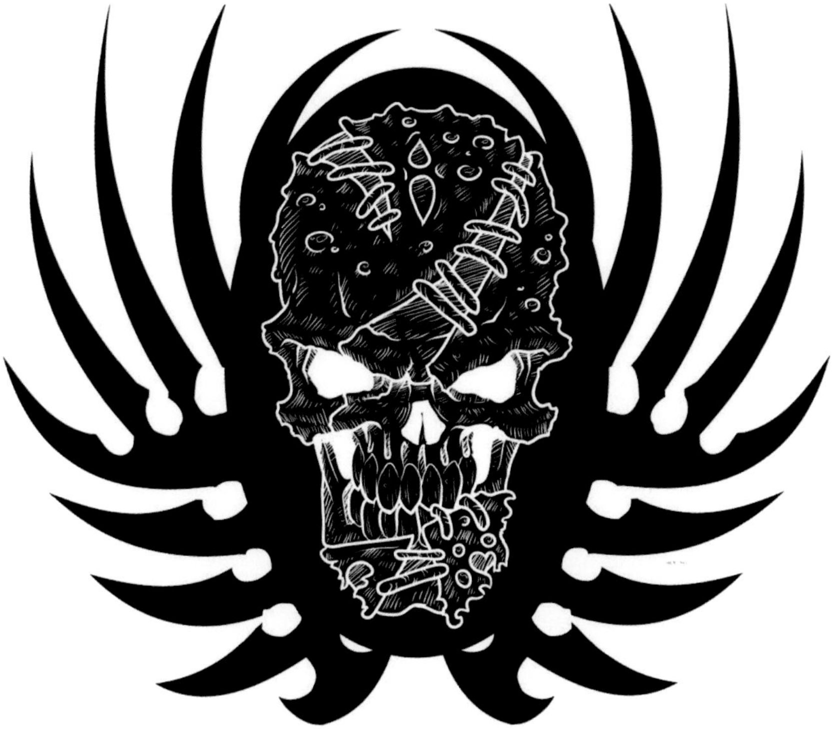

Dash Skull by ribano2

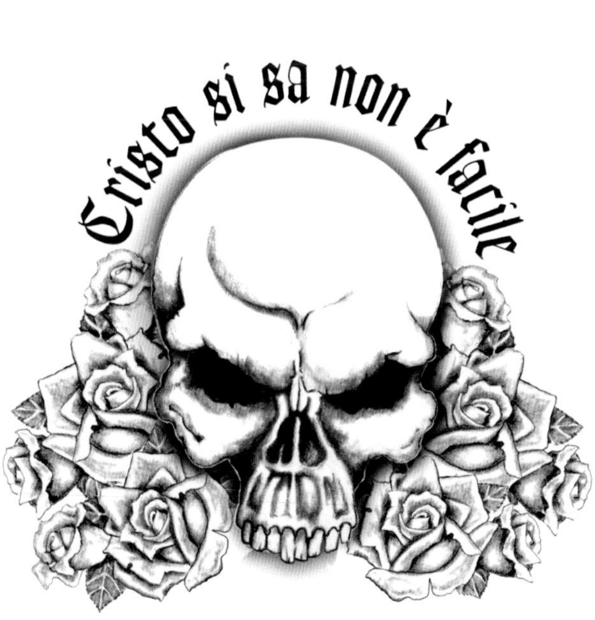

"Christ You Know It Ain't Easy" Skull by Picasso2b

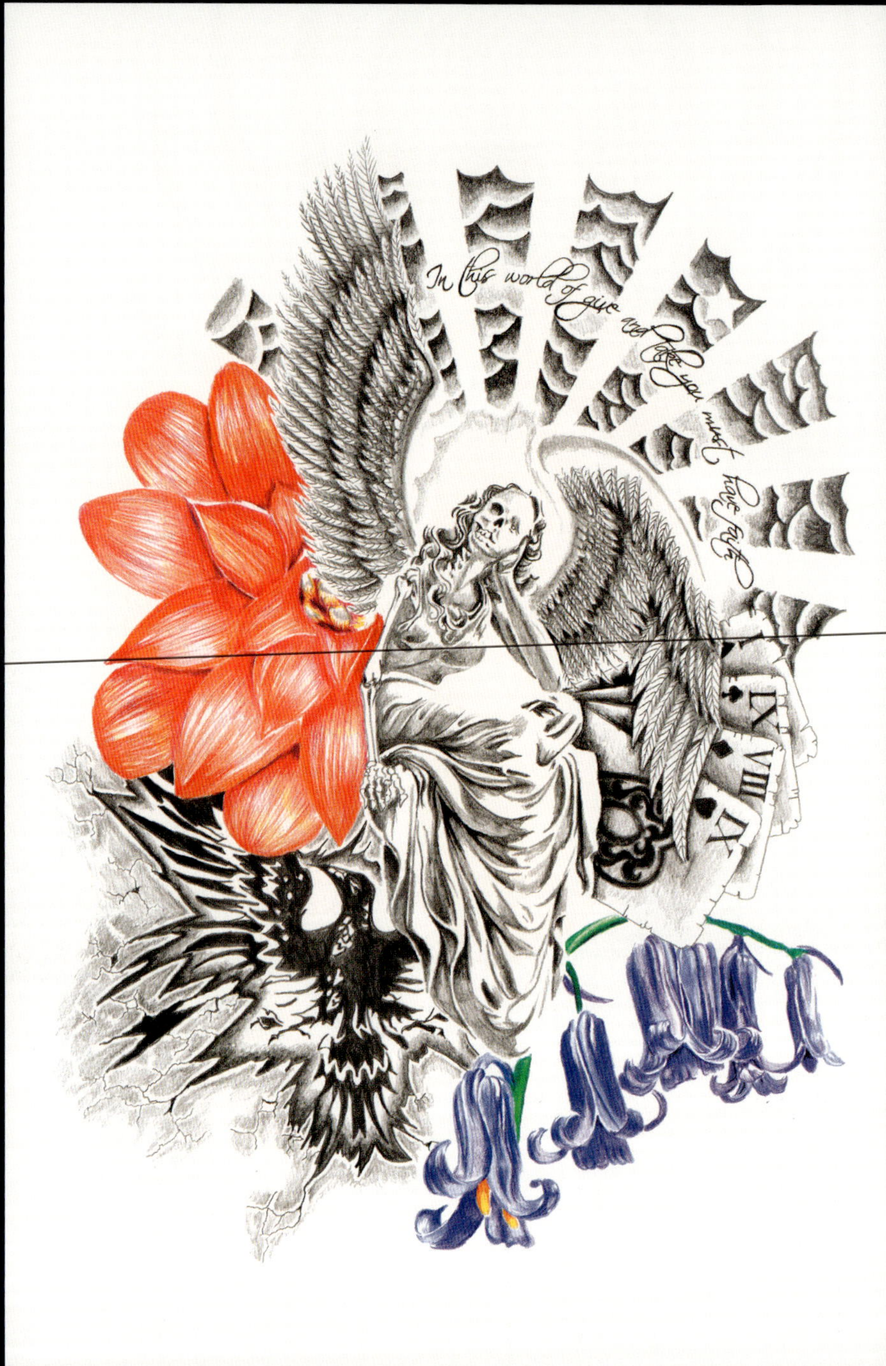

This spread: Alasdair Roy from Dark Design Graphics in Newcastle upon Tyne, England, drew these skull tattoos.

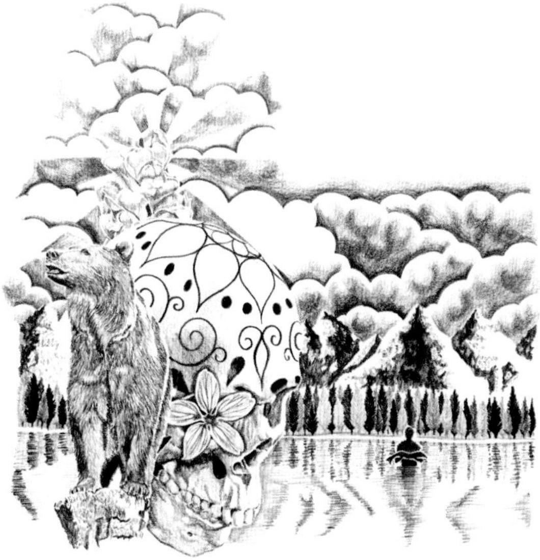

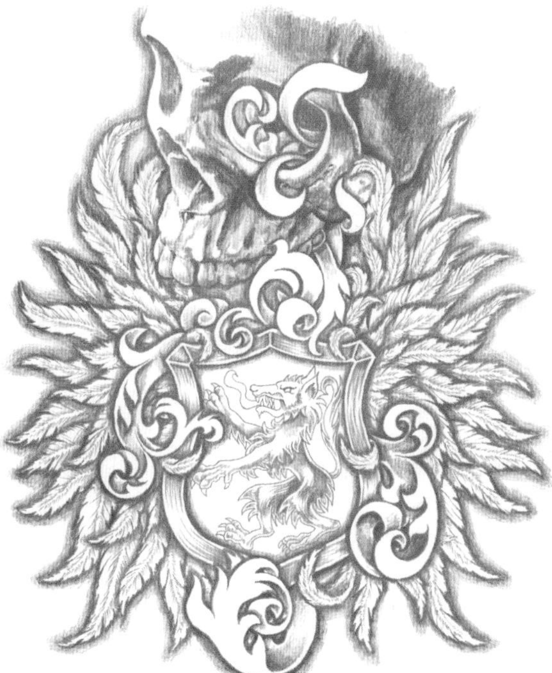

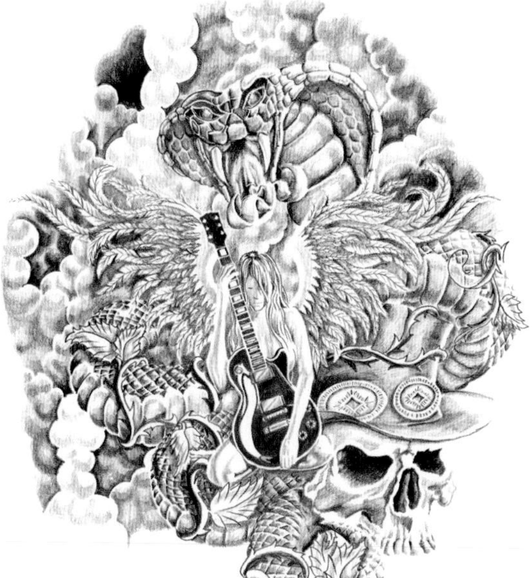

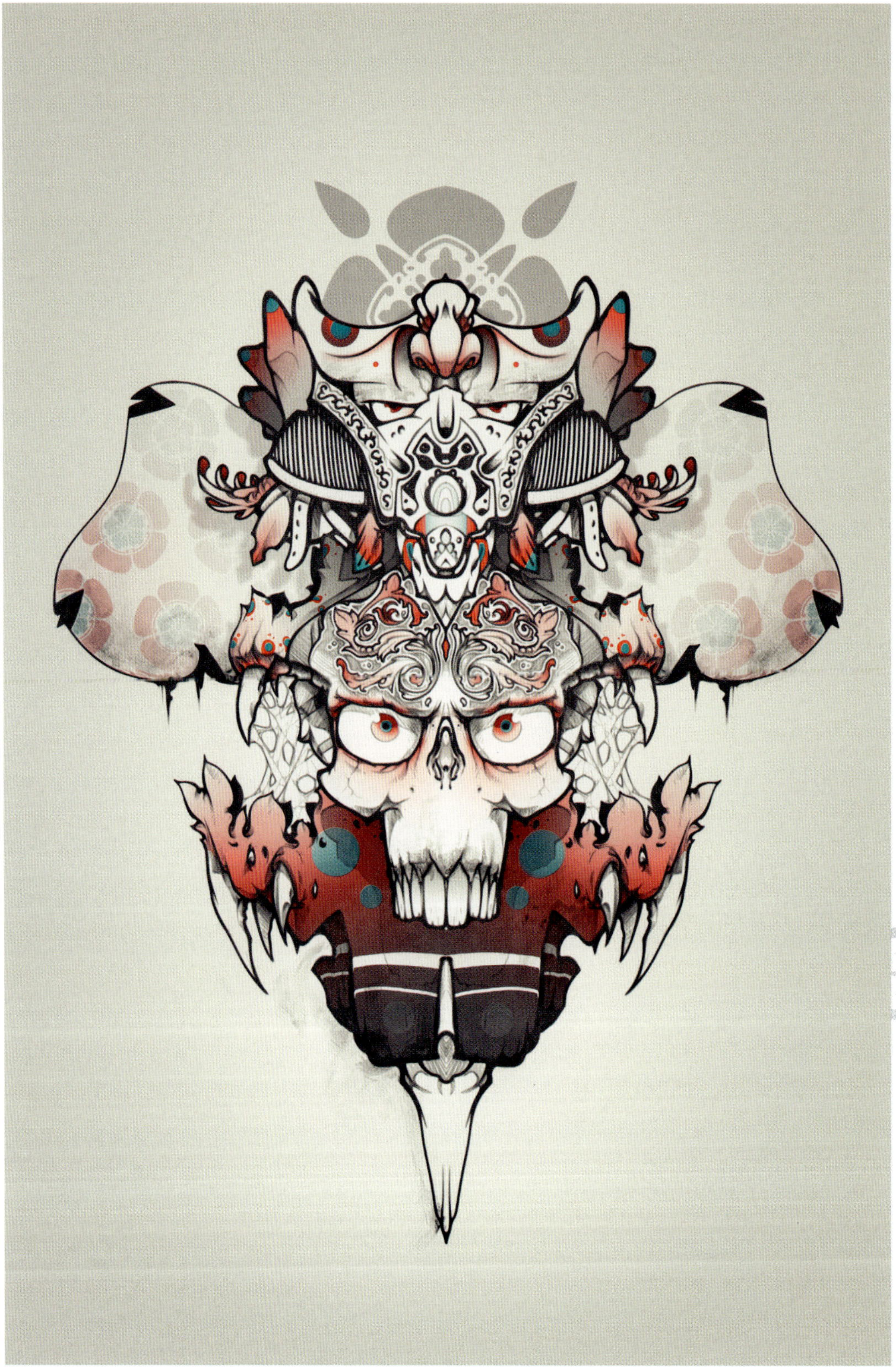

This spread and pages 108–109: These original tattoos were designed by Remi Larroque of Montreal, Canada.

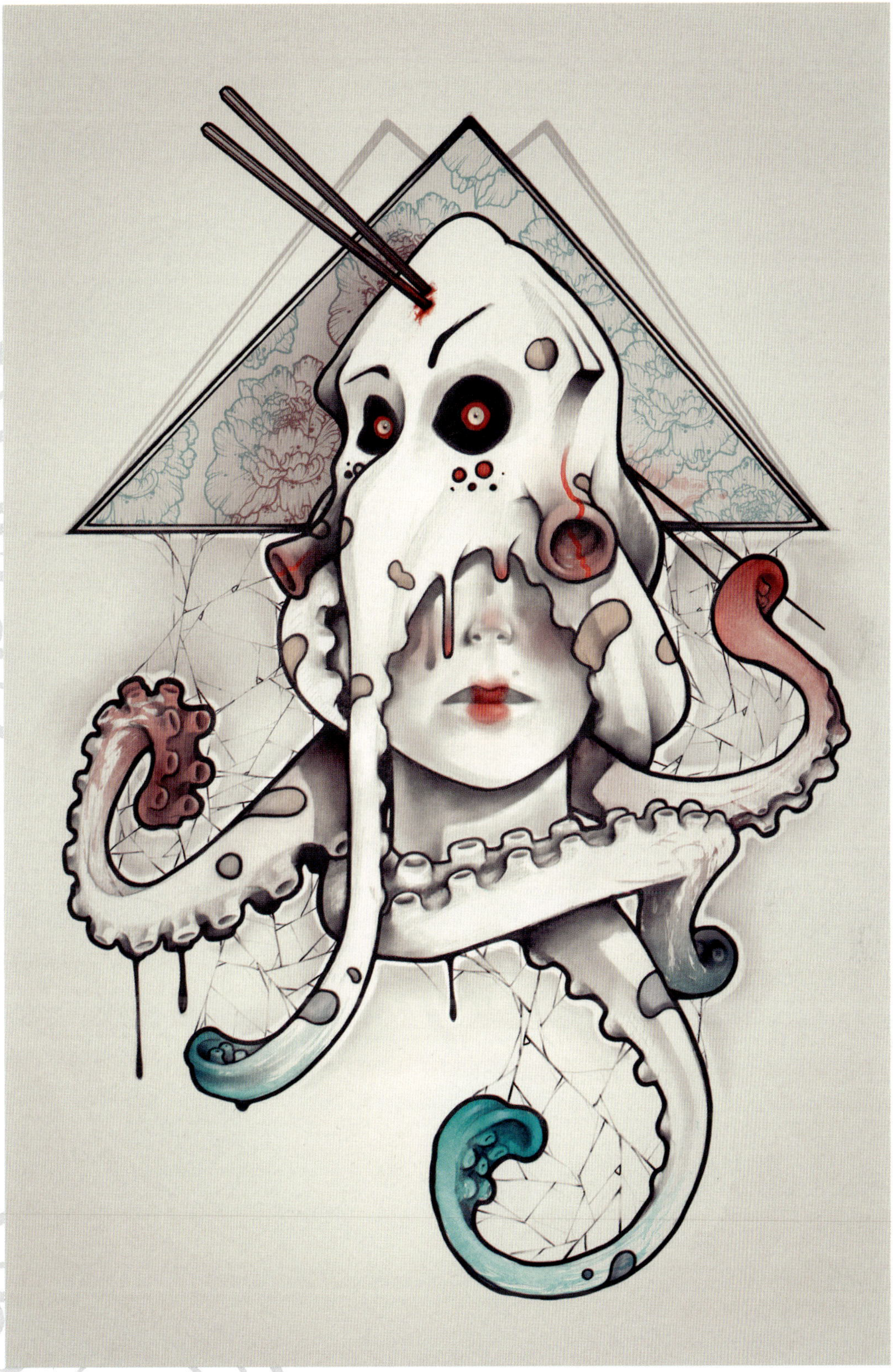

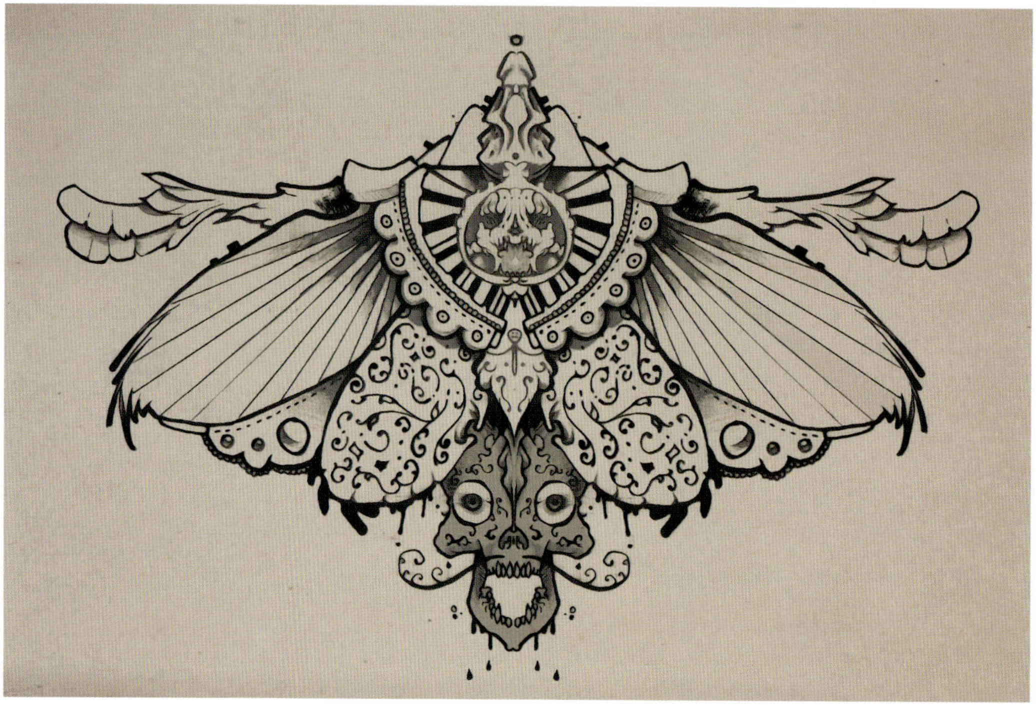

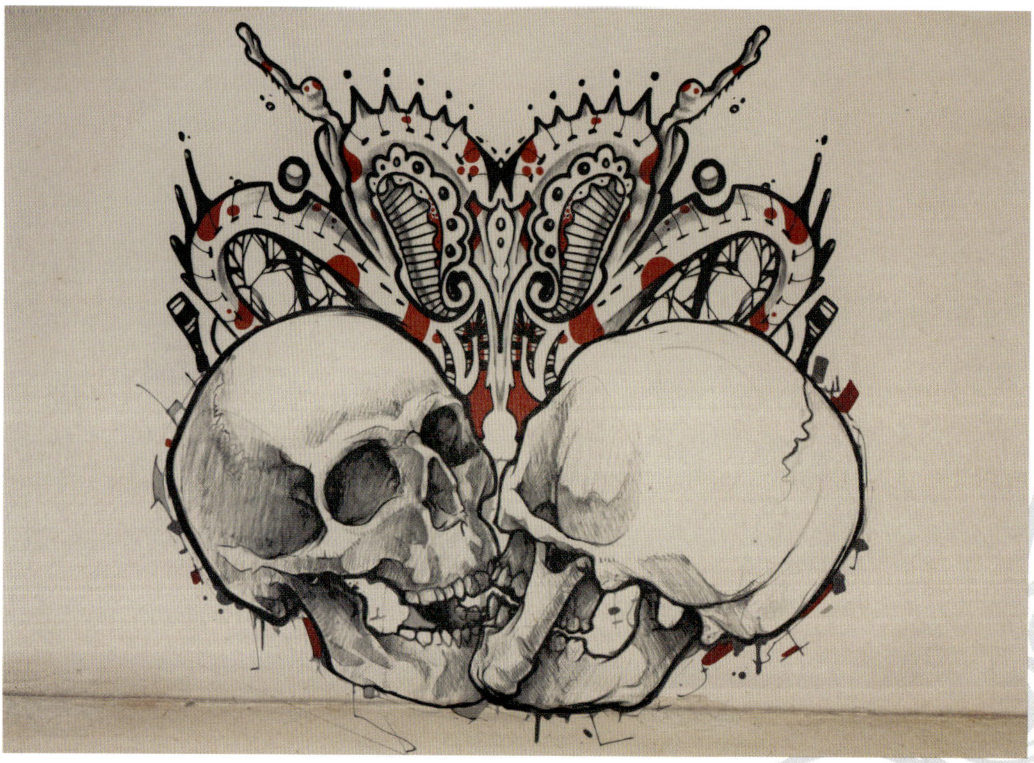

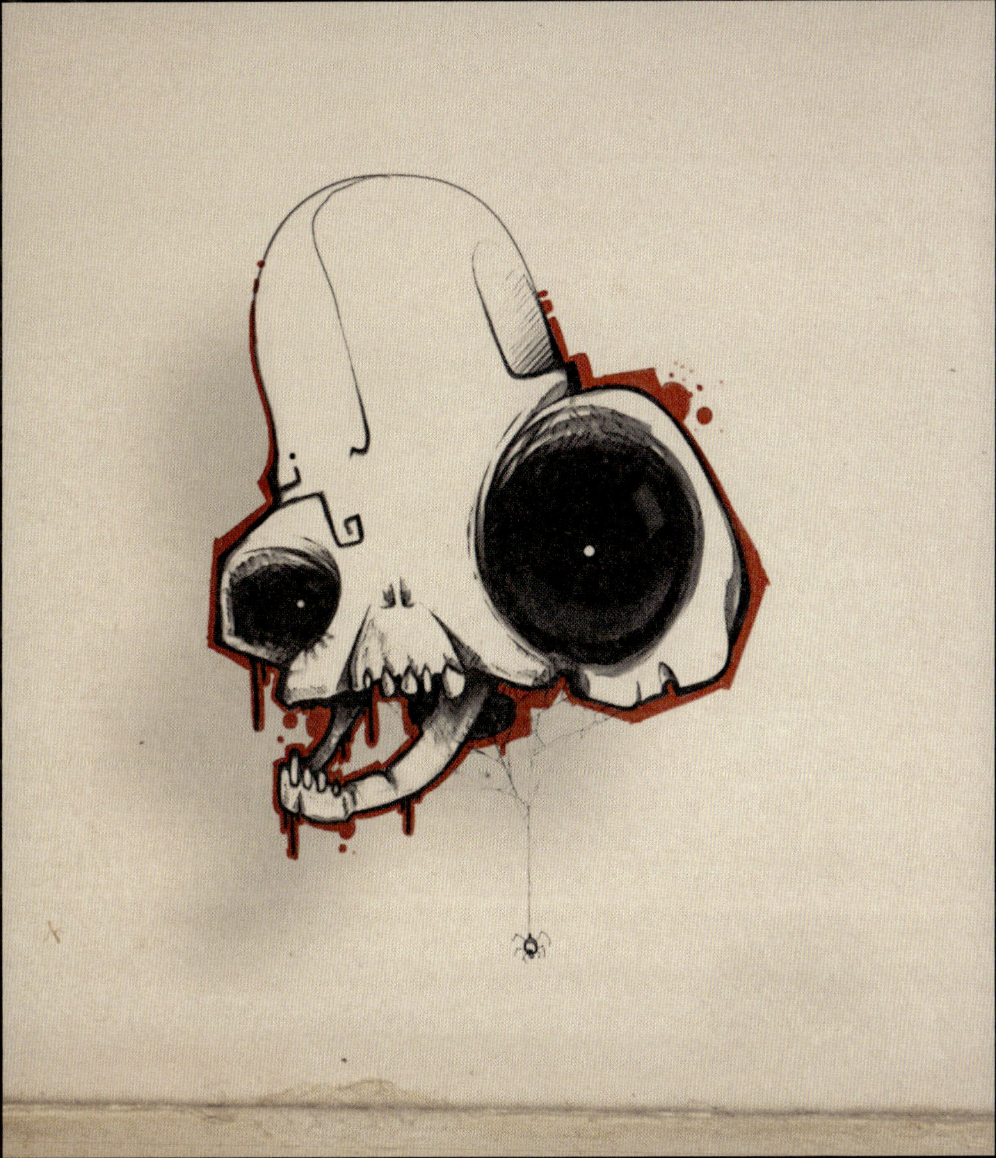

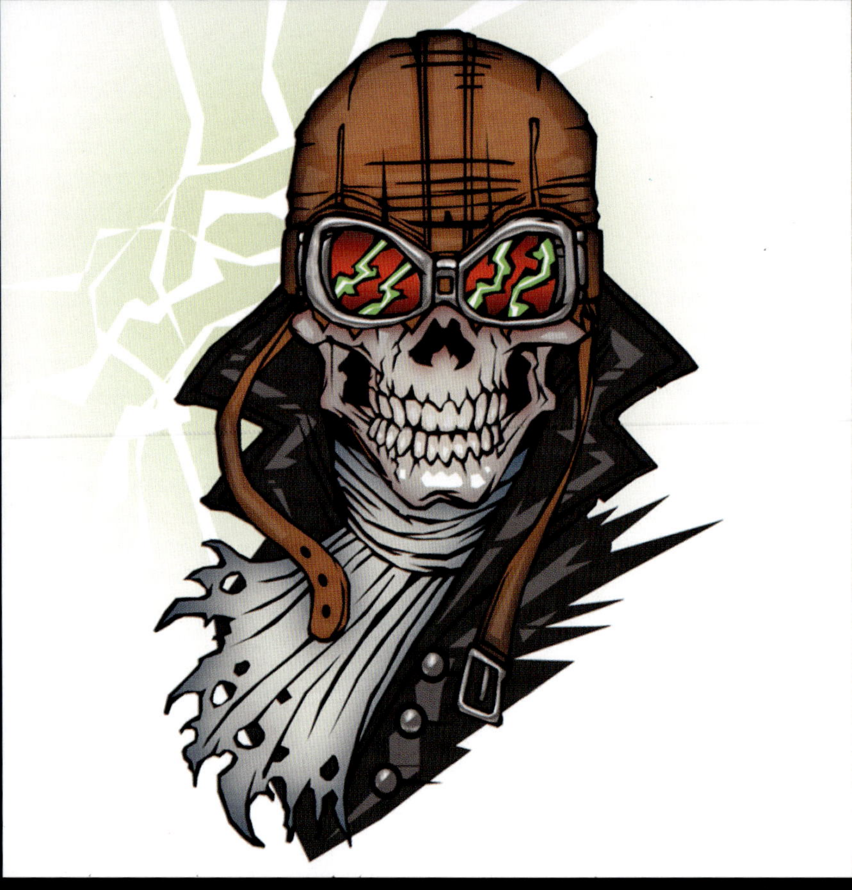

The Dead Baron by munkywrench

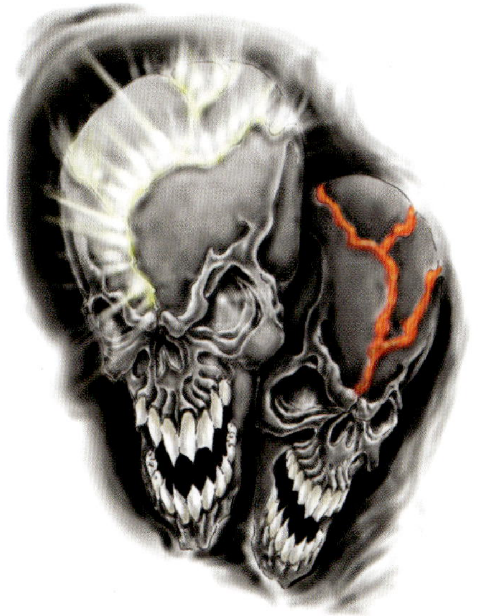

Evil Skulls by mrsinister

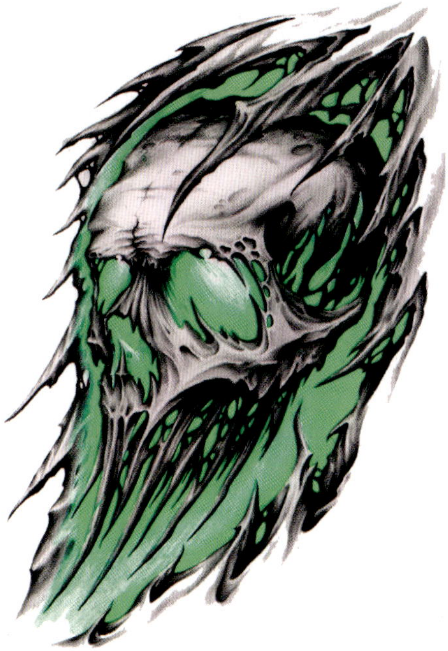

Biomech Skull by WhiteWhale

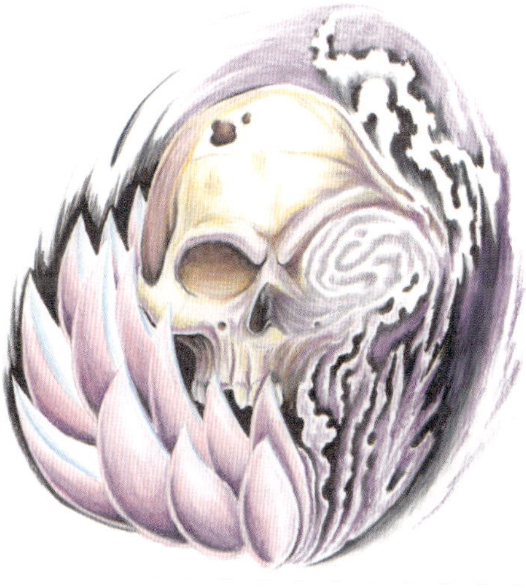

Skull and Lotus by inkaholick

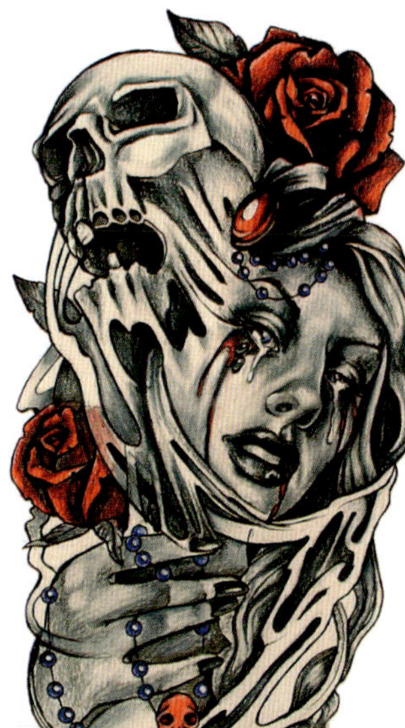

Gothic Girl by eraek

SKULLS IN ART

SKULLS ARE USED IN A SIMILARLY SYMBOLIC WAY IN VANITAS PICTURES. VANITAS IS LATIN FOR "VANITY" (THE ORIGINAL MEANING OF WHICH IS "FUTILITY"). IN AN ARTISTIC SENSE, IT REFERS TO A STYLE OF PAINTING THAT CAME INTO prominence during the 16th and 17th centuries, mainly amongst Northern European artists, particularly those from the Netherlands. Frans Hals' *Young Man Holding a Skull* (painted between 1626–8) is a good example. A handsome, ruddy-faced youth, evidently in excellent health, holds a wax-colored skull in one hand, while gesticulating animatedly with the other. The paintings provide a pictorial statement as to the meaningless and transitory nature of earthly life, with its material possessions and pointless pursuits. The paintings are rich in symbols that represent earthly obsessions, counterpointed by images of death—hence the prominence of the skull in these dramatic works of art.

In Elizabethan England, a particular style of skull—the *Death's Head Skull,* defined by its missing lower jawbone—was popular among rakes, prostitutes, and sexual adventurers of all kinds. Worn as a ring, the Death's Head signaled those in the know as to the nature of the wearer. In polite company, the ring could be swiveled around to show a plain, innocent, band. This is one of the earliest instances of skulls in somewhat mainstream fashion, but it is certainly not the last.

In 1900, French artist Paul Cézanne painted three skulls, dubbed *Still Life, Three Skulls.* Five years later, in 1905, and nearing the end of his life, Cézanne almost prophetically painted another still-life entitled *Pyramid of Skulls.* The skulls in both paintings look remarkably similar and art critics believe that he may have used the same models for each painting.

Mexican José Guadalupe Posada used skull imagery as a way to communicate the importance of art to Mexican culture during the early part of the twentieth century. His most notable works are his engravings of calaveras, or cartoonish images of skeleton heads wearing costumes. These often satirized the life of the Mexican upper class during the time that he was alive.

Artist Otto Dix's images of skulls wearing gas masks pre-empted the Second World War. Dix, a German painter, was an enthusiastic volunteer during the First World War, where he experienced the horrors of war. His etching of *Stormtroopers Advancing Under Gas* was done in 1924, just six years after he was discharged from military service and fifteen years before the outbreak of the Second World War.

Georgia O'Keeffe was an American photographer known for her images of flowers. O'Keeffe also painted images of skulls with flowers, a trend she said started when someone visited her house and she stuck a rose she was holding into a skull's eye socket so she could answer the door. The beauty of the flowers adds a soft note to the otherwise-macabre skull portraits. In 1931, she produced a piece entitled *Horse's Skull with White Rose.*

Spanish surrealist Salvador Dalí found that his unique take on the world suited itself perfectly to skulls. His *Ballerina in a Death's Head*, painted in 1939 is a perfect example of this.

Nancy Glenn-Nieto, an American-Mexican actor and artist, is a major influence in Skull Art. In 1985, for the first Day of the Dead after the death of her husband, artist Rodolfo Nieto, she started work on a series of paintings featuring skulls, for which she has become renowned.

British artist George Ioannou started depicting skulls in 2013. His *Dead Famous* collection of work gives strips celebrities' heads to the bone. The pop-artist depicted thirteen popular stars in the collection, including Madonna, Audrey Hepburn, and Amy Winehouse.

Renowned British artist Damien Hirst famously studded an adult human skull with diamonds. Created in 2007, *For the Love of God* used £50 million worth of diamonds to cover the skull. The diamond-studded skull demonstrated the dichotomy of riches and mortality, providing a poignant modern illustration for the vanitas idea that wealth is meaningless beyond the grave. *For Heaven's Sake* is a 2008 repeat of the idea, except this time, Hirst used a baby's skull as a background for over 8,000 pink and white diamonds. As you can imagine, *For the Love of God* and *For Heaven's Sake* were met with a great deal of controversy, both for the use of human skulls and for the value of the jewels he used. However, Hirst's modern skull art emulated the art of turquoise-encrusted pre-Columbian American skulls that archaeologists have discovered.

The current love for skulls as a decorative motif is due in no small part to the late British designer, Alexander McQueen. Known for his innovative and original designs, McQueen initially used skull images on scarves—an idea which rapidly caught on. It made a swift transition from catwalk to sidewalk and even market stalls. He followed the skull scarves with elaborately designed hand-knits in the cable-knit style, using large skulls to adorn what looked to be traditional knitwear. Since McQueen's early skull experiments, the floodgates of skulls in fashion have opened, with skulls appearing on tee shirts, bags, belts and buckles, boots, bags, baby clothes, fabric prints, wallpaper, bedclothes, and other home furnishings.

A European furniture design company, Acrila, has taken the skull-motif and made elegant, baroque-style chairs. Acrila's chairs feature large, graphically-designed skulls on the backrest. The chairs are constructed from acrylic and are named the "Let's Rock" chair, a nod to the use of the skull imagery amongst the Hells Angels and rock bands.

British chocolatiers and cake makers Choccywoccydoodah construct elaborate wedding cakes with skulls artfully rendered in fondant. One such gothic masterpiece, resplendent with a stack of pale skulls and black roses, features the phrase "Till Death do us Part." Their elaborate designs, combining chocolate, cake, and skulls, are an original and tasty take on old motifs.

Noah Scalin, an American artist, is known for his "Skull-A-Day" project, in which he presents one skull image a day to the world. One of those images, dubbed Dead Media 2, shows an anamorphic skull created from used books. The image of the skull is apparent only when viewed from one specific angle. In this, Scalin was emulating the anamorphic techniques of "hidden skull" artists, such as Holbein's The Ambassadors, except in a three-dimensional format.

Julien Lemoine is a French graphic designer. He has a vast portfolio of work, much of which features skulls. Lemoine is known for highly detailed skulls, typically drawn in black and white. Some of his art portrays instantly recognizable people, such as Audrey Hepburn.

Russian artist Sasha Vinogradova uses skulls as a canvass for a collection of traditional folk art paintings. Vinogradova describes the project as a fusion of Russian and Mexican artwork through ornately designed Day of the Dead-themed skulls. The paint sometimes spills away from the skull onto the surrounding area.

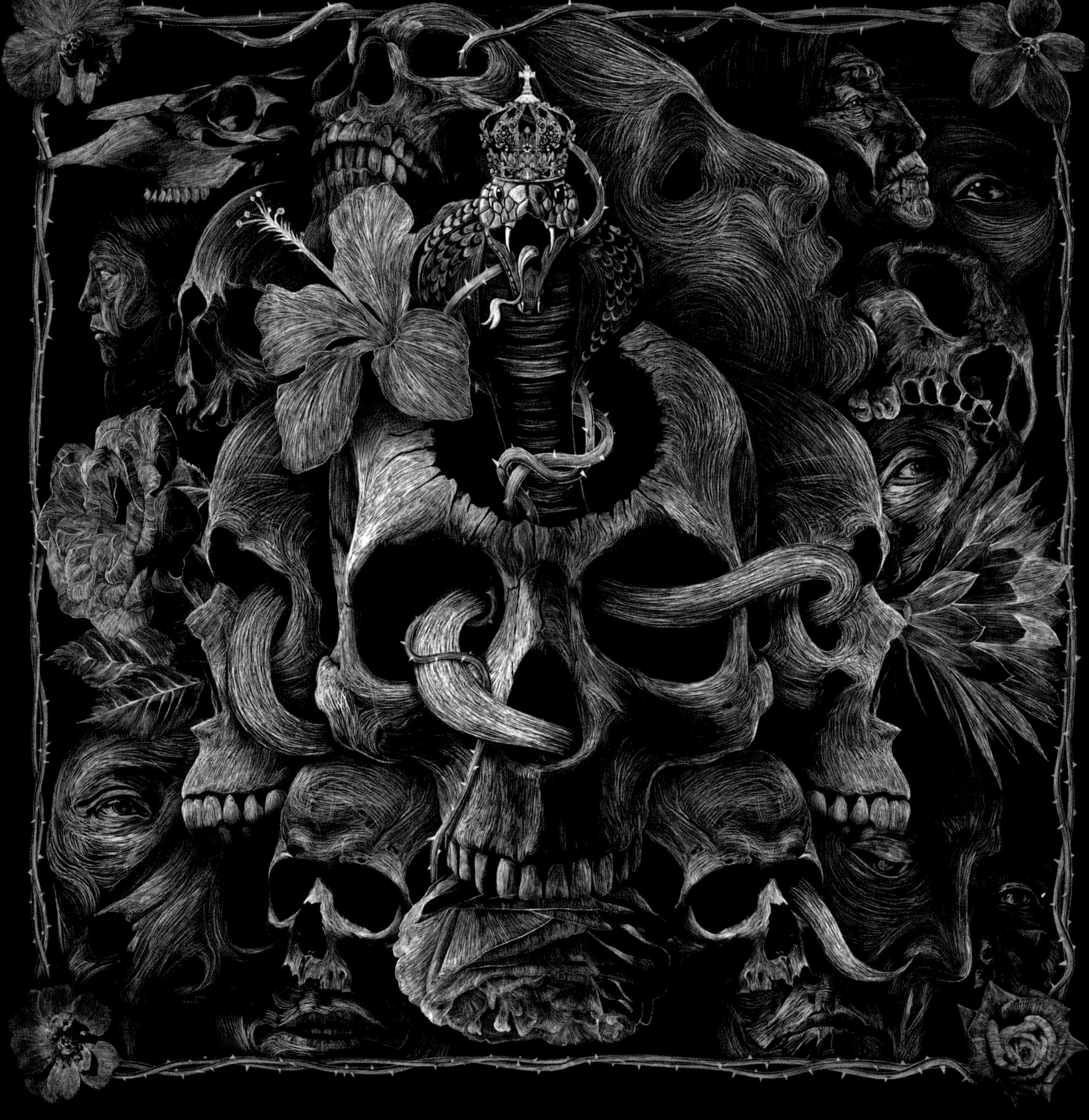

This spread and pages 118–120: Skull artwork and illustrations by Julien Lemoine.

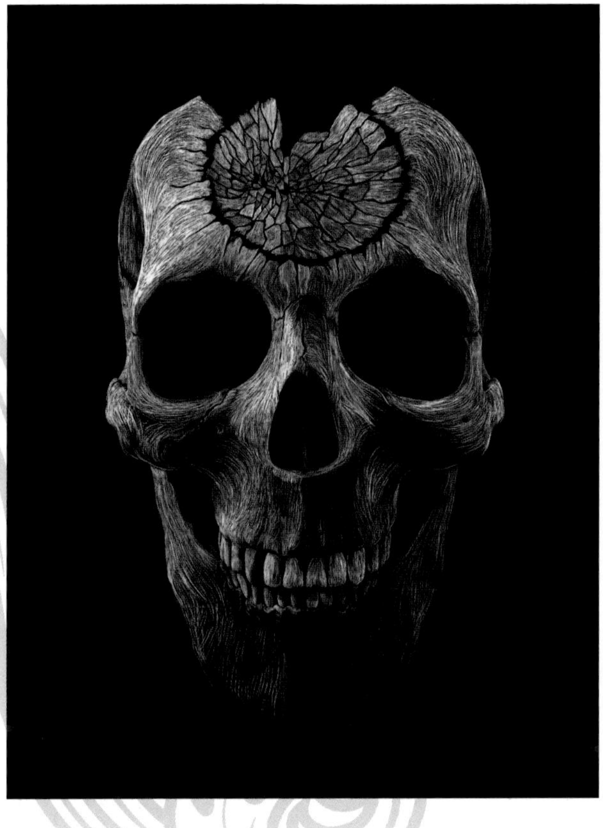

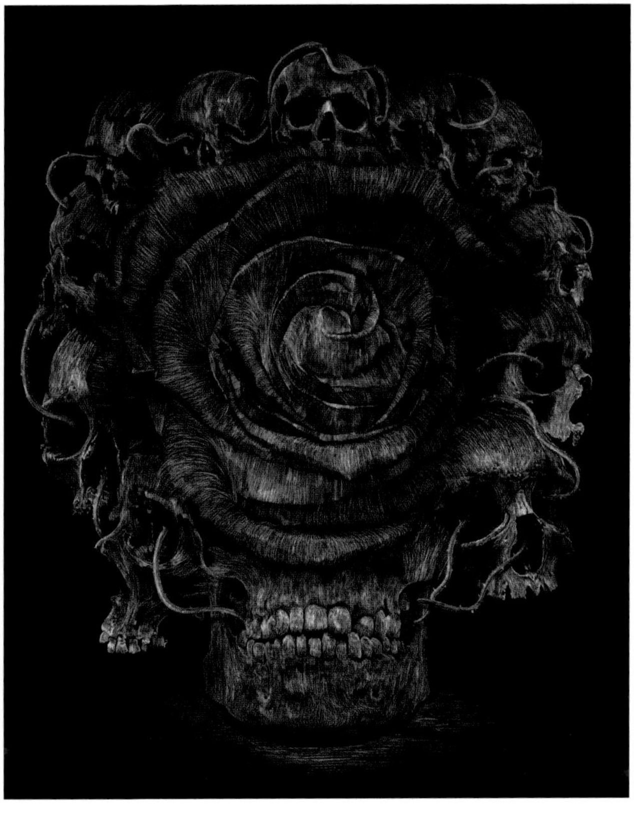

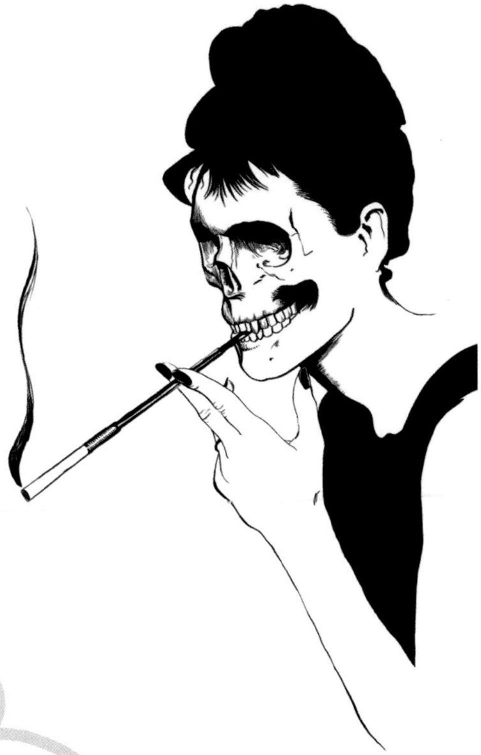

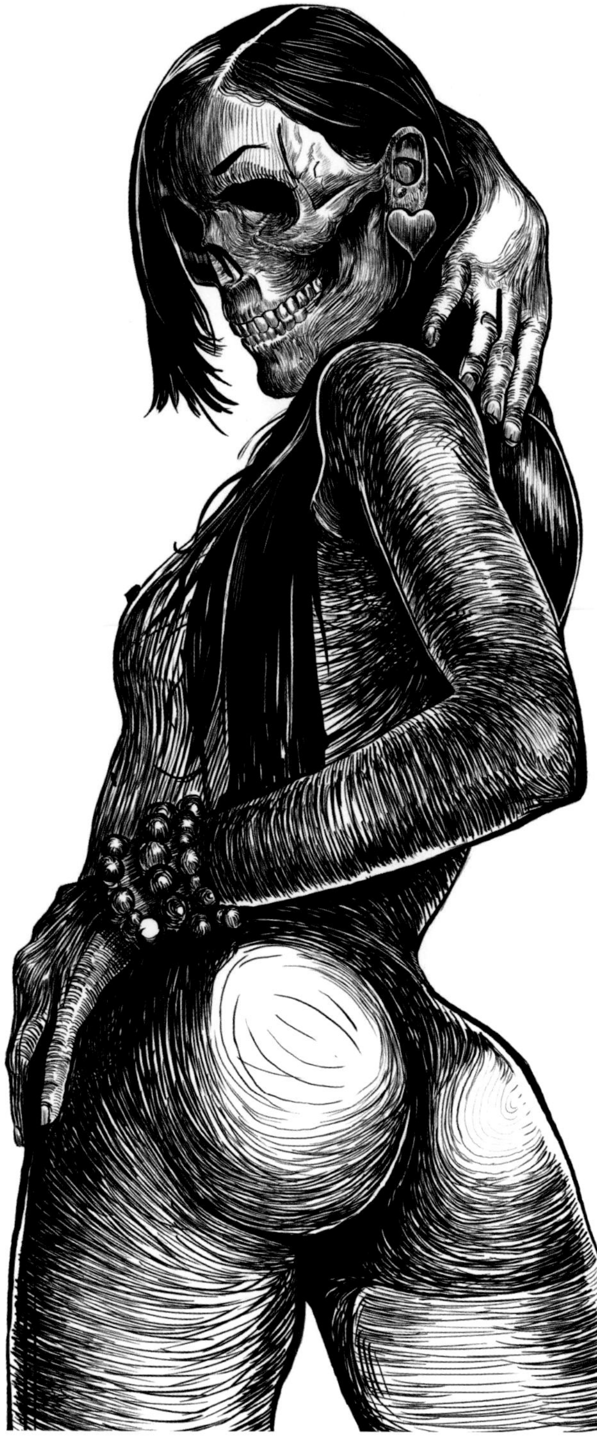

SKULLS IN ART

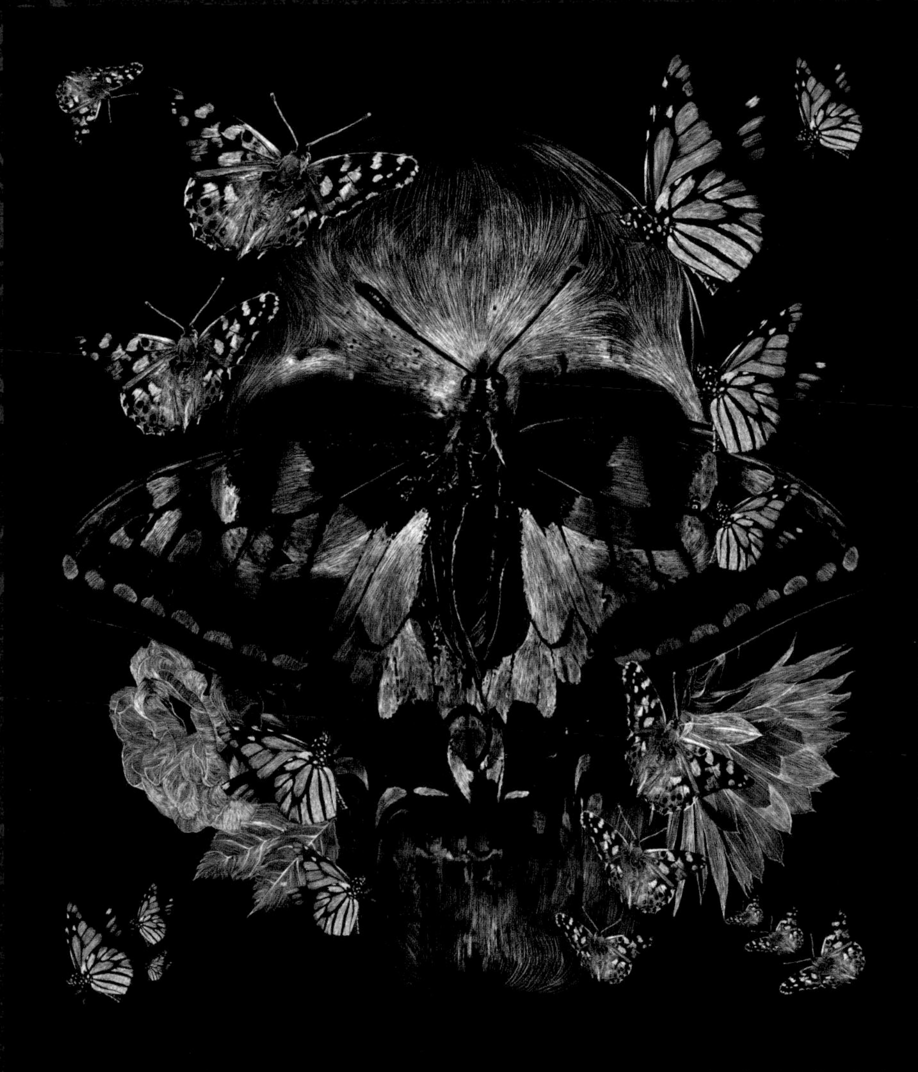

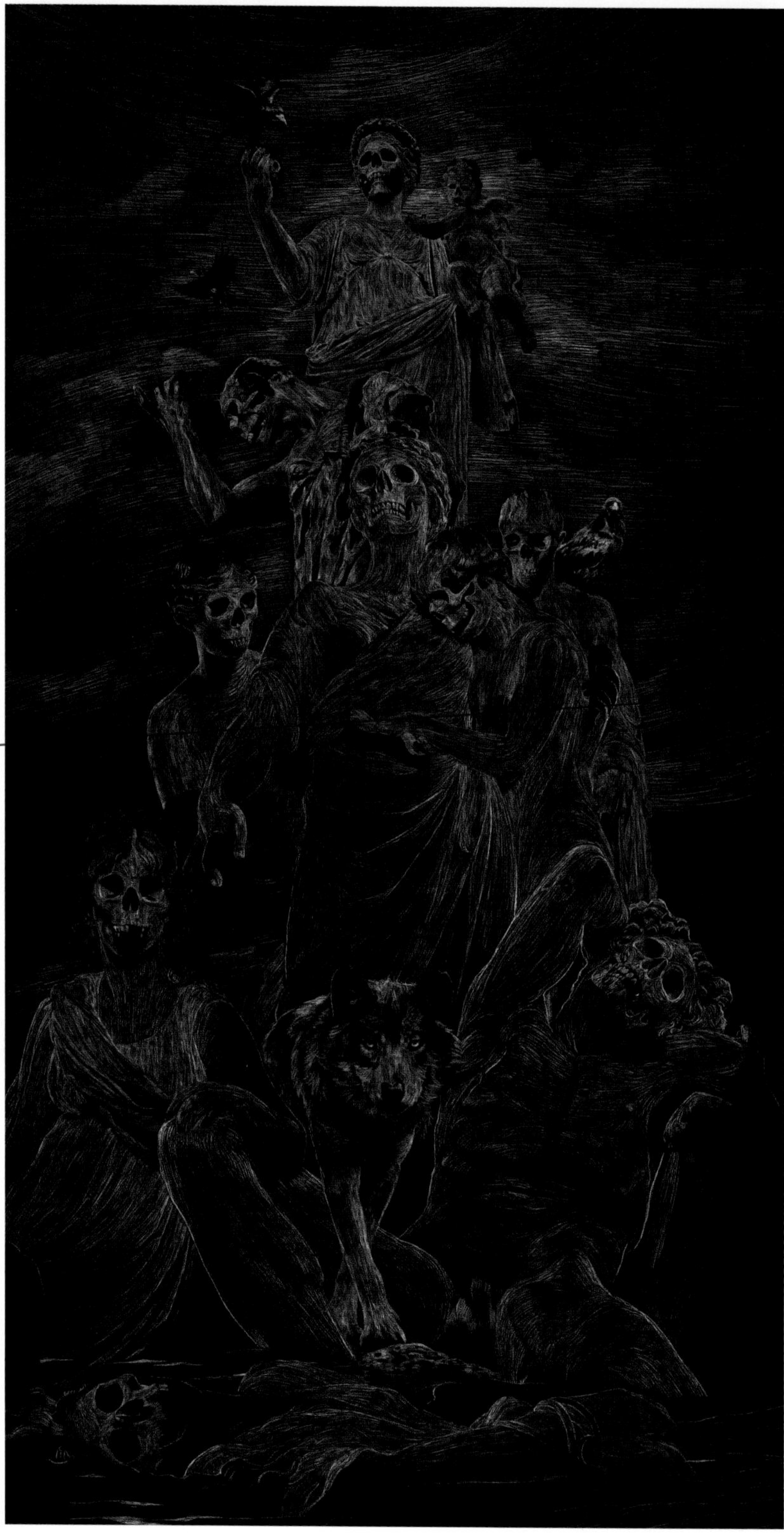

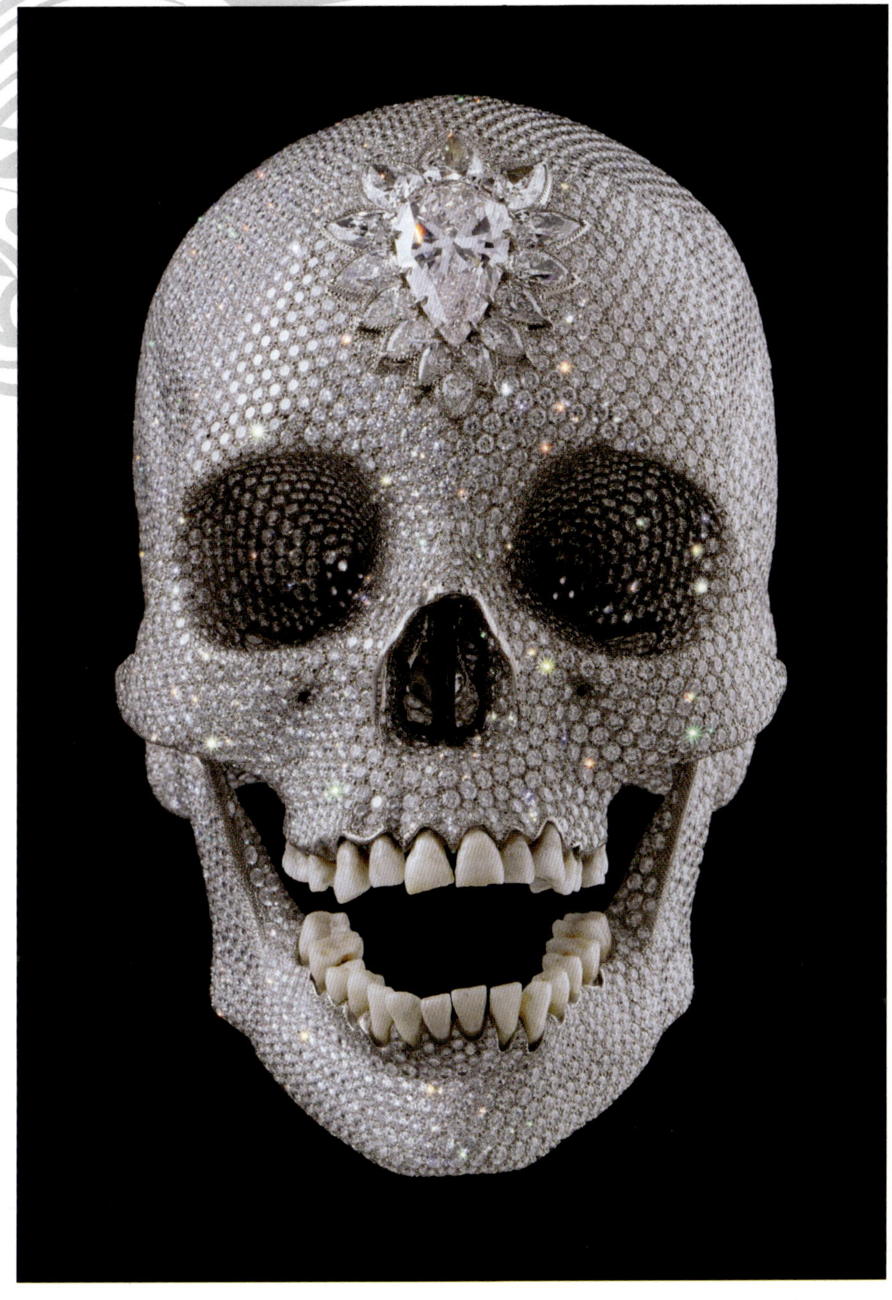

Damien Hirst *For the Love of God* 2007, Platinum, diamonds and human teeth.

Photo: Prudence Cuming Associates Ltd

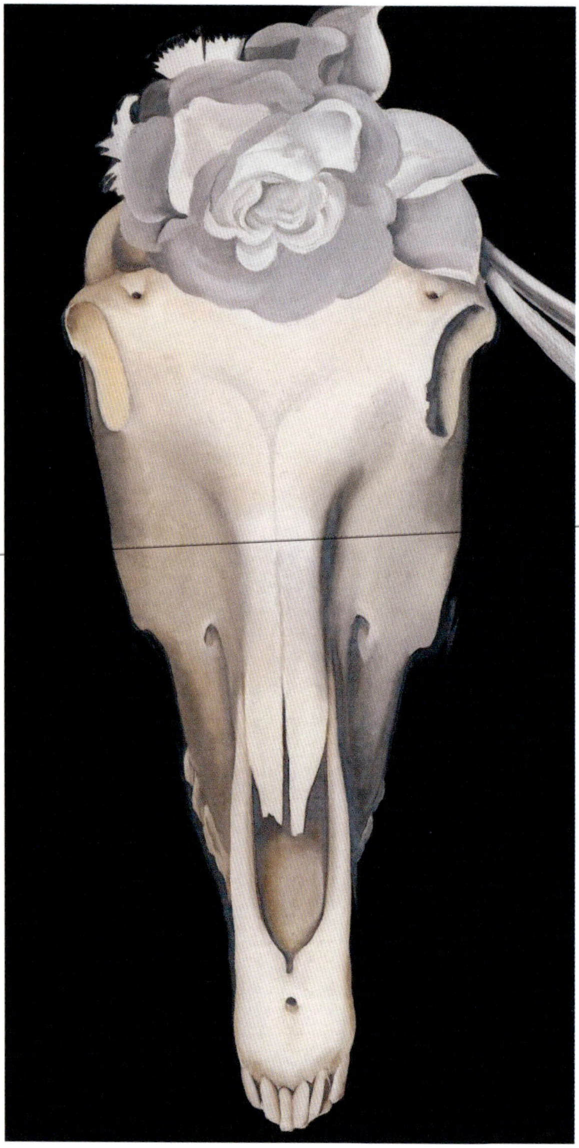

This 1931 painting by Georgia O'Keeffe is called *Horse's Skull with White Rose.*

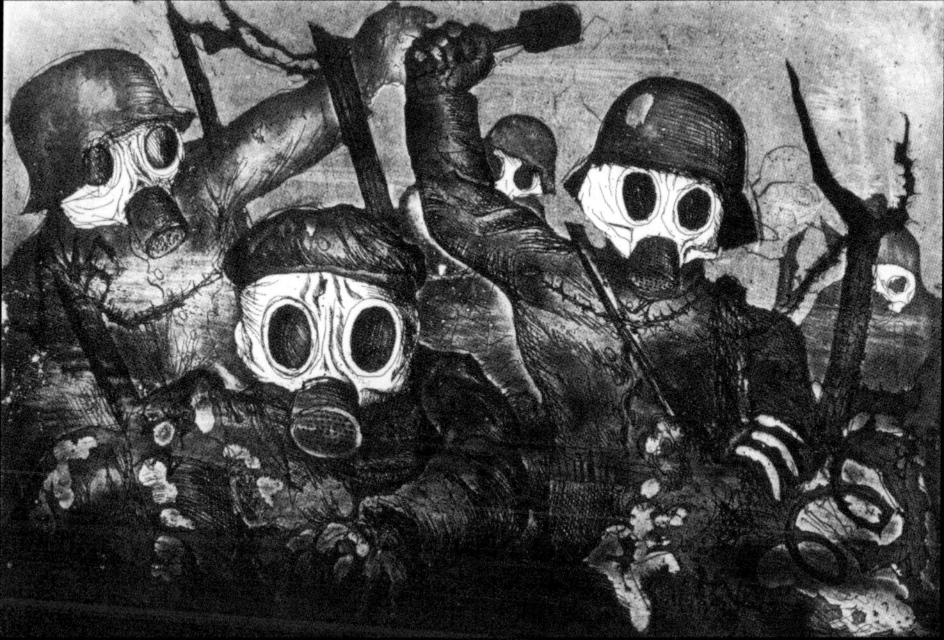

Otto Dix's 1924 artwork from the series *The War* shows stormtroopers advancing under a gas attack.

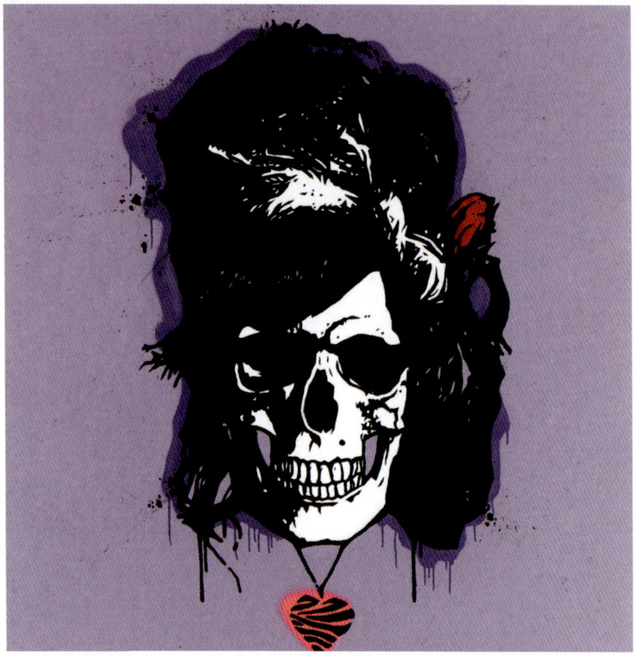

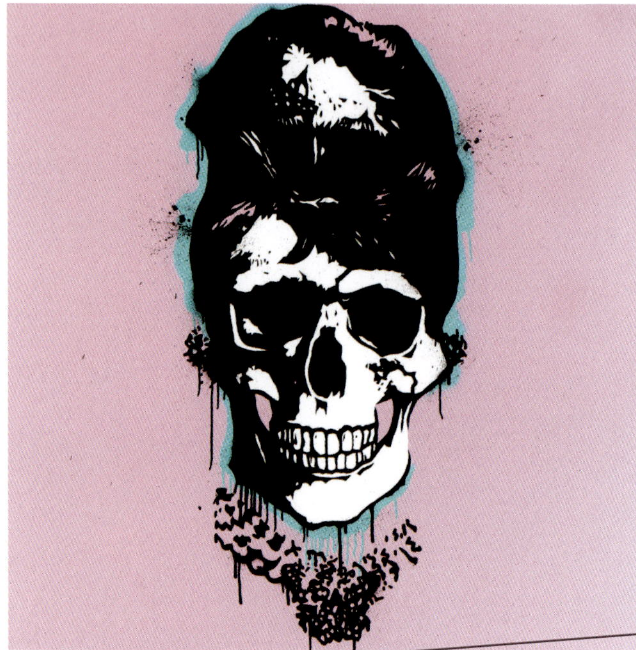

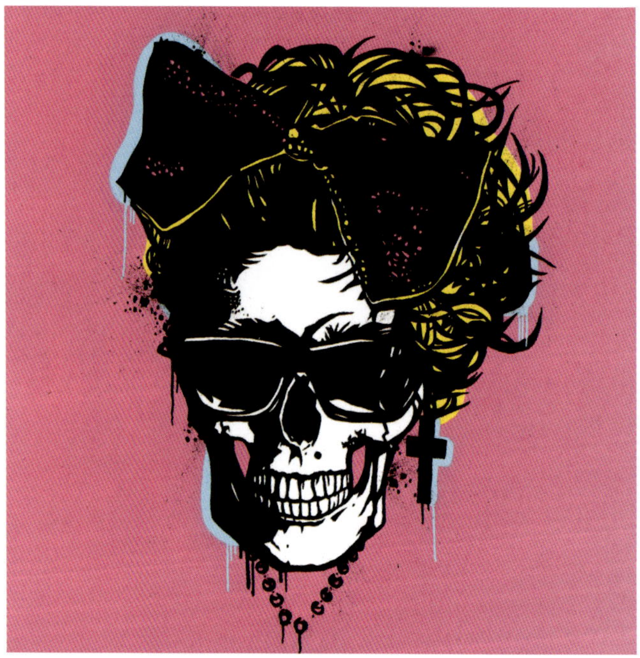

George Ioannou's *Dead Famous Collection* depicts celebrities as skulls and includes such stars as Amy Winehouse (above left), Audrey Hepburn (above right), and Madonna.

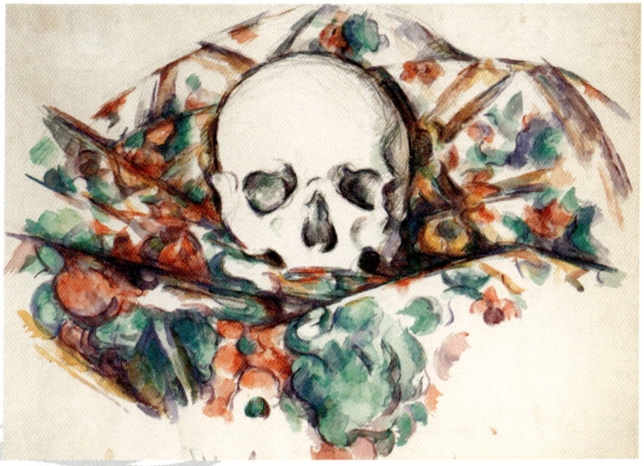

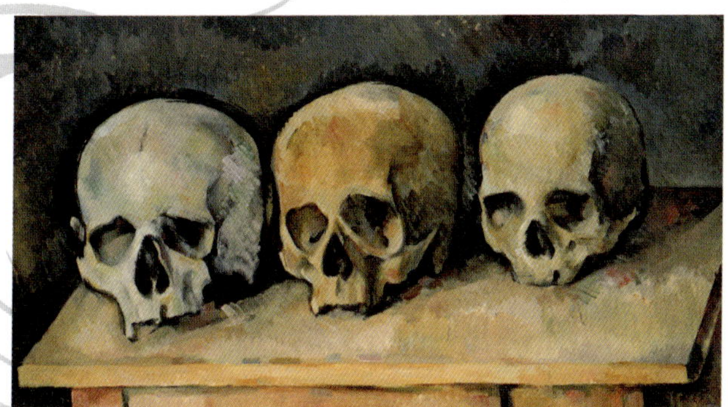

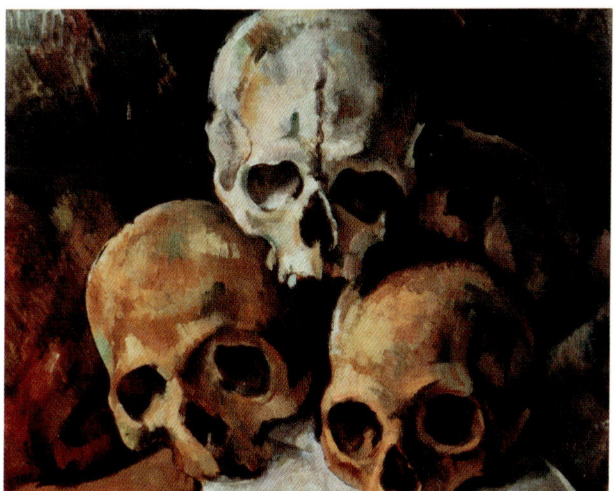

Paul Cézanne painted many skulls, including these three; from top: *Skull on a Curtain* (c. 1902–1906), *The Three Skulls* (c. 1900), and *Pyramid of Skulls* (c. 1898–1900).

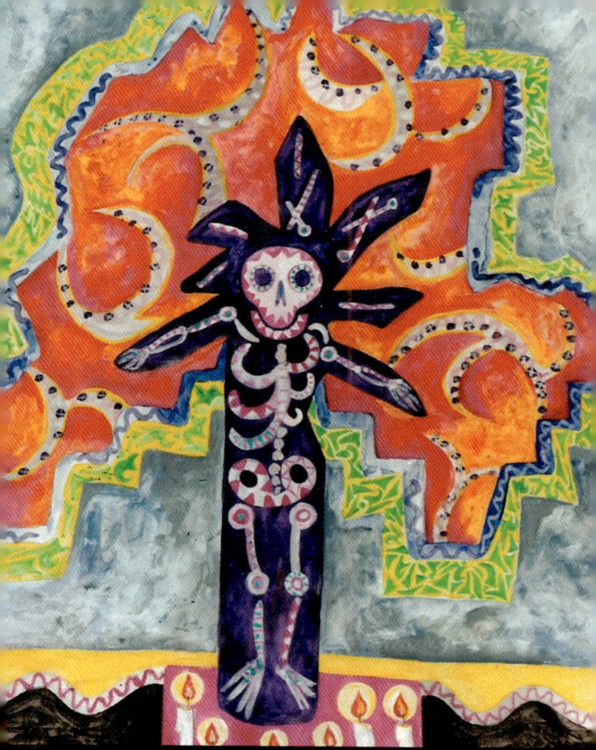

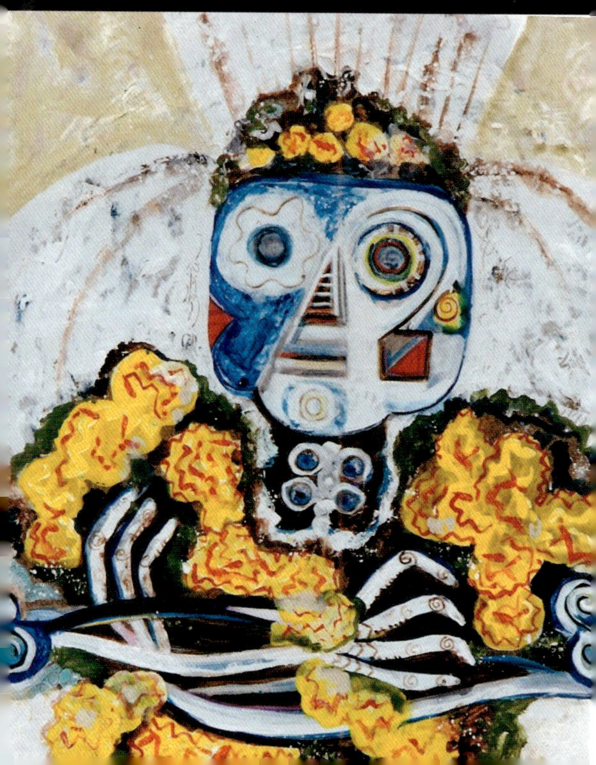

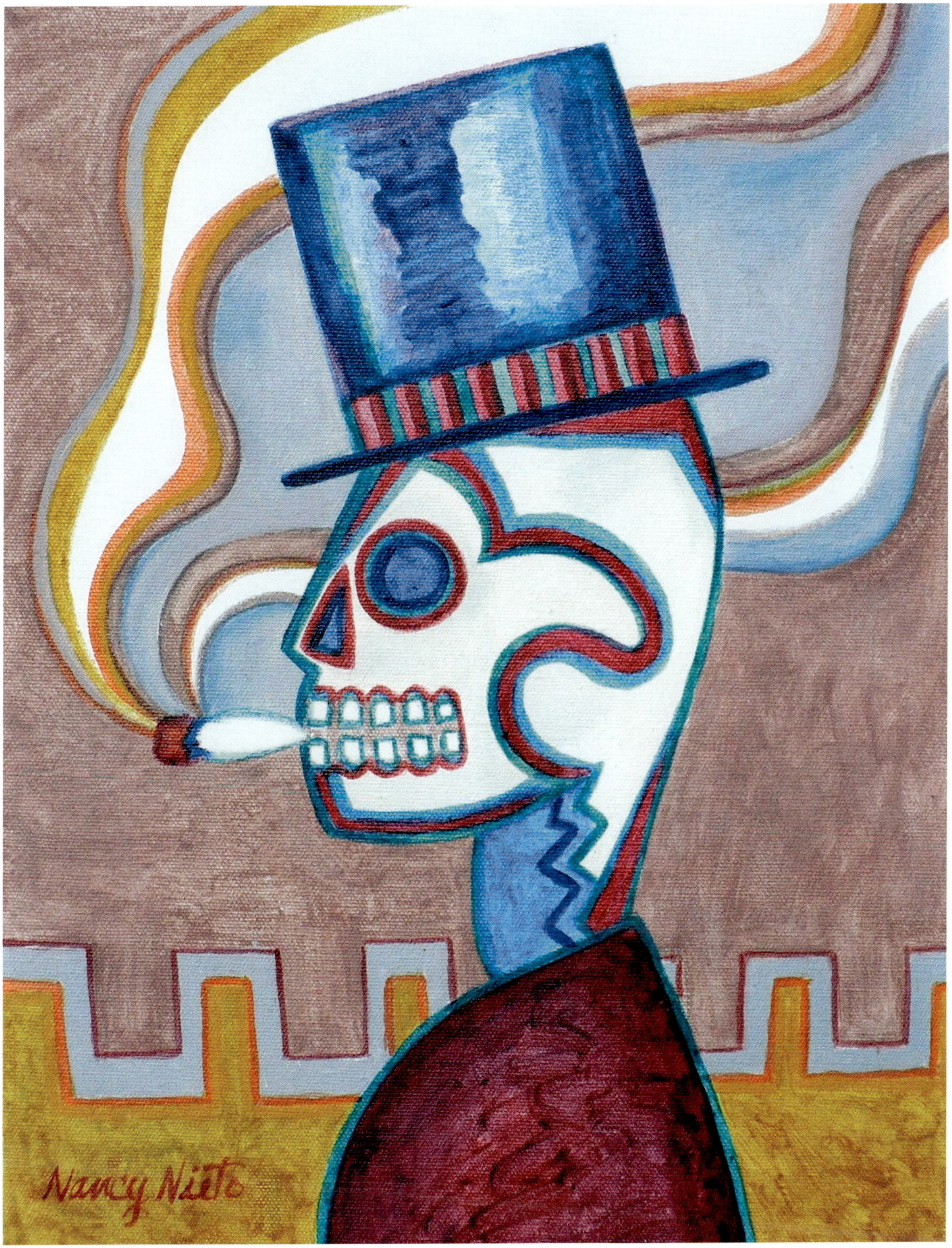

This spread: Paintings by artist Nancy Glenn-Nieto named *Red Skull Tree* (opposite top), *Skull Bride* (opposite bottom), and *Monte Alban* (above).

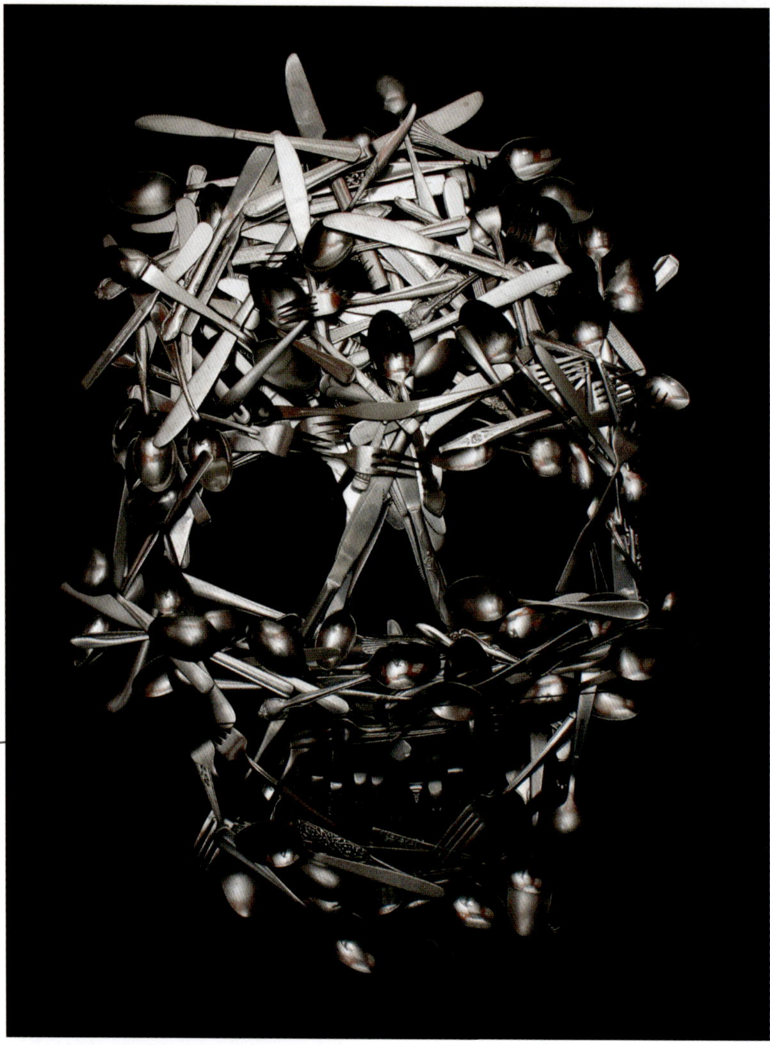

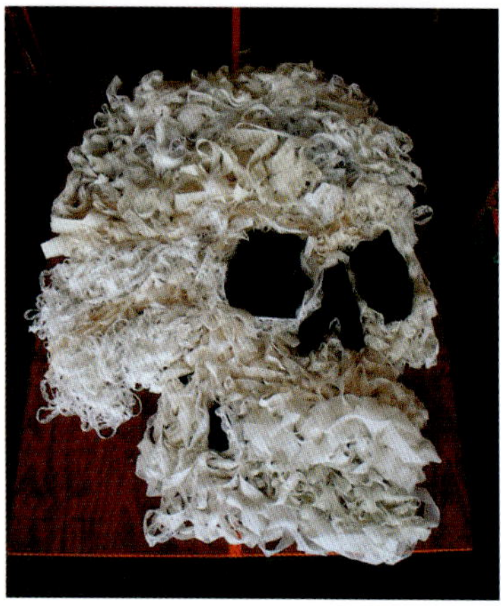

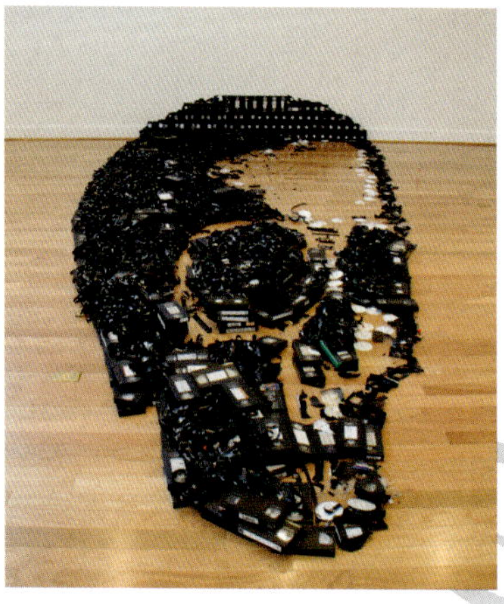

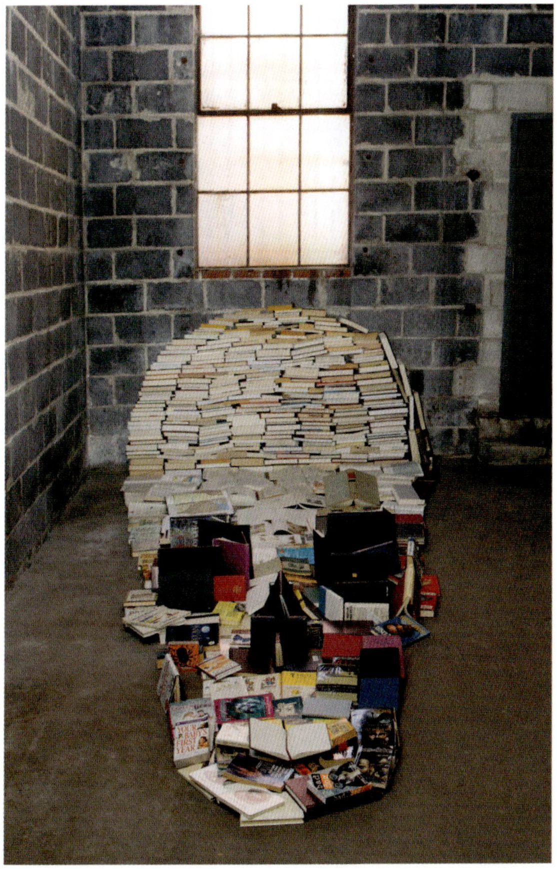

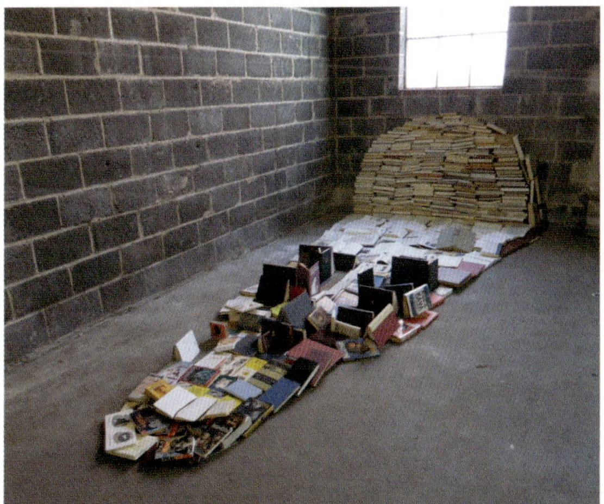

This spread: Selections from Noah Scalin's Skull-A-Day project, including *Silverware* (opposite top), an anamorphic skull of books shown in two views (above), *Dead Media* (opposite right), and *Toilet Paper* (opposite left).

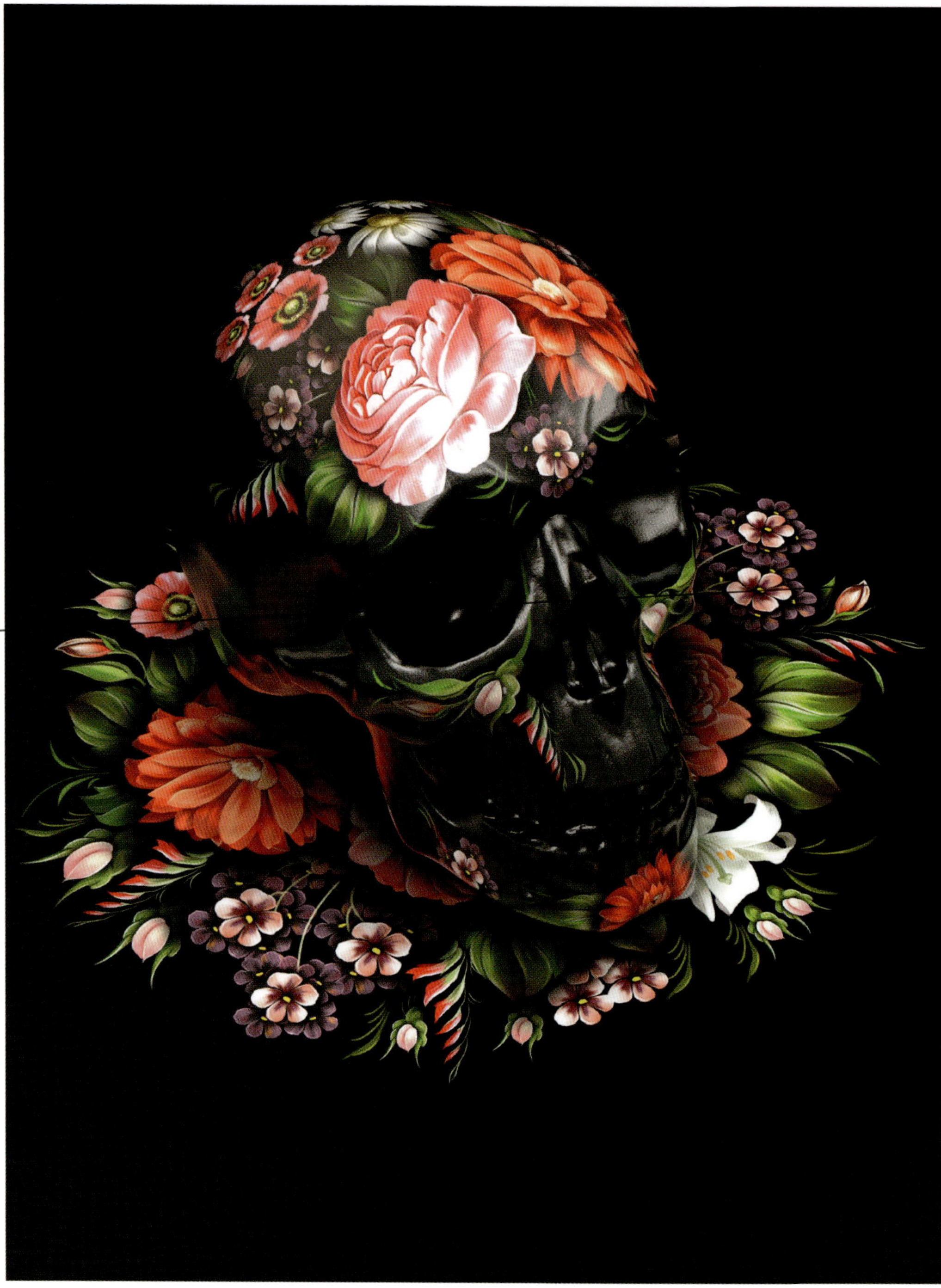

This spread: A fusion of Russian and Mexican artwork through ornate Dia de los Muertos–themed skulls, designed by Sasha Vinogradova.

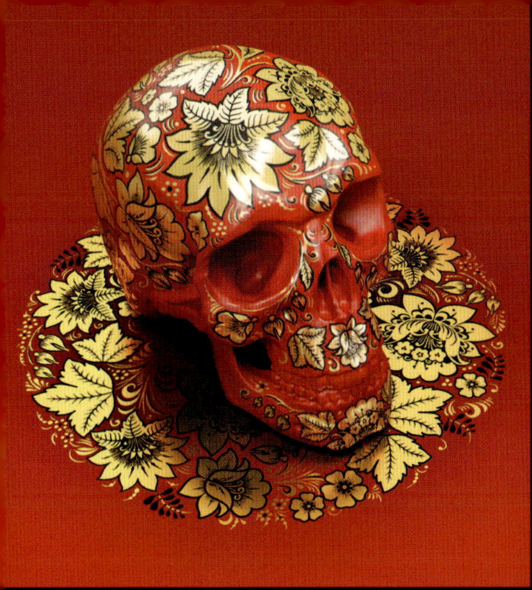

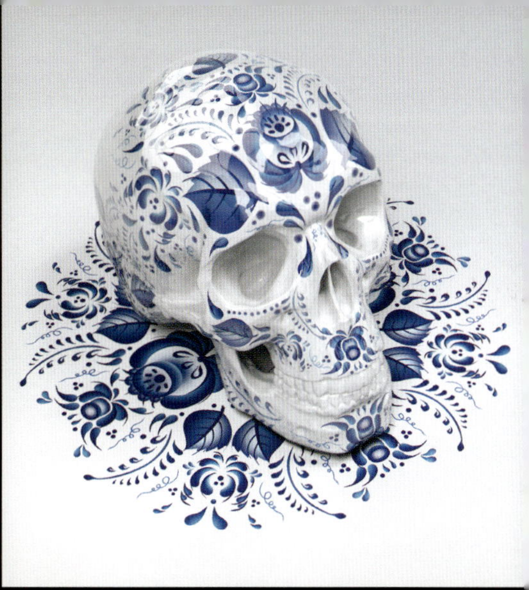

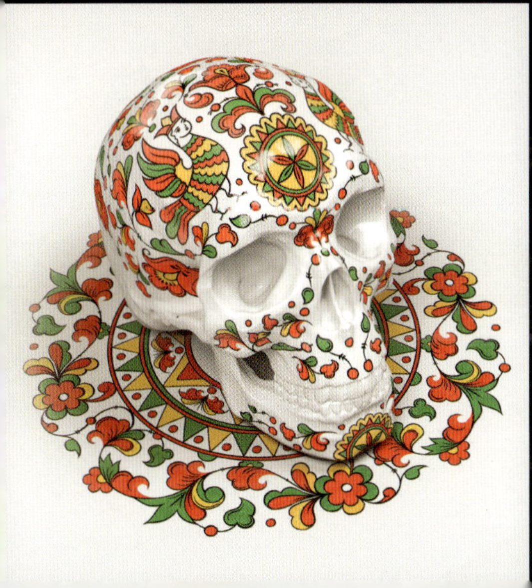

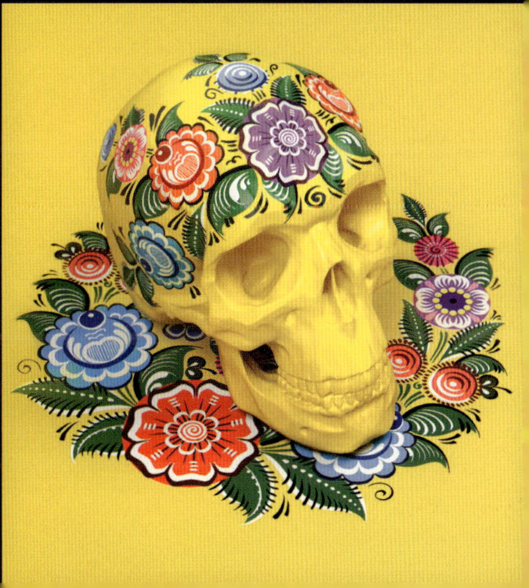

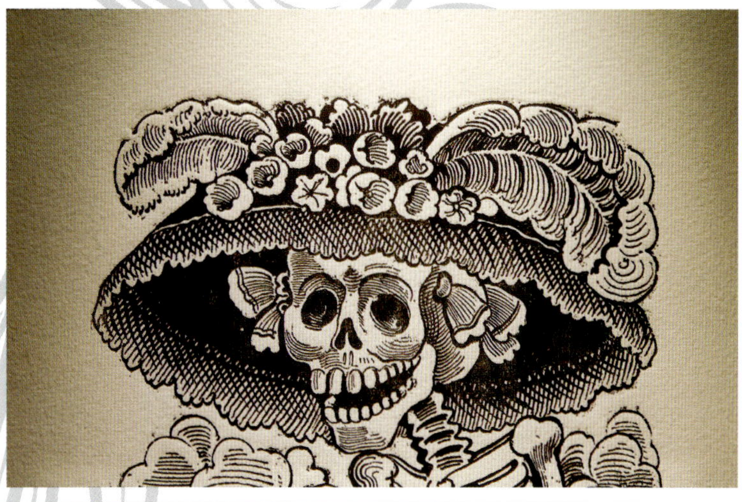

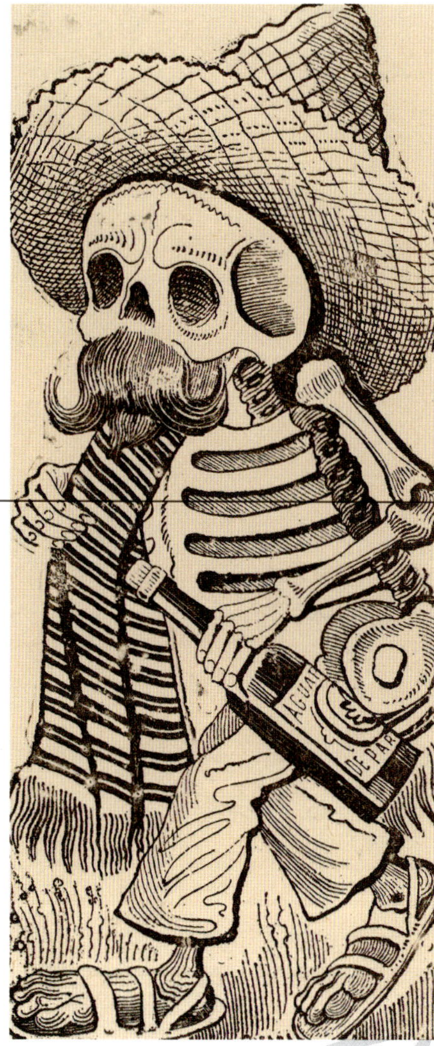

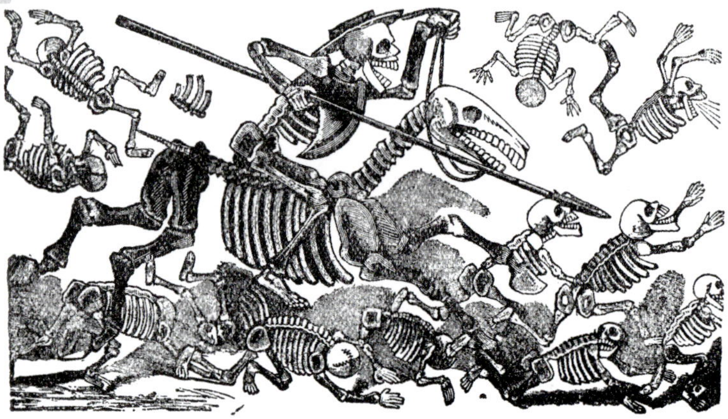

Engravings by José Guadalupe Posada (clockwise from top left): *Don Quixote*, *Calavera Maderista*, and *The Catrina*.

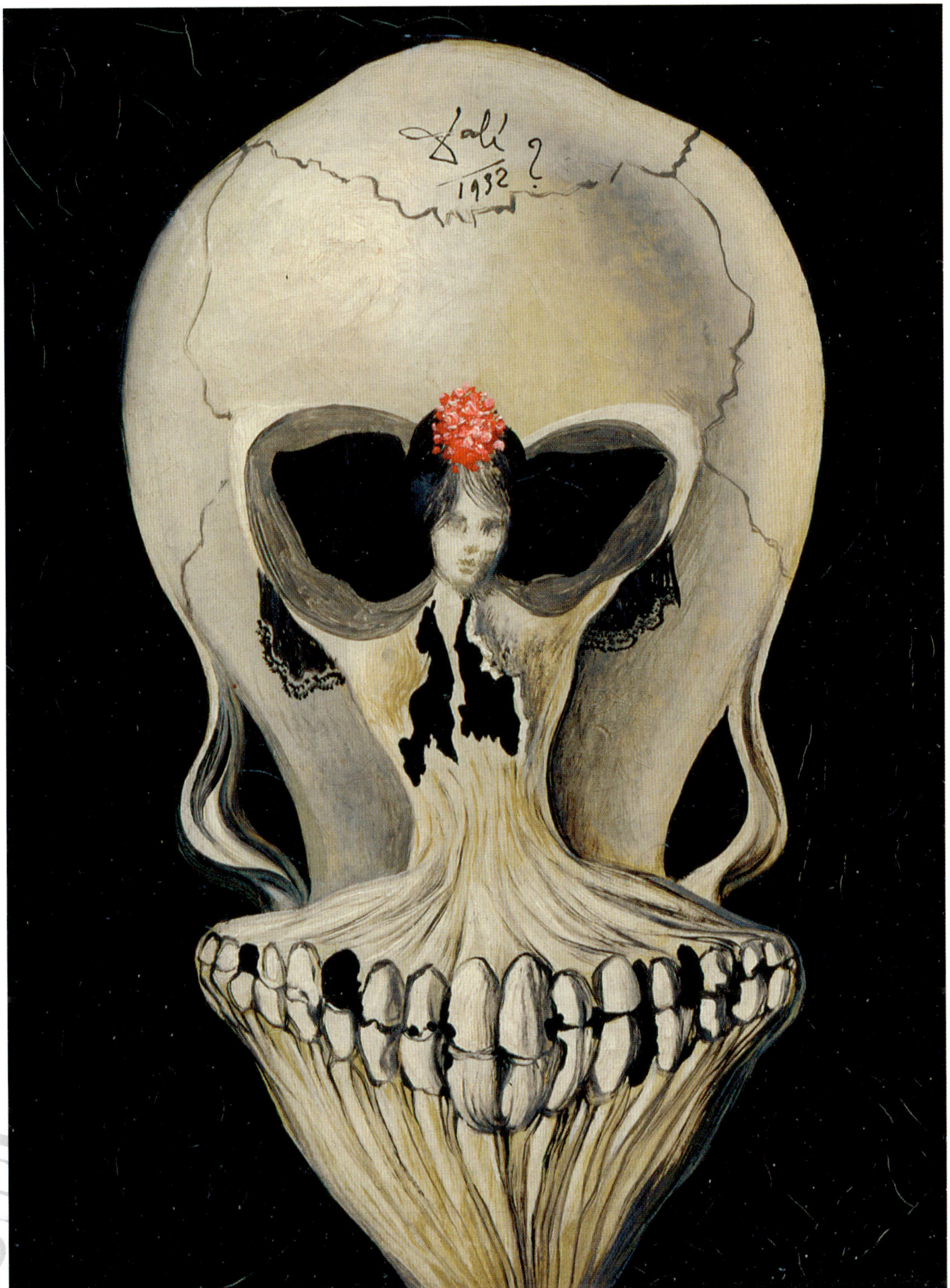

This 1939 painting by Salvador Dalí is named *Ballerina in a Death's Head.*

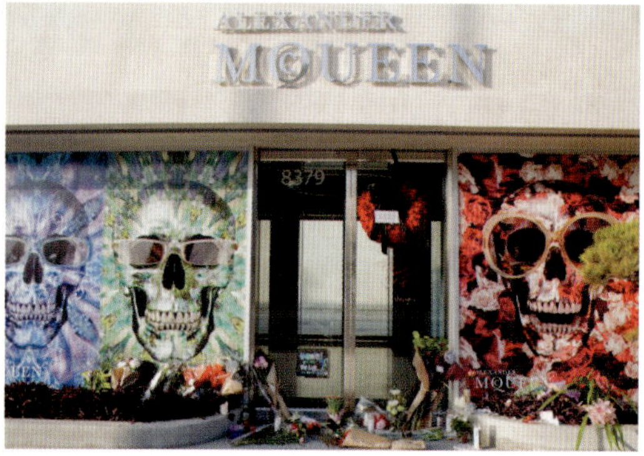

Top: Flowers left outside the skull-decorated Alexander McQueen store in Los Angeles, after the designer's death in 2011.

Bottom: An Alexander McQueen purse photographed in London, England, in September 2011.

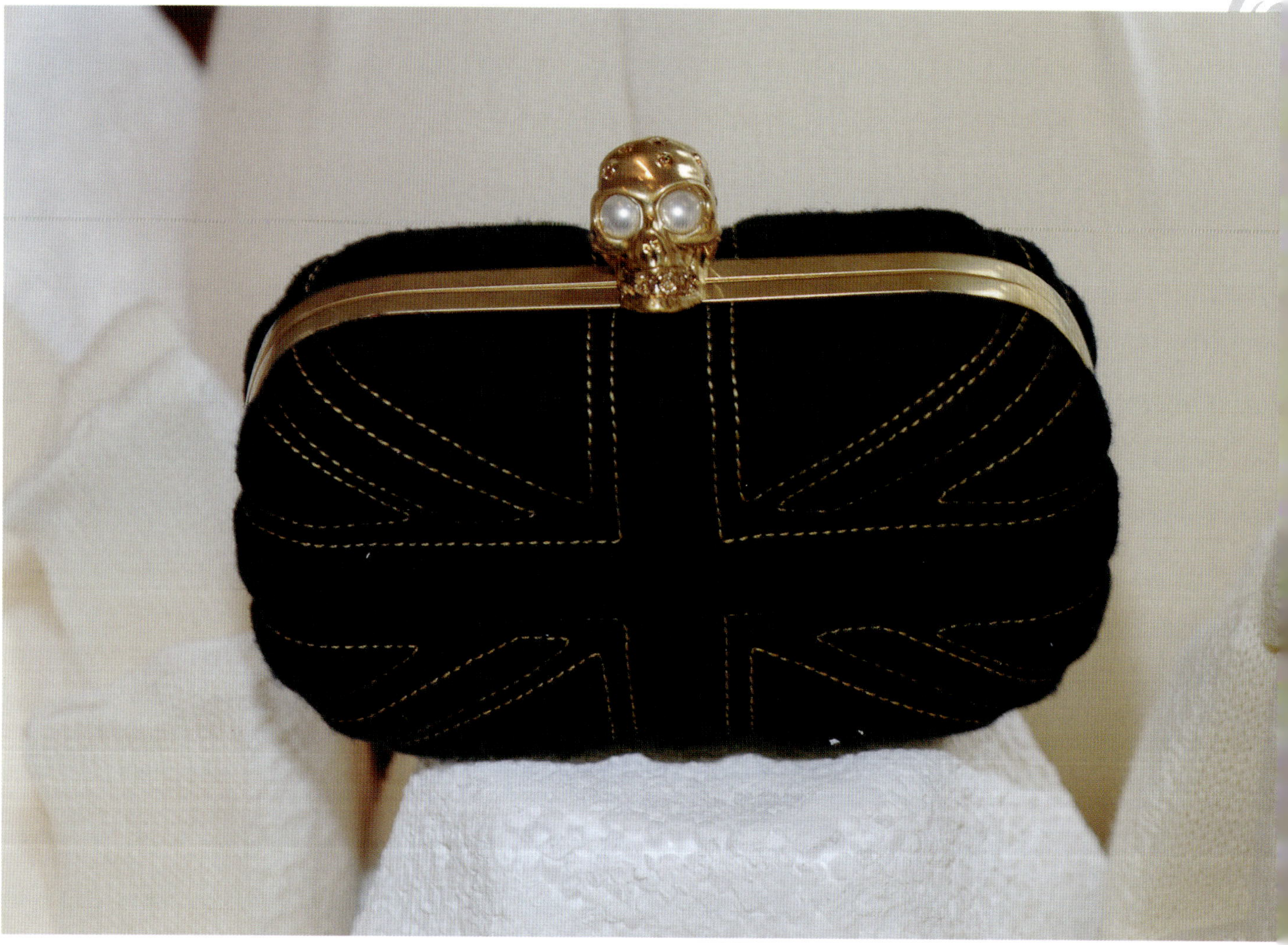

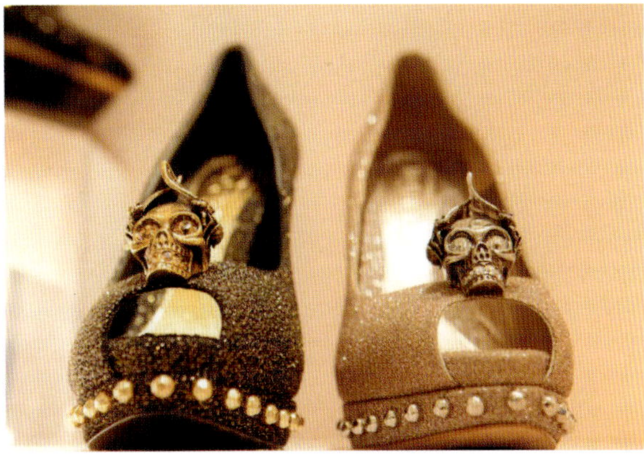

Designer shoes by Alexander McQueen, photographed in Selfridges Department Store, London, England.

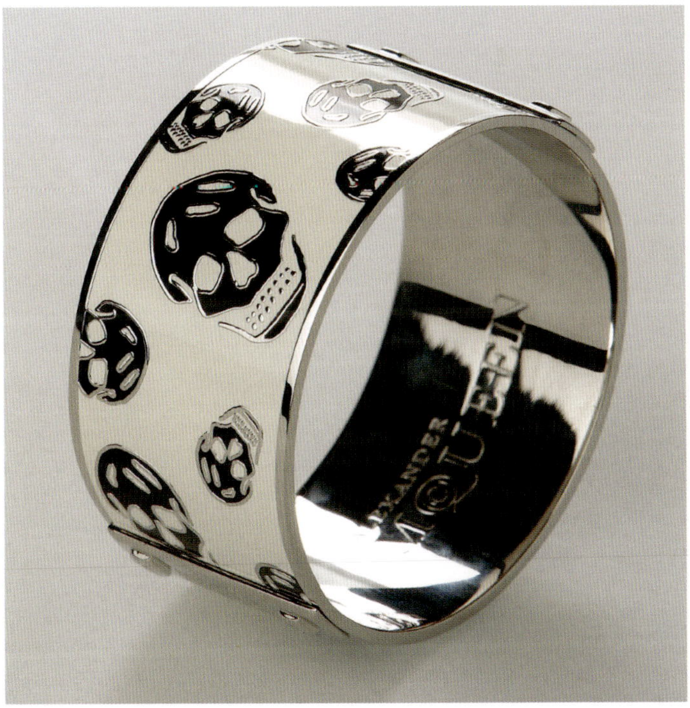

Alexander McQueen bracelet decorated with skulls.

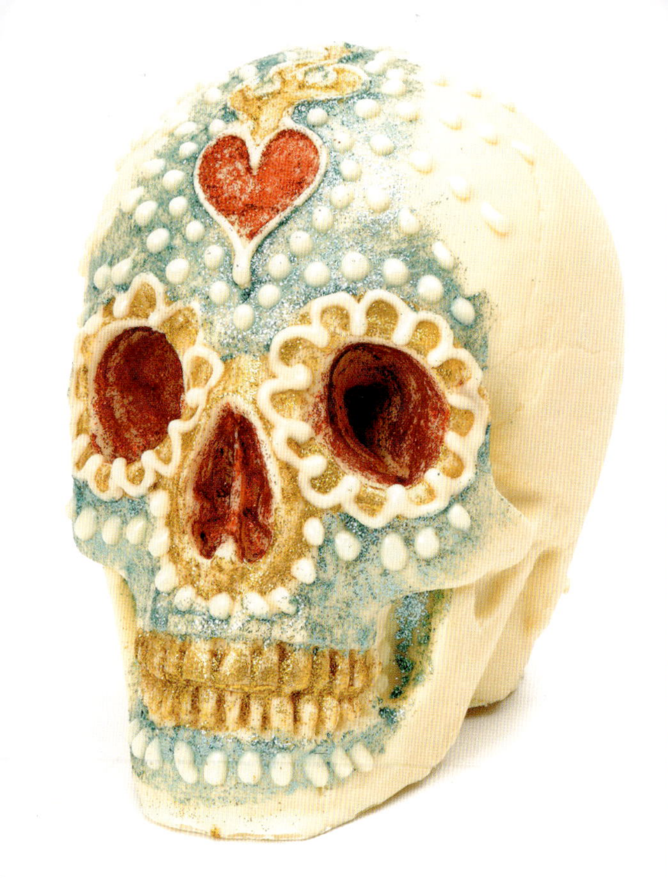

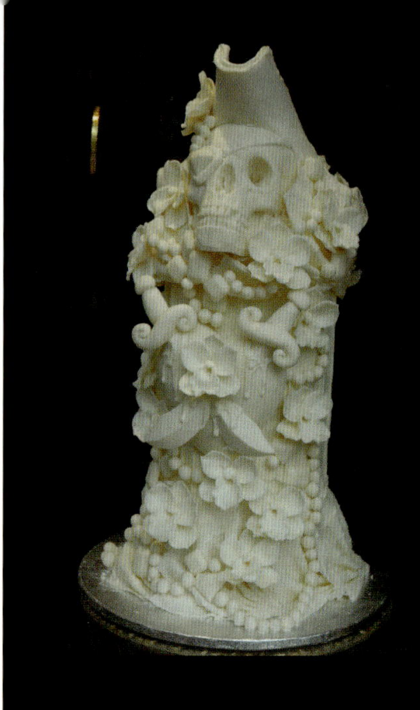

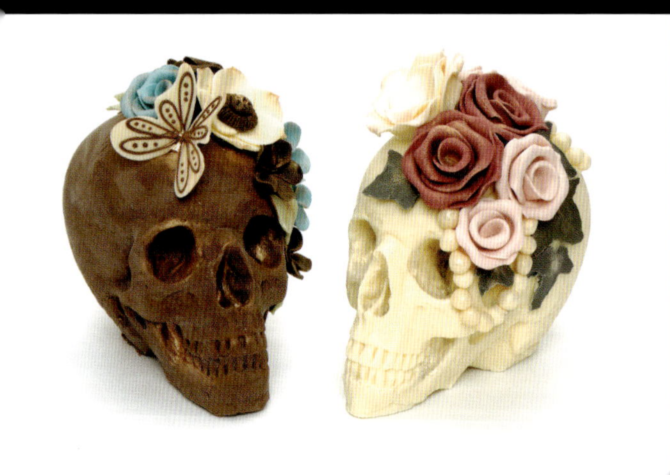

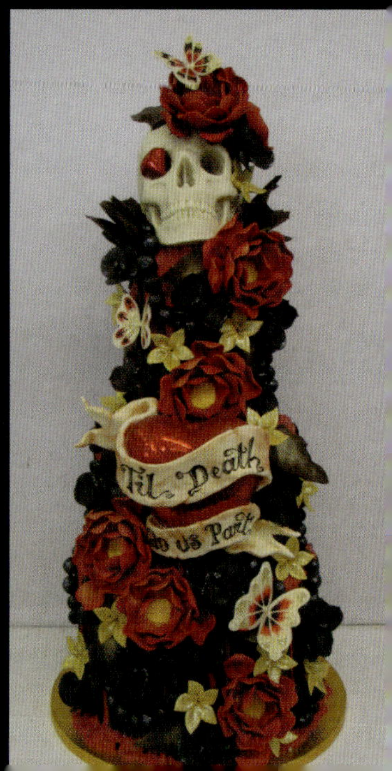
Til Death
us Part

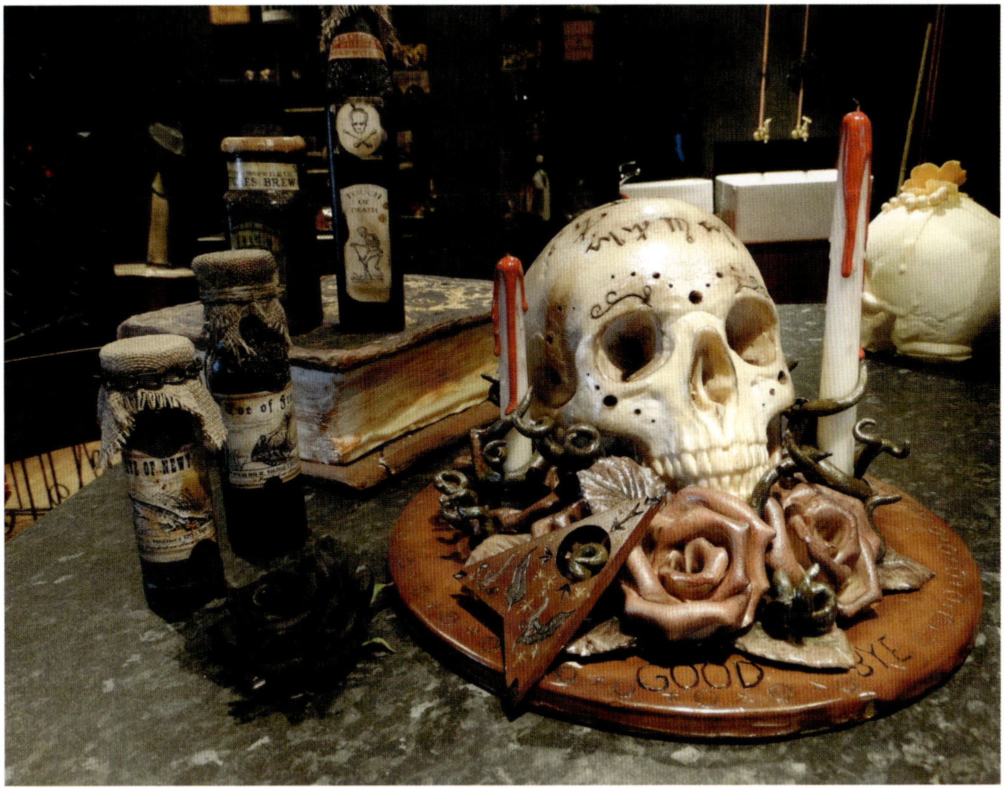
GOOD BYE

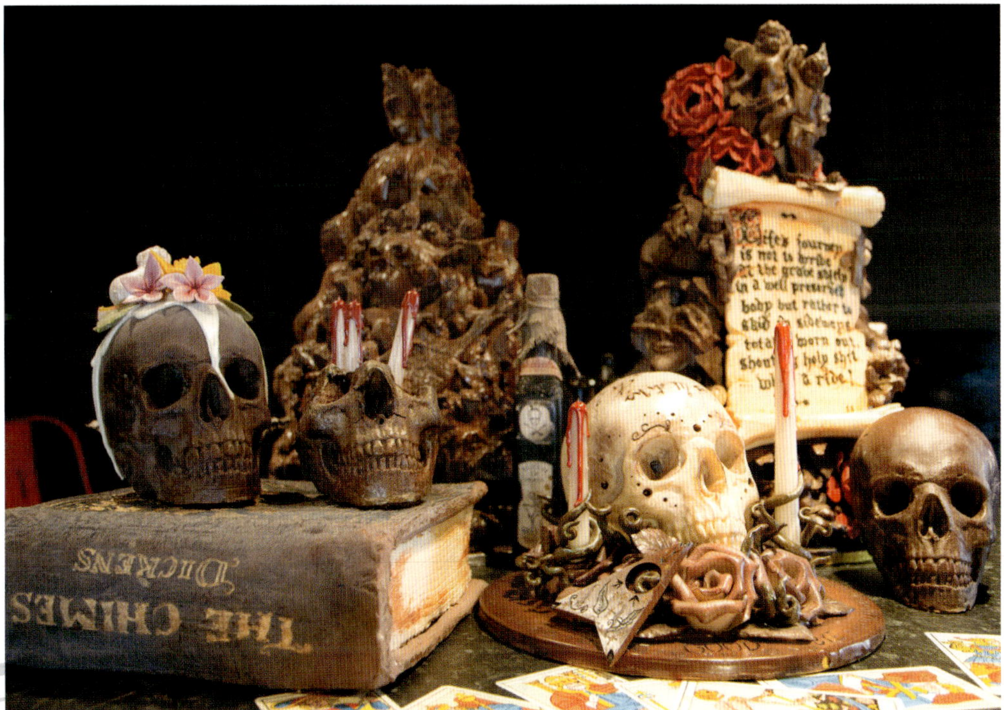
THE CHIMES
DICKENS

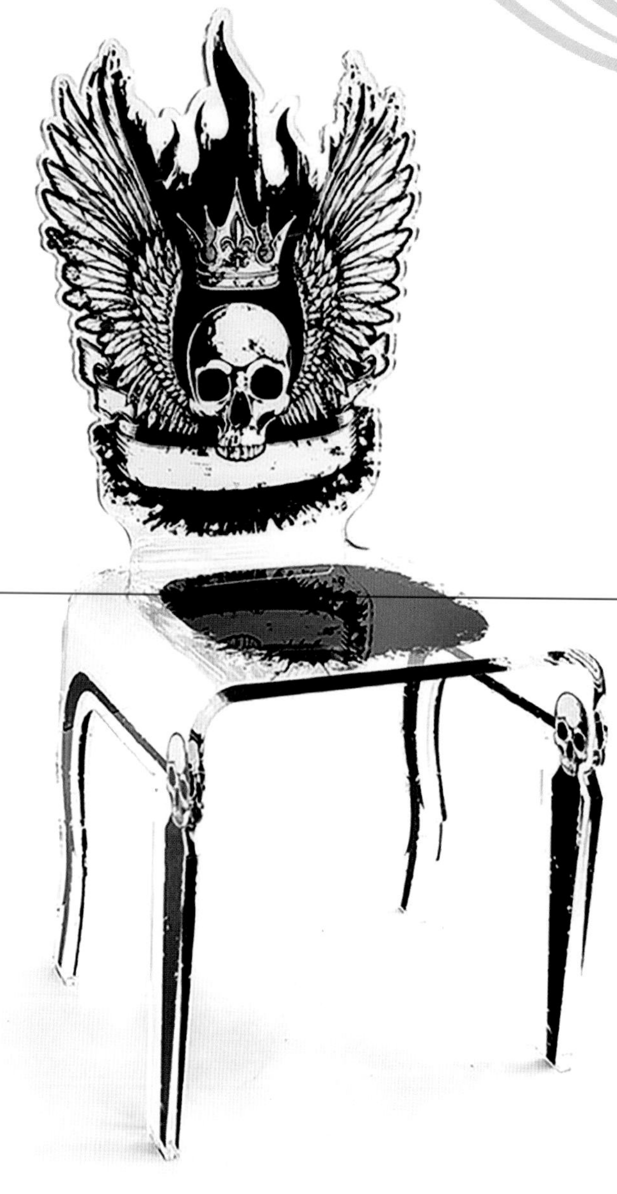

The "Let's Rock" skull chair from Acrila Furniture.

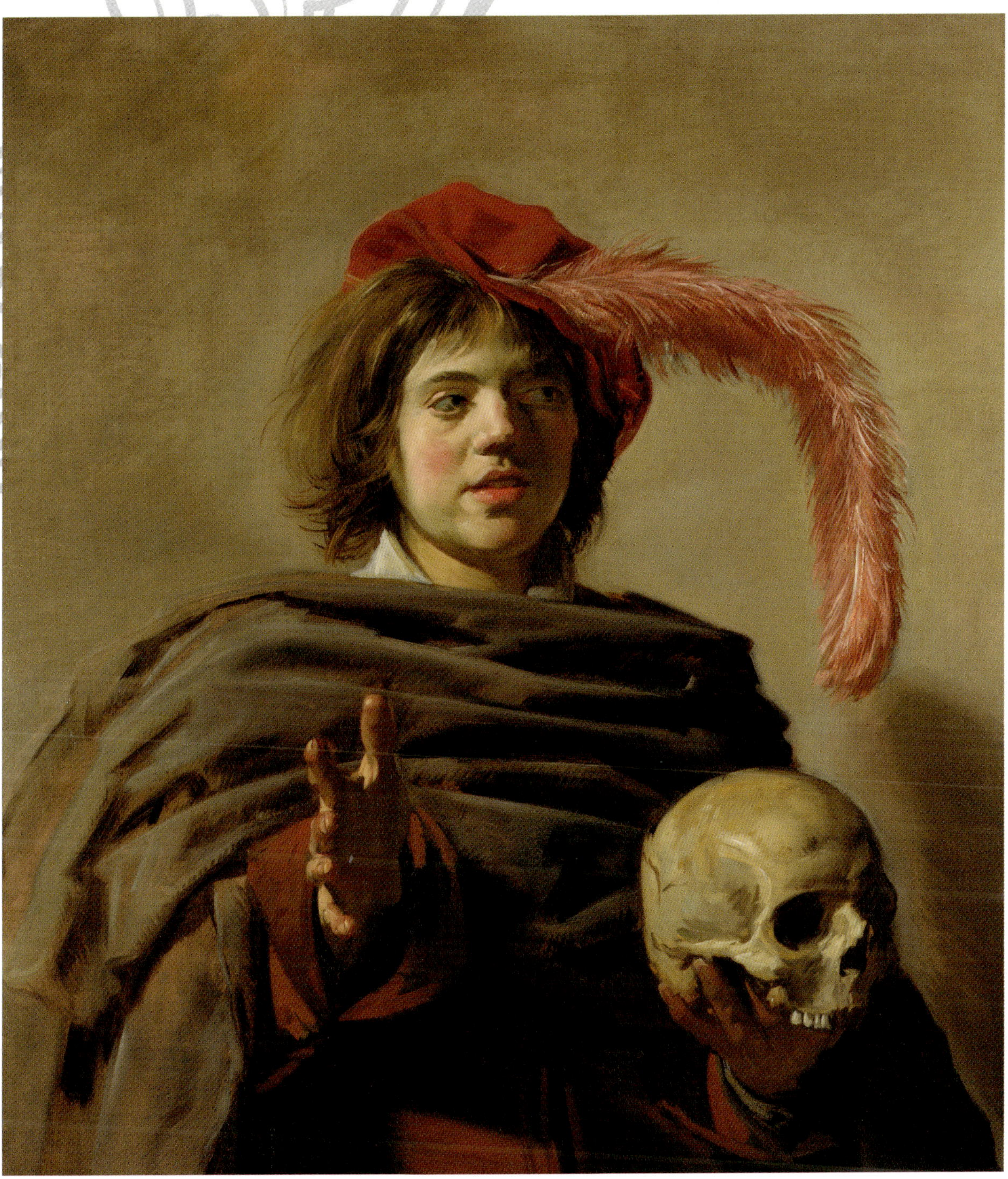

A 1627 painting by Frans Hals named *Young Man Holding a Skull*.

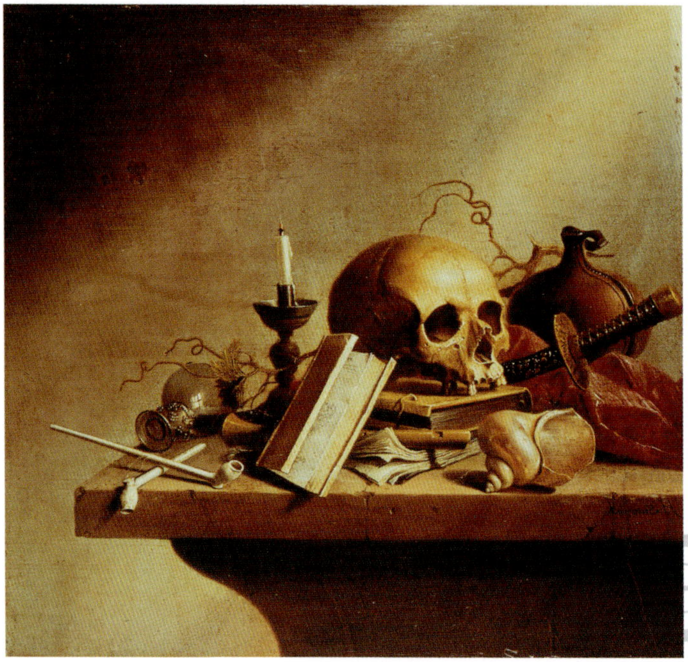

Vanitas artwork by Harmen Steenwijck, painted in 1640.

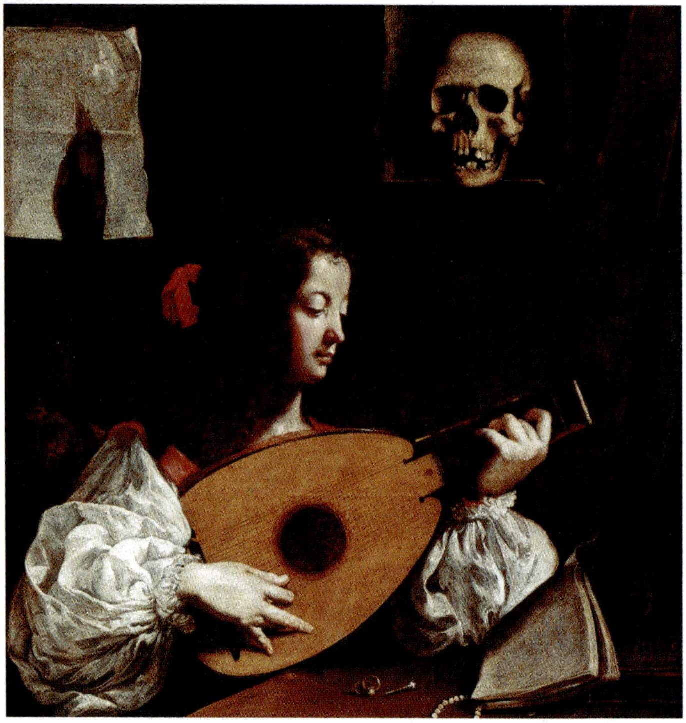

Vanity, a painting by Italian Luigi Miradori, also known as Genovesino.

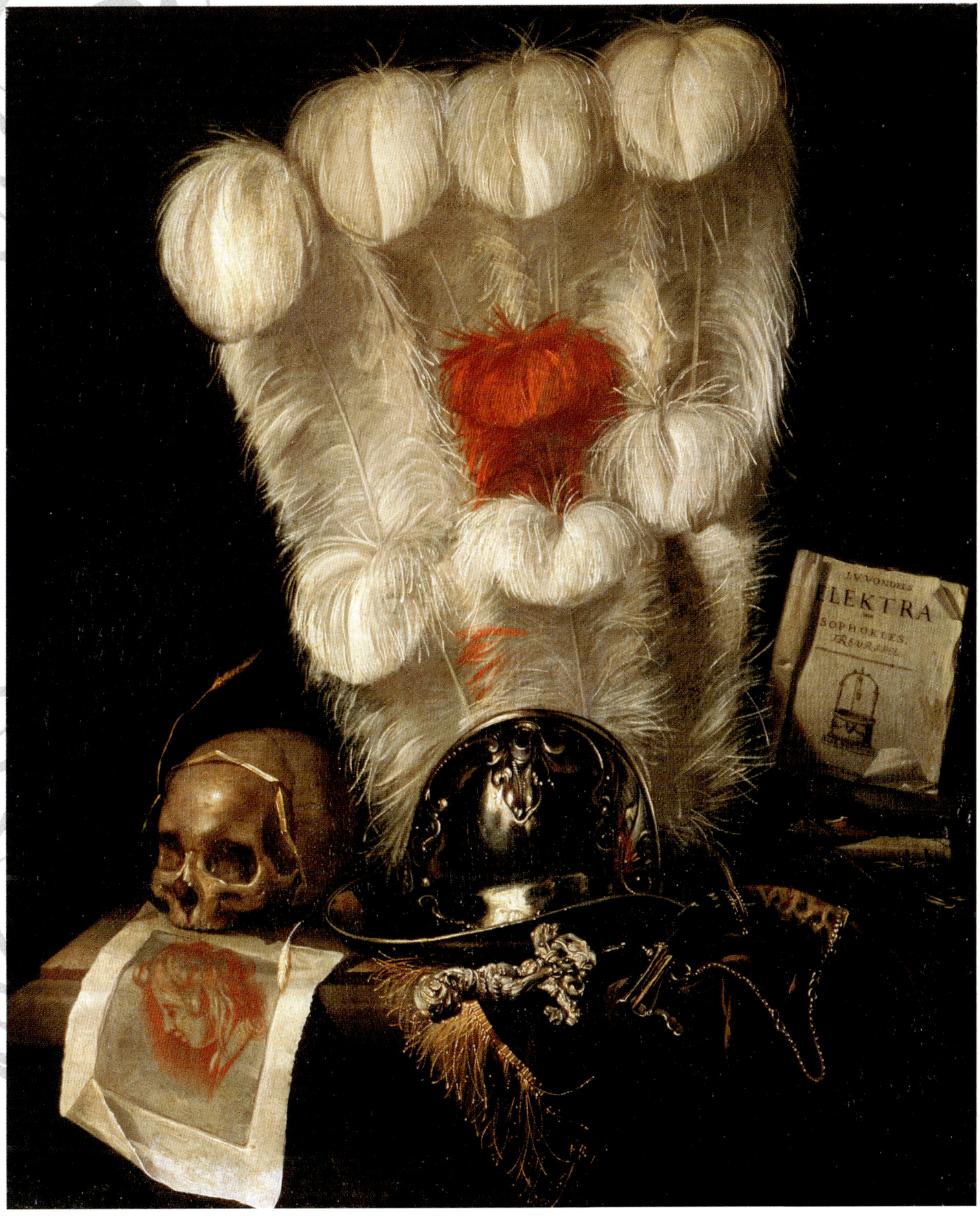

Vanitas, by Juriaen van Streeck, painted in the seventeenth century.

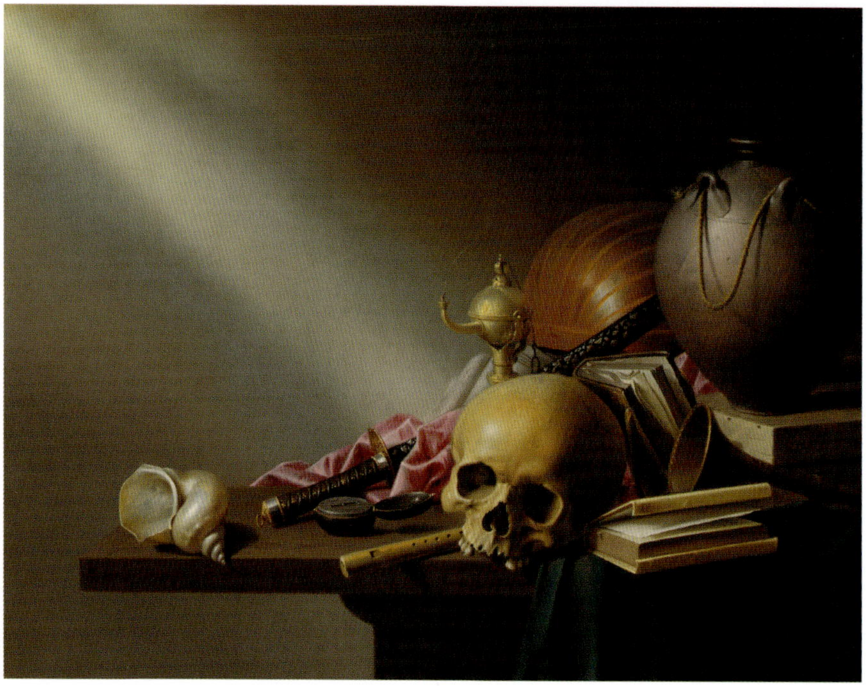

A c. 1640 painting by Harmen Steenwijck named *An Allegory of the Vanities of Human Life*.

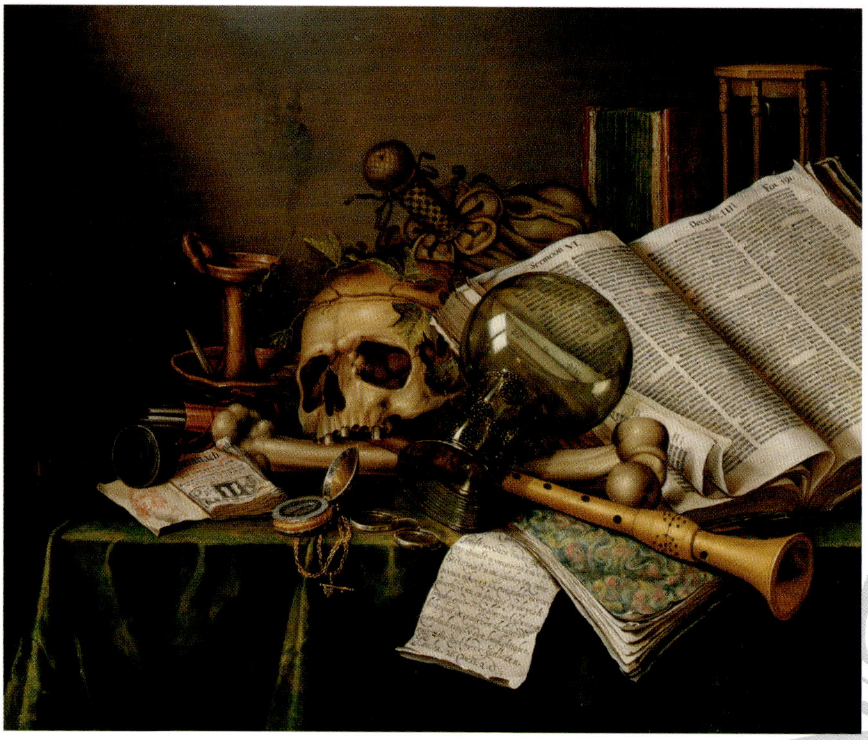

A painting by Edward Collier named *Still Life with Books, Manuscripts, and a Skull*.

SKULLS IN ART

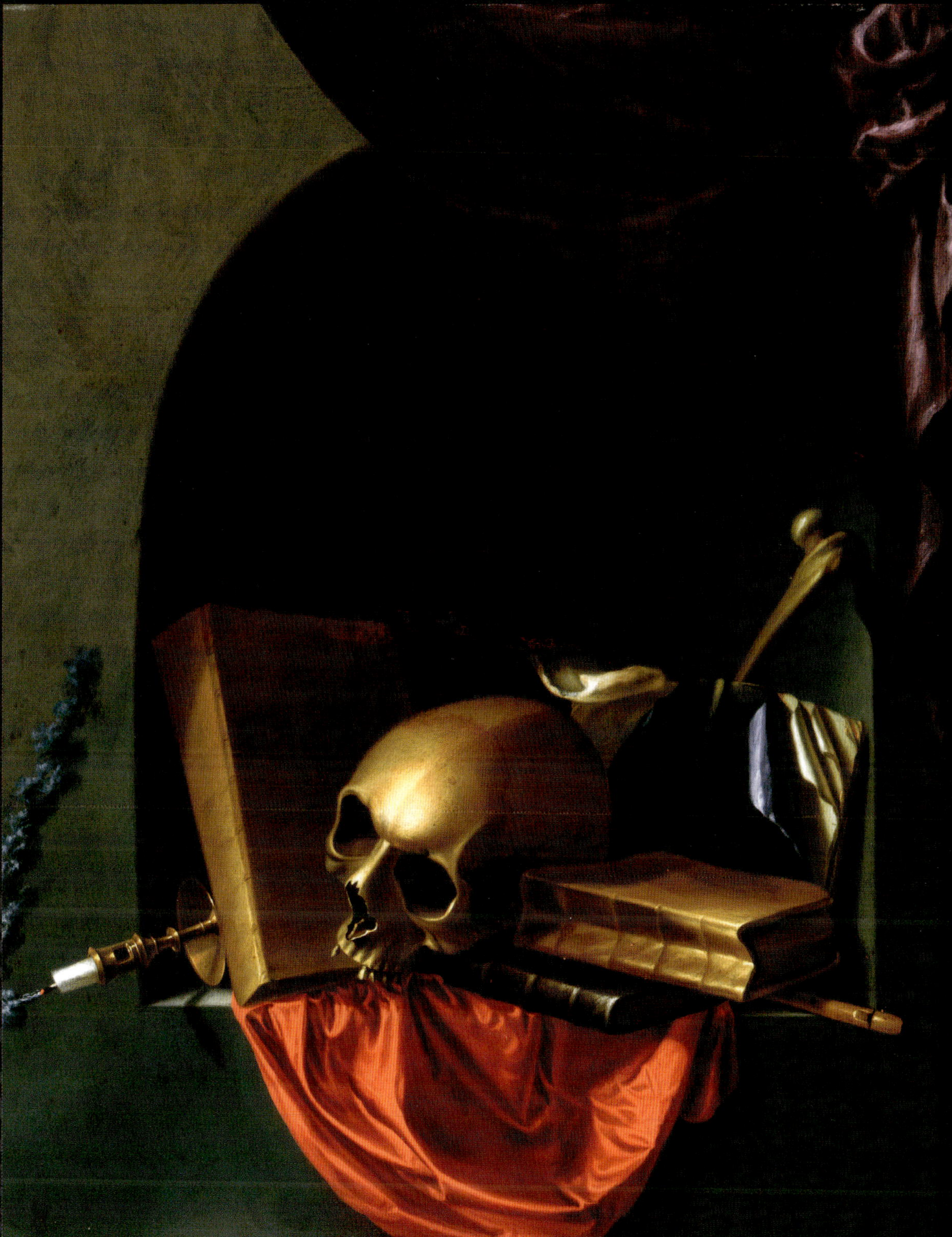

A vanitas painting by Francois Van Daellen.

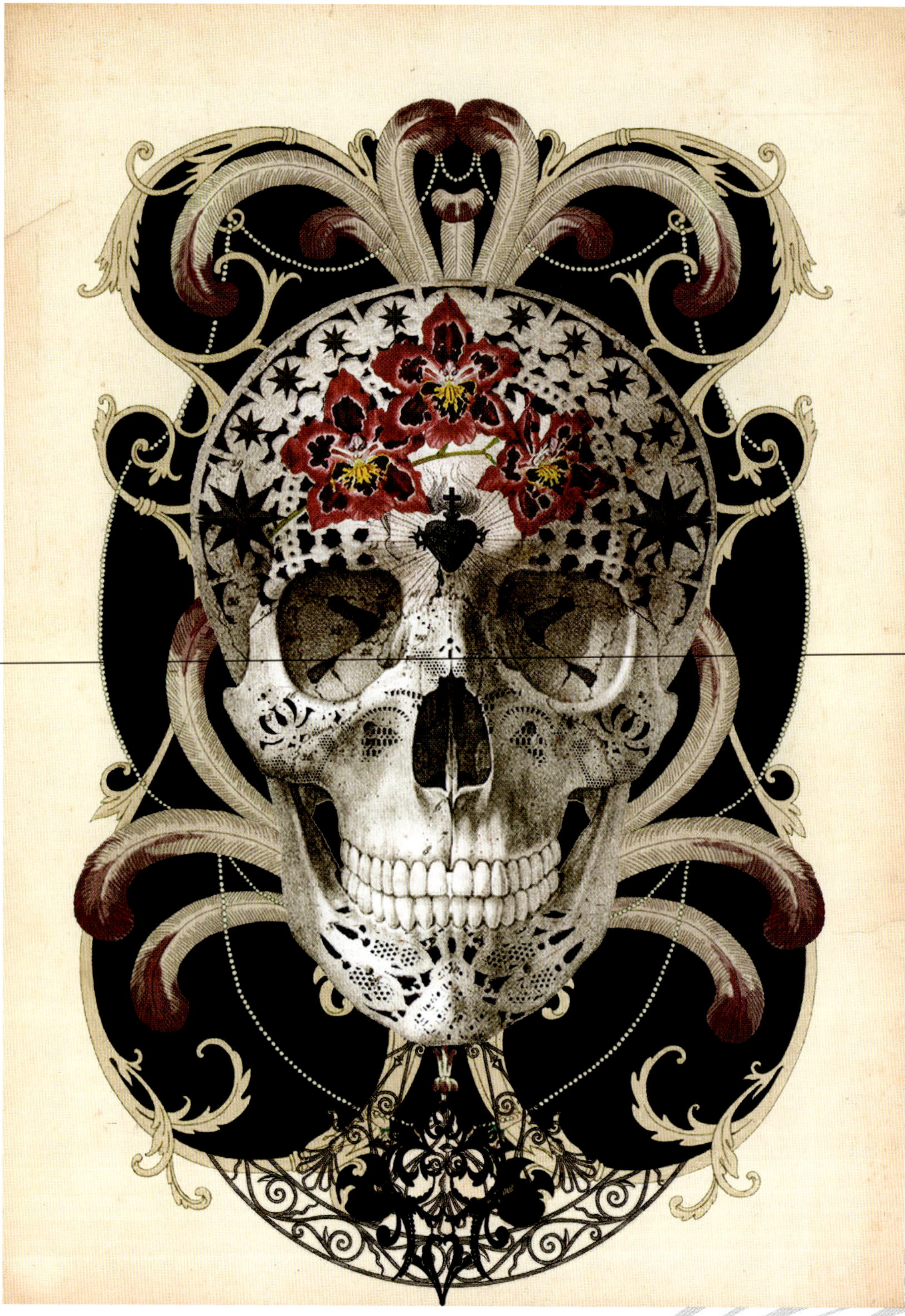

This spread and page 146: Skull art contributions from Isabelle Dalle.

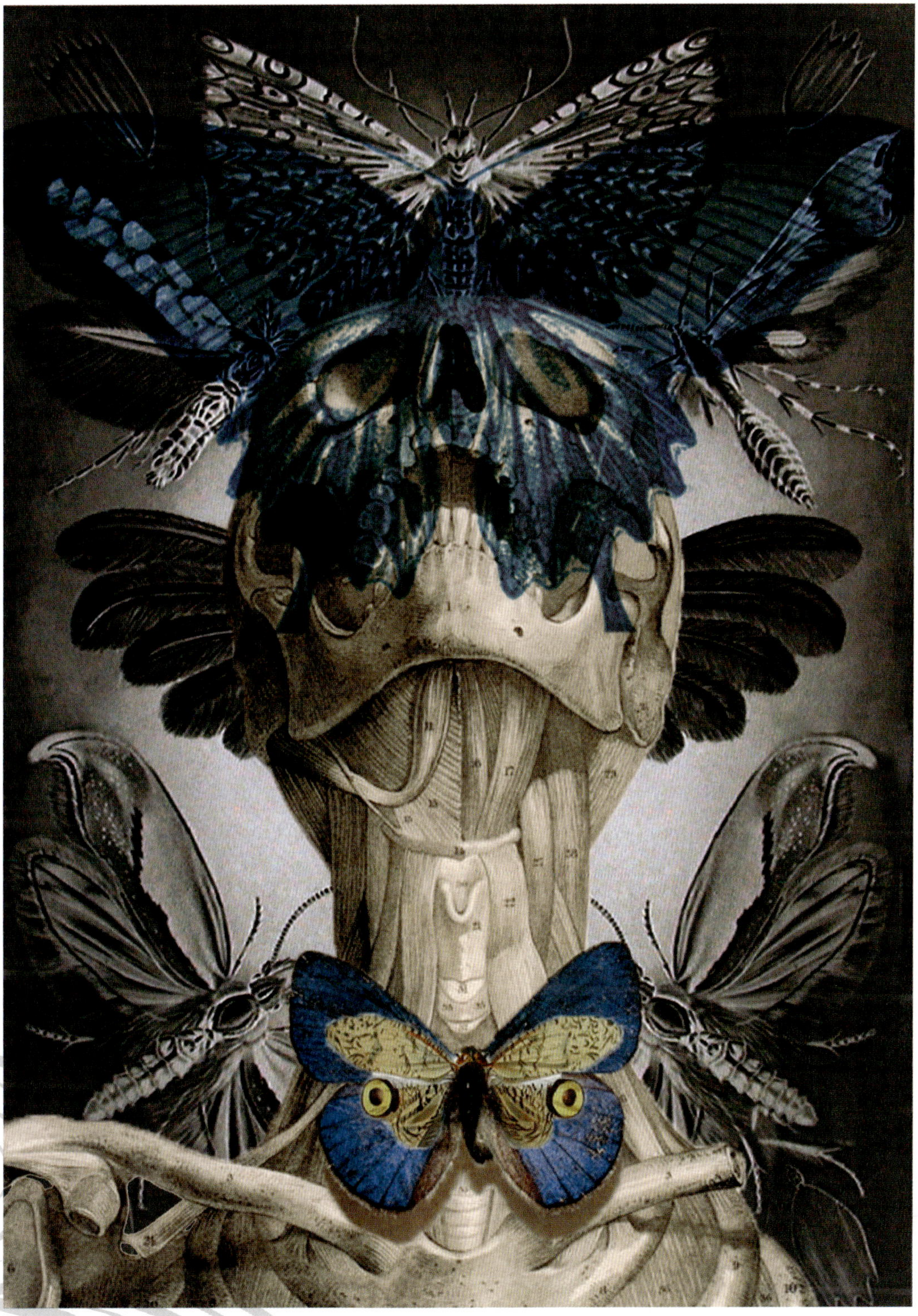

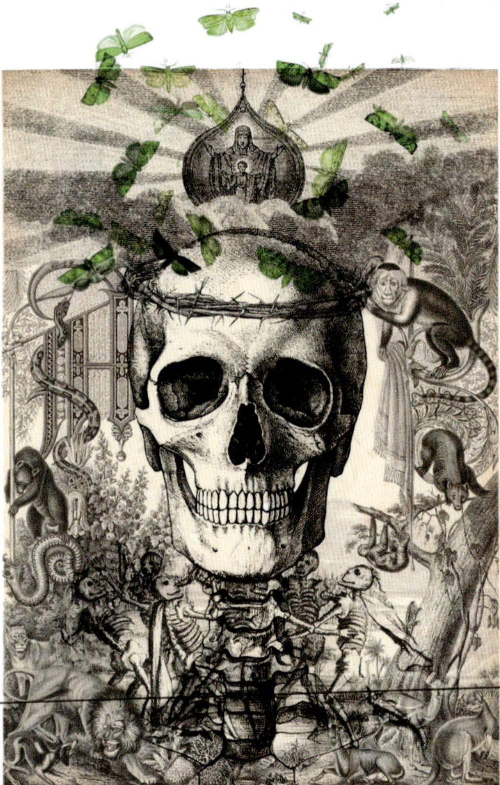

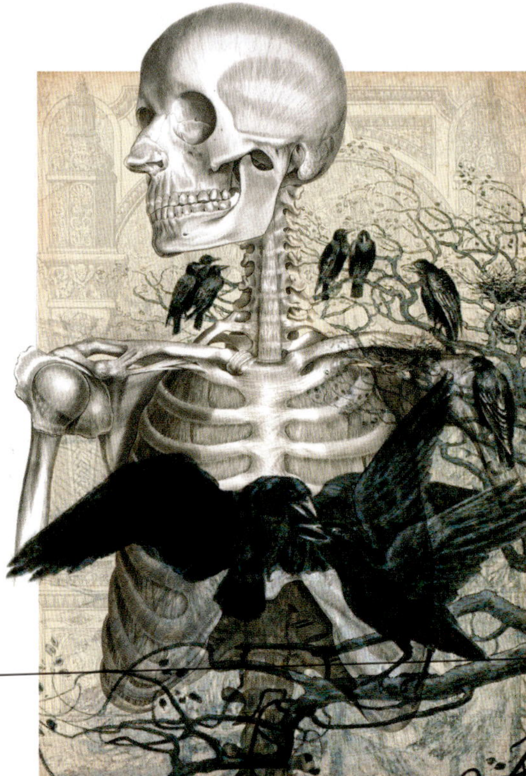

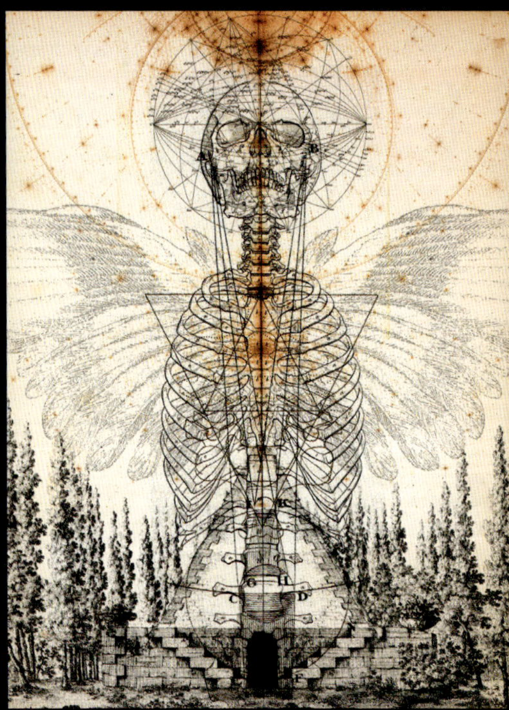

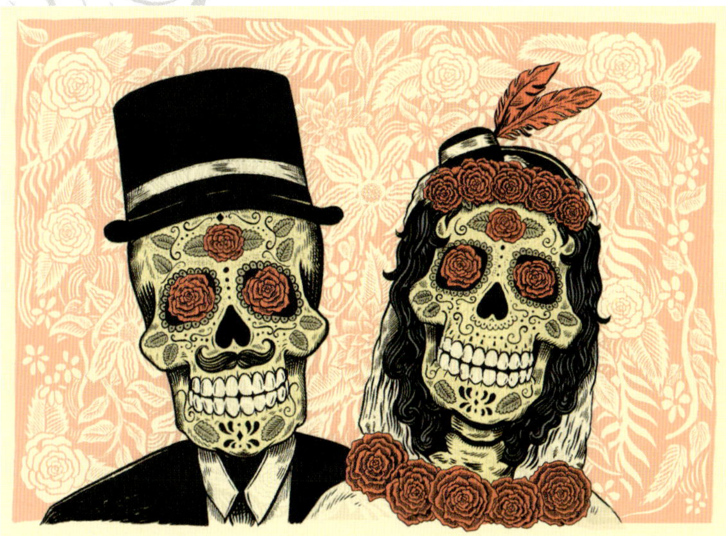

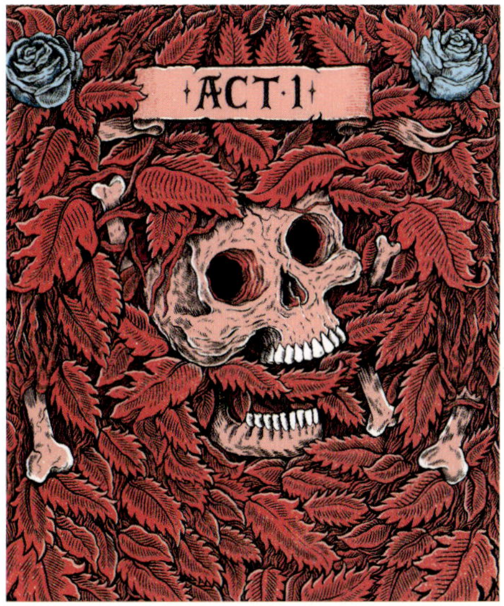

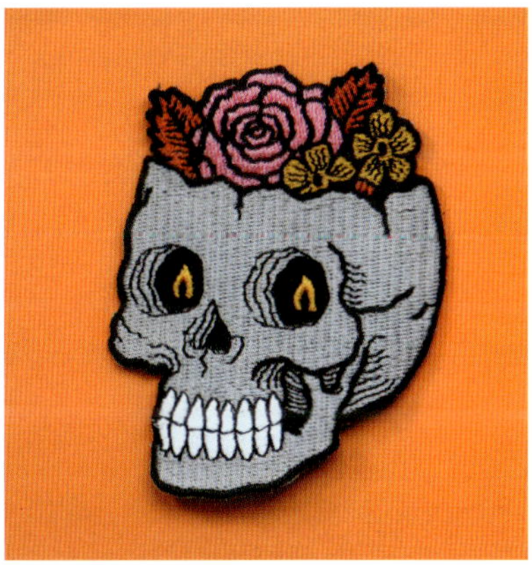

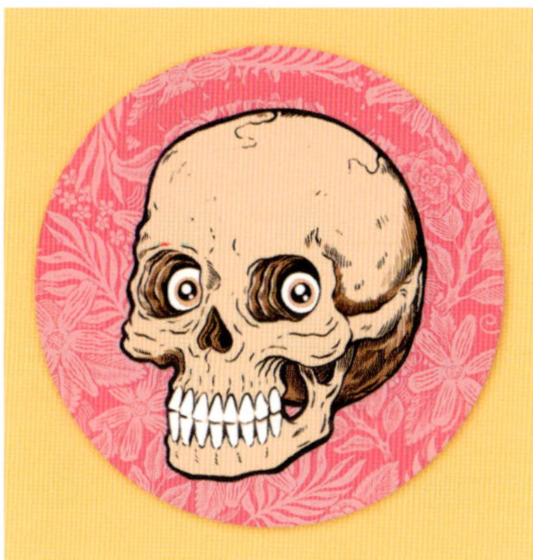

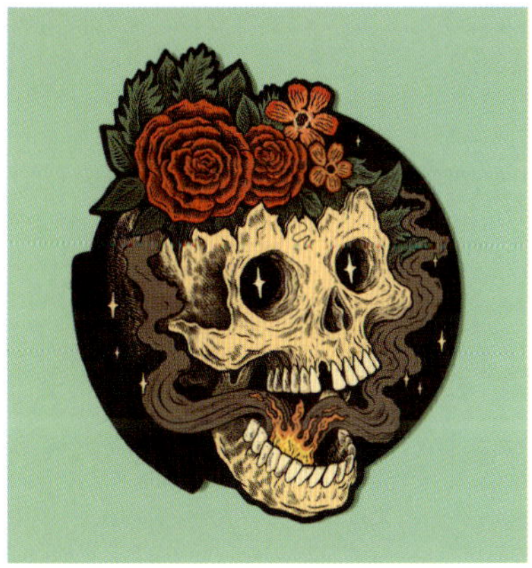

Skull-related artwork by Sam Dunn.

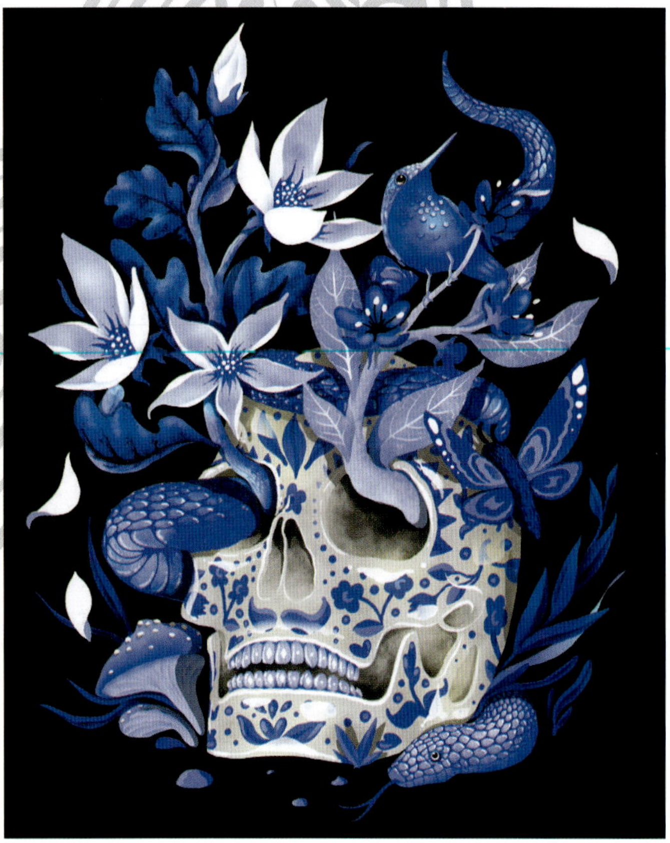

Beauty Immortal, an illustration by artists Lidija Paradinovic and Kimby Dawson.

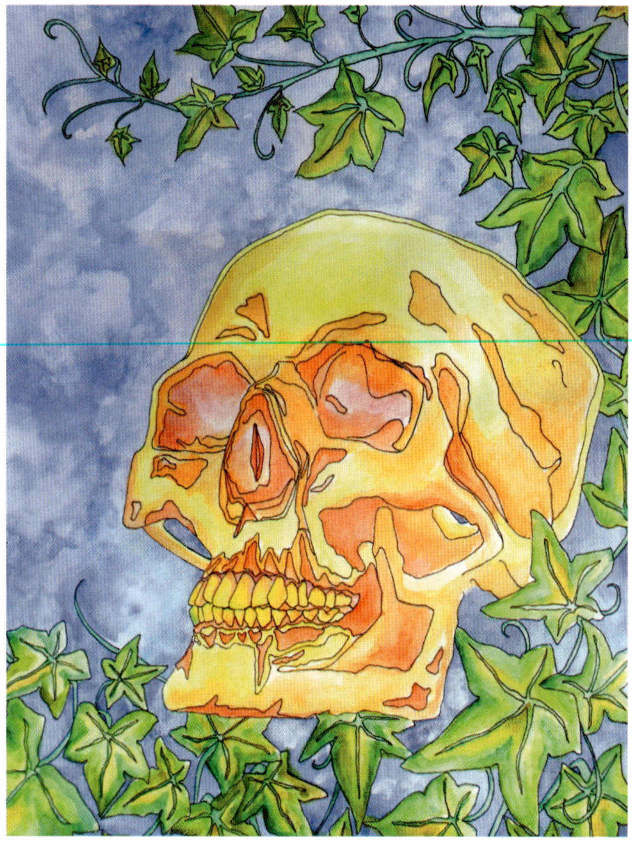

Artist Emily Marchesiello painted this work, named *Ivy & Skull.*

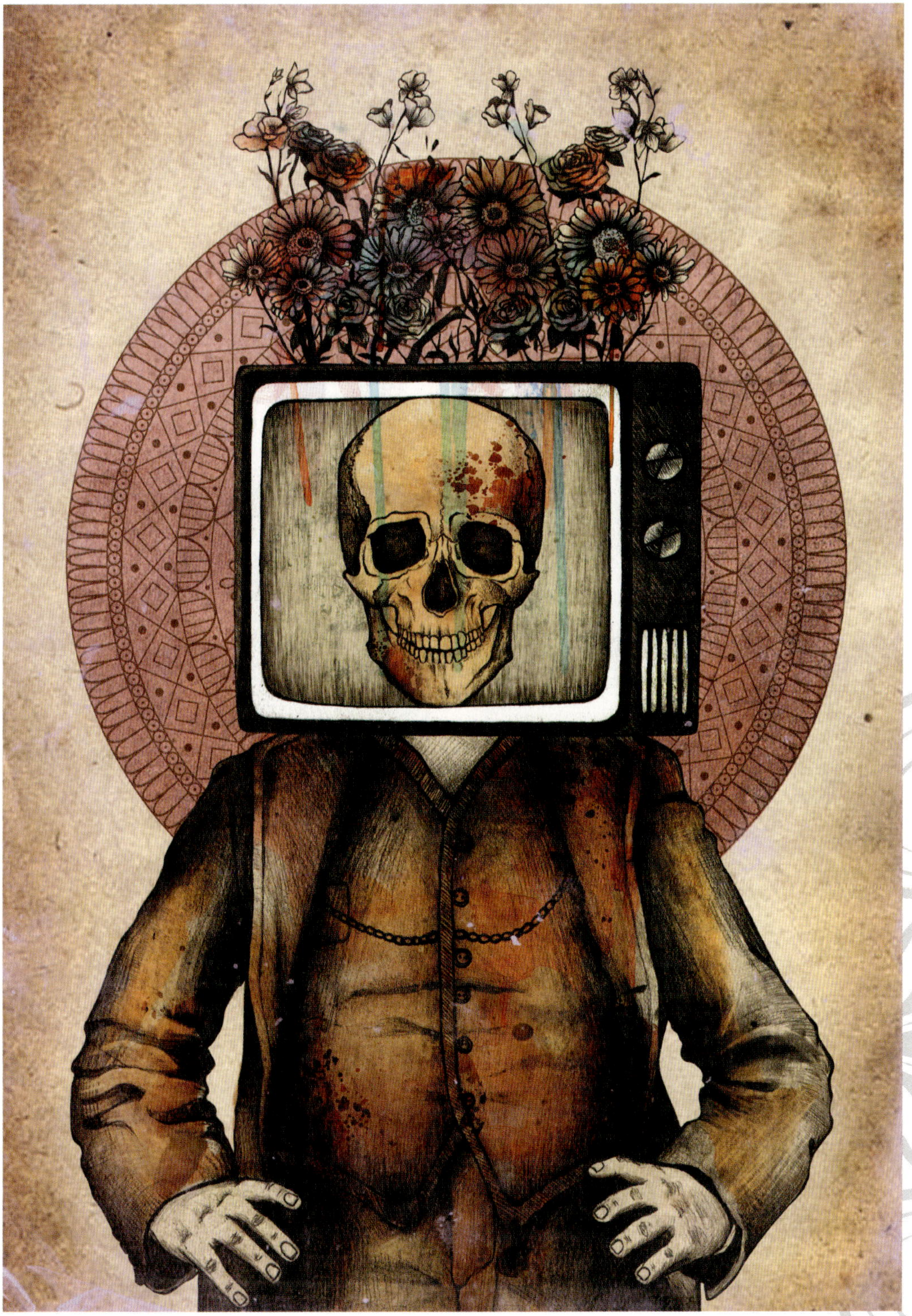

A digital illustration by Alisa Grace.

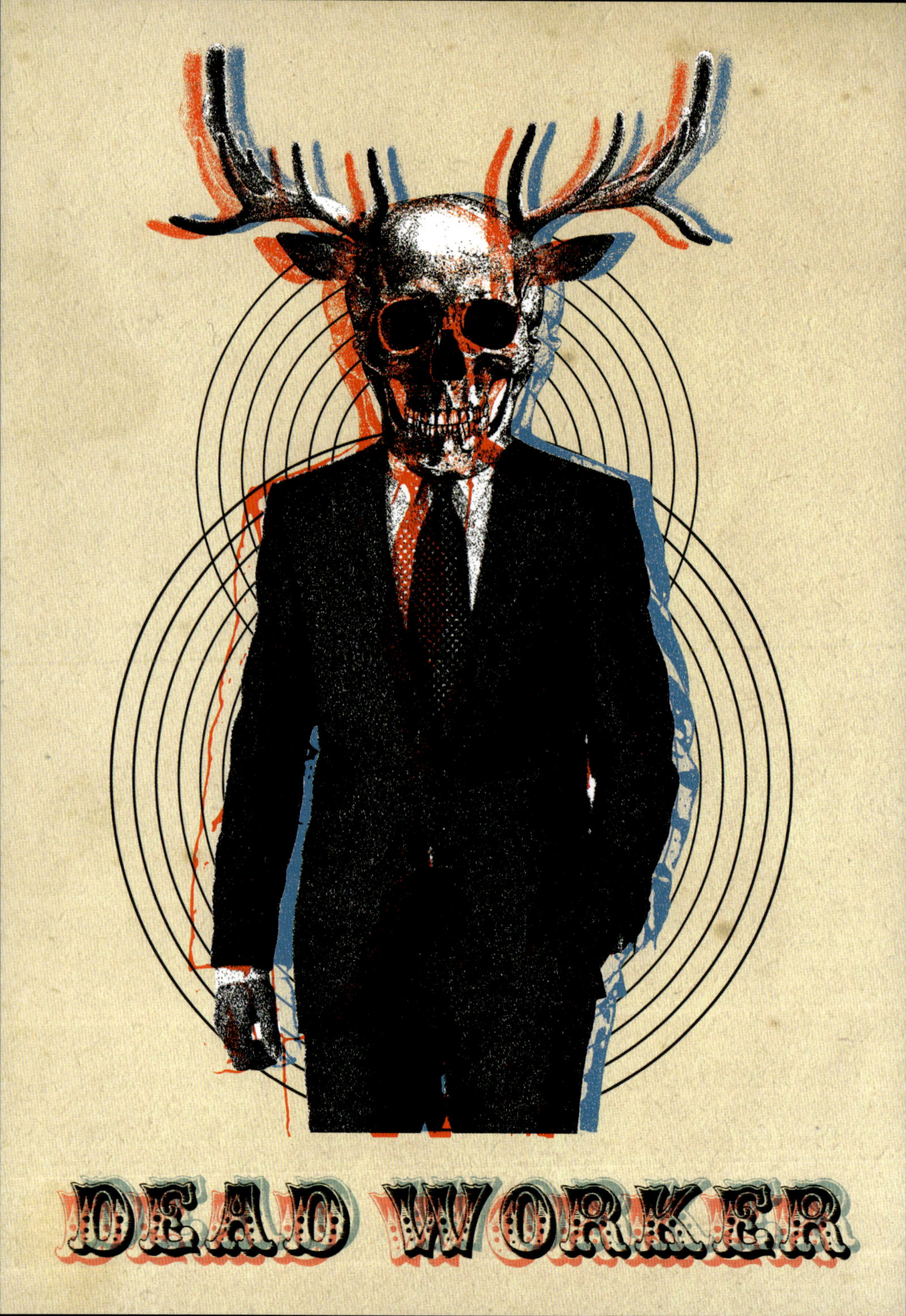

Dead Worker, a 3-D illustration by Alice Dodard.

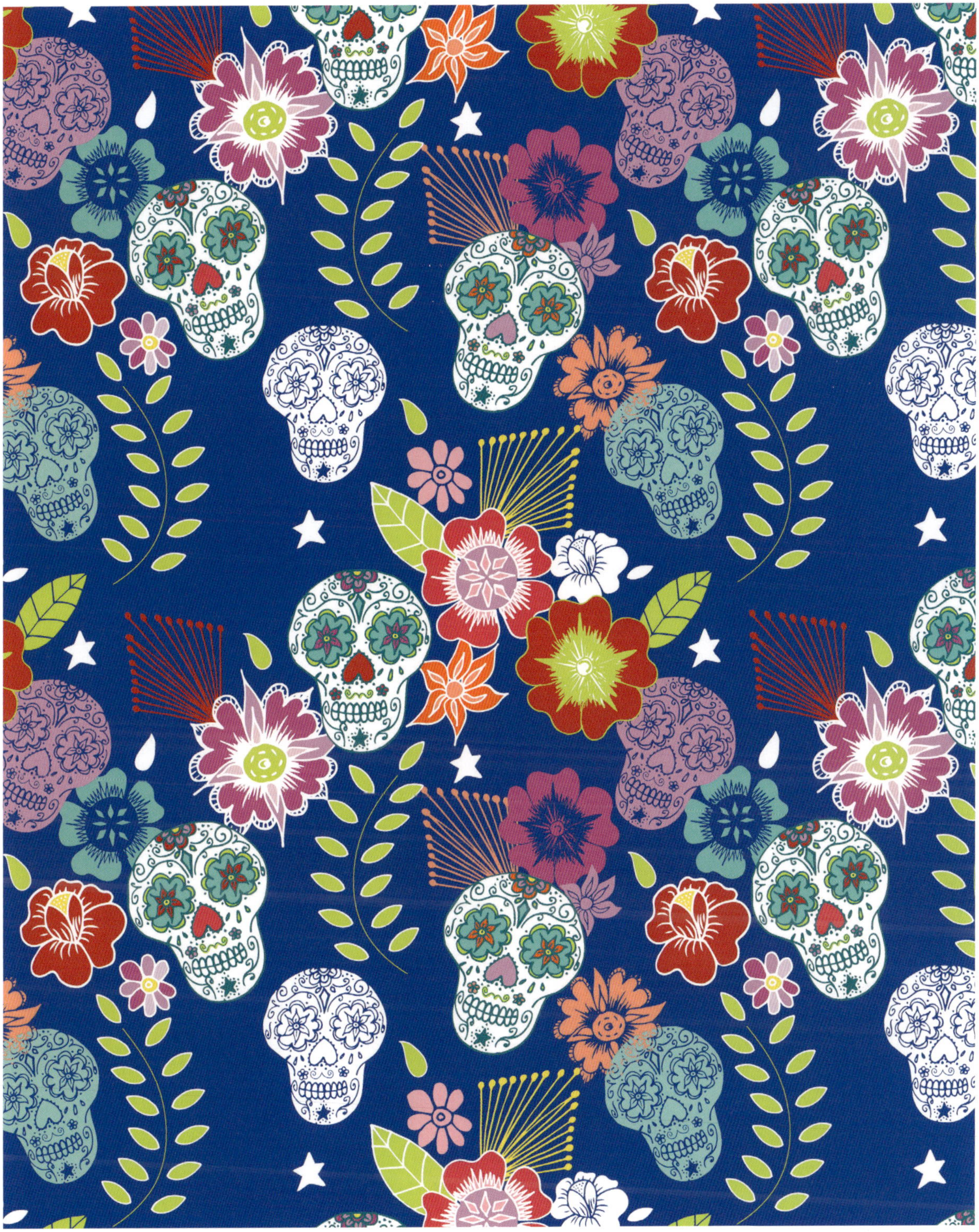

A Day of the Dead–themed collage by artist Anneline Sophia.

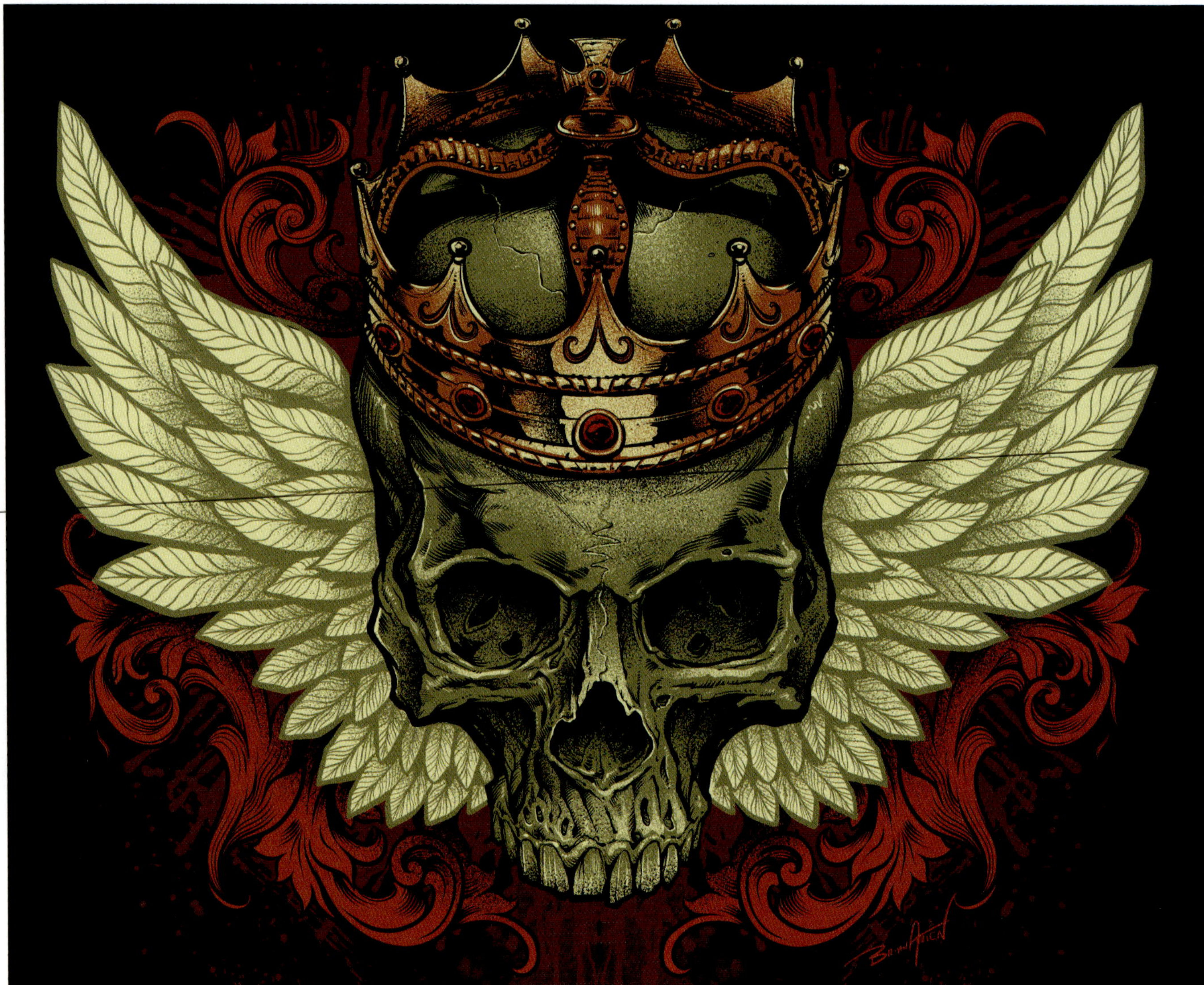

This skull wearing a crown was designed by artist Brian Allen.

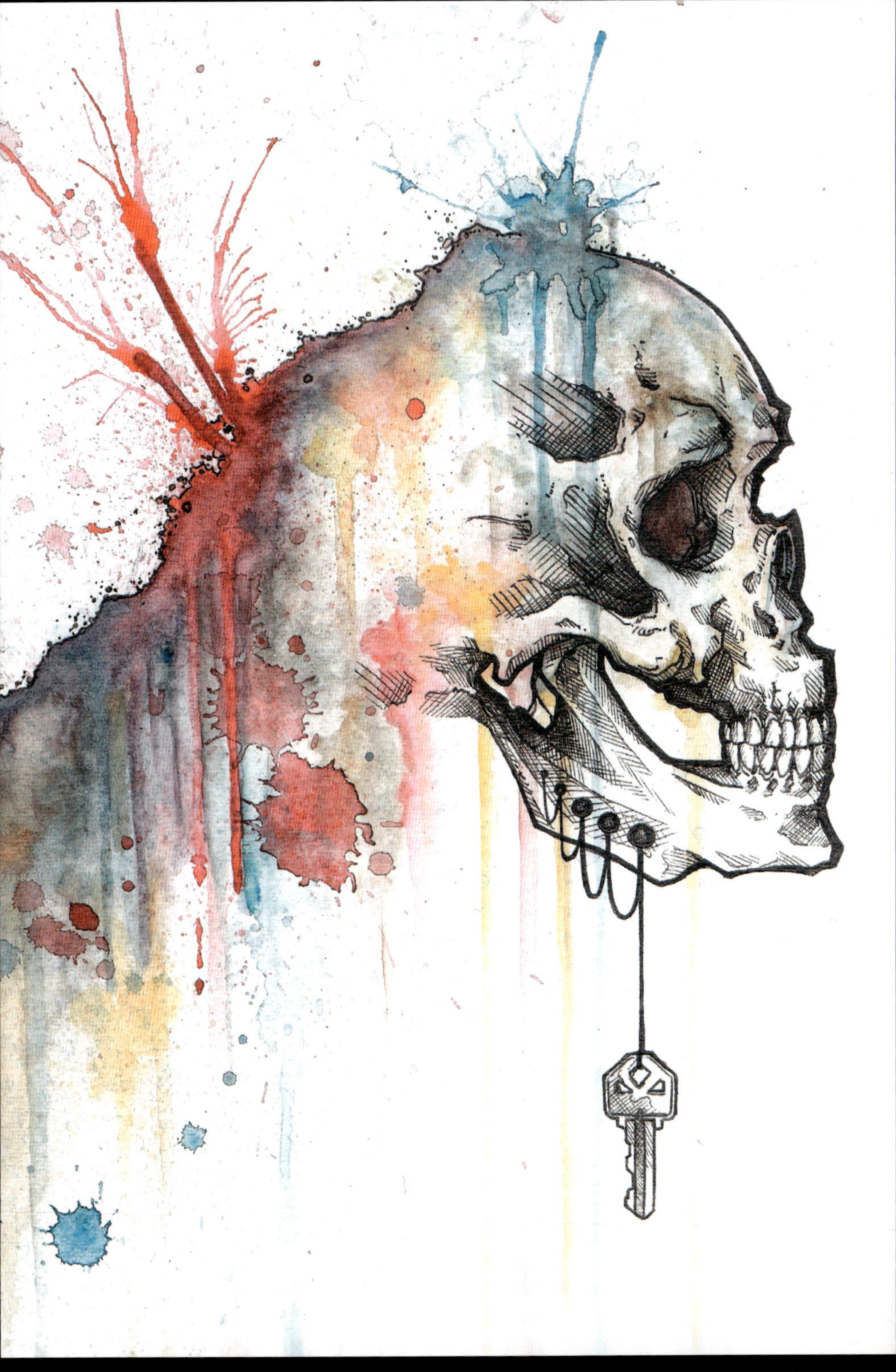

Headrush and Lockjaw by artist Ian Lawrence.

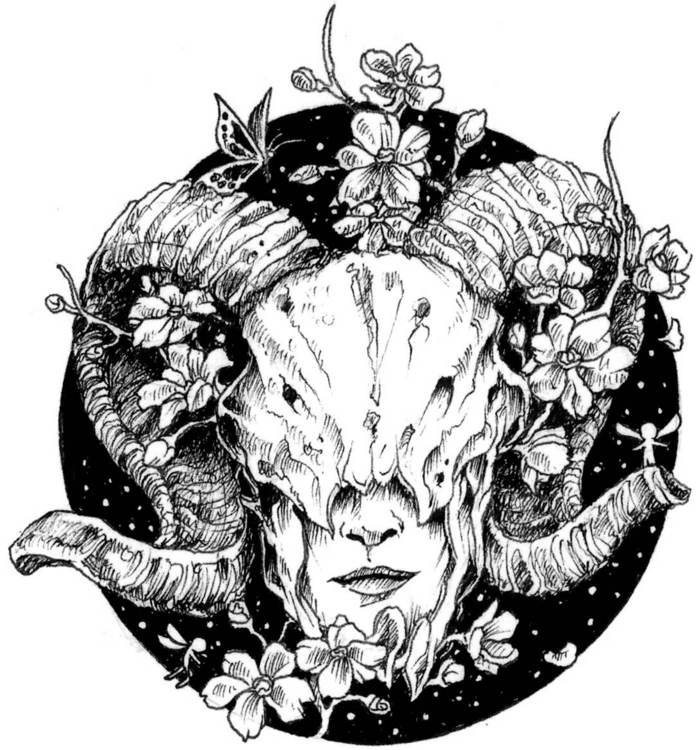

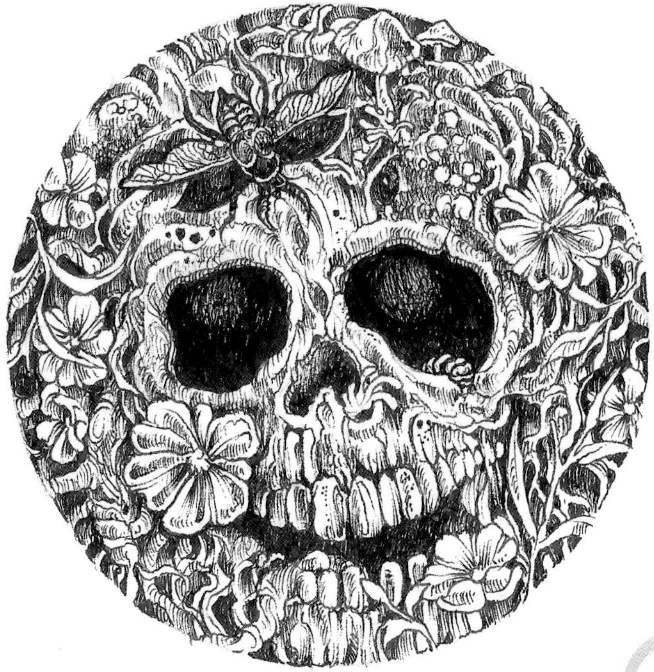

This spread: Illustrated by Kerby Rosanes (www.kerbyrosanes.com) using pen and ink on paper.

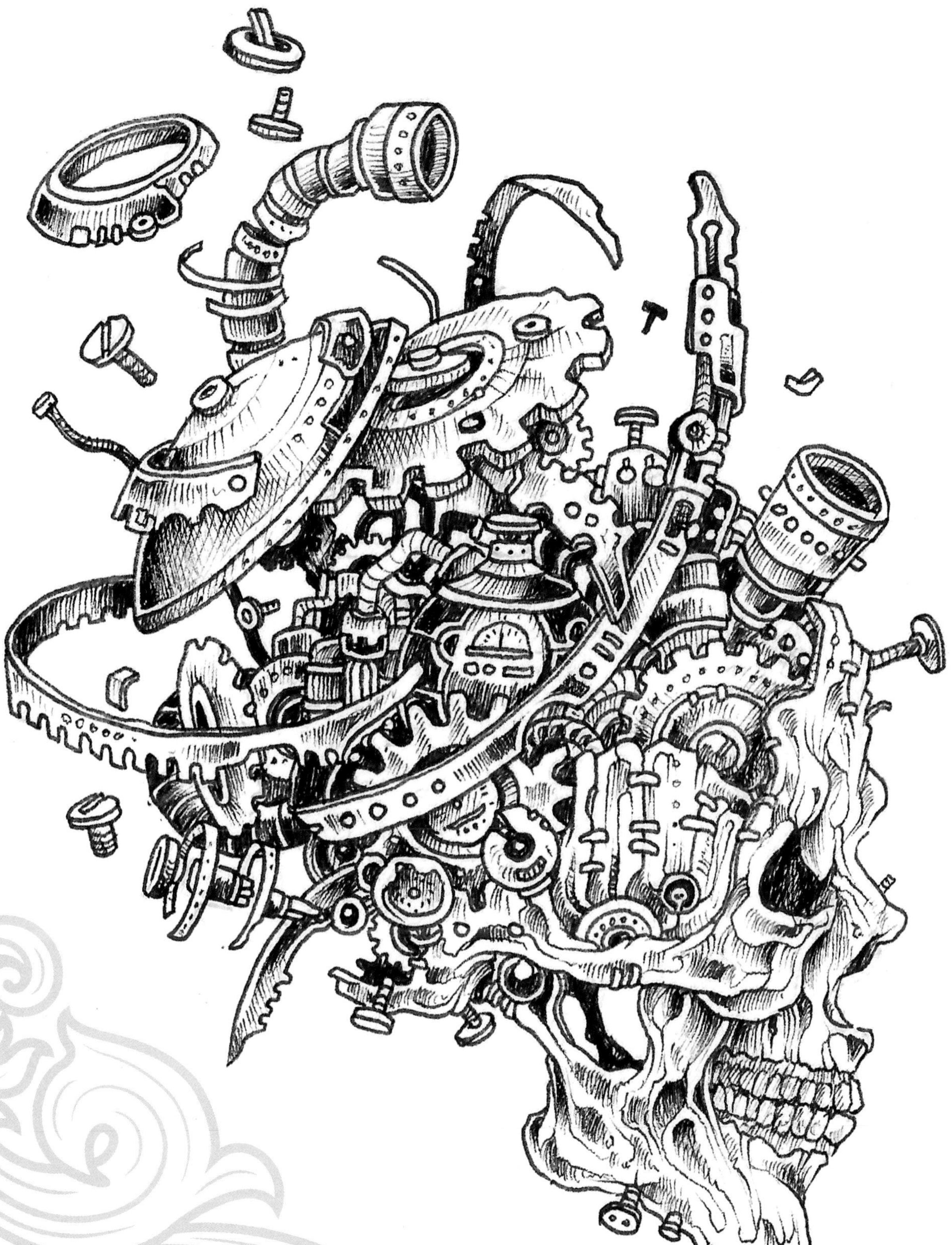

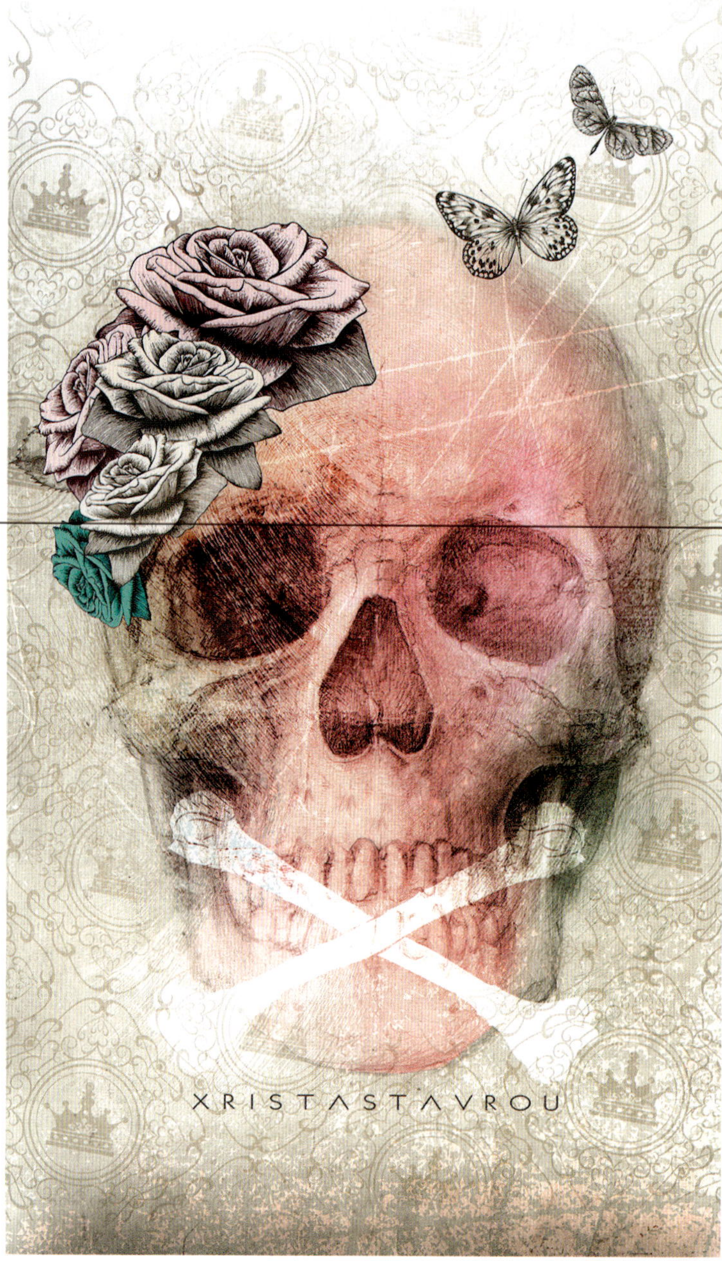

Skull Queen by artist Xrista Stavrou.

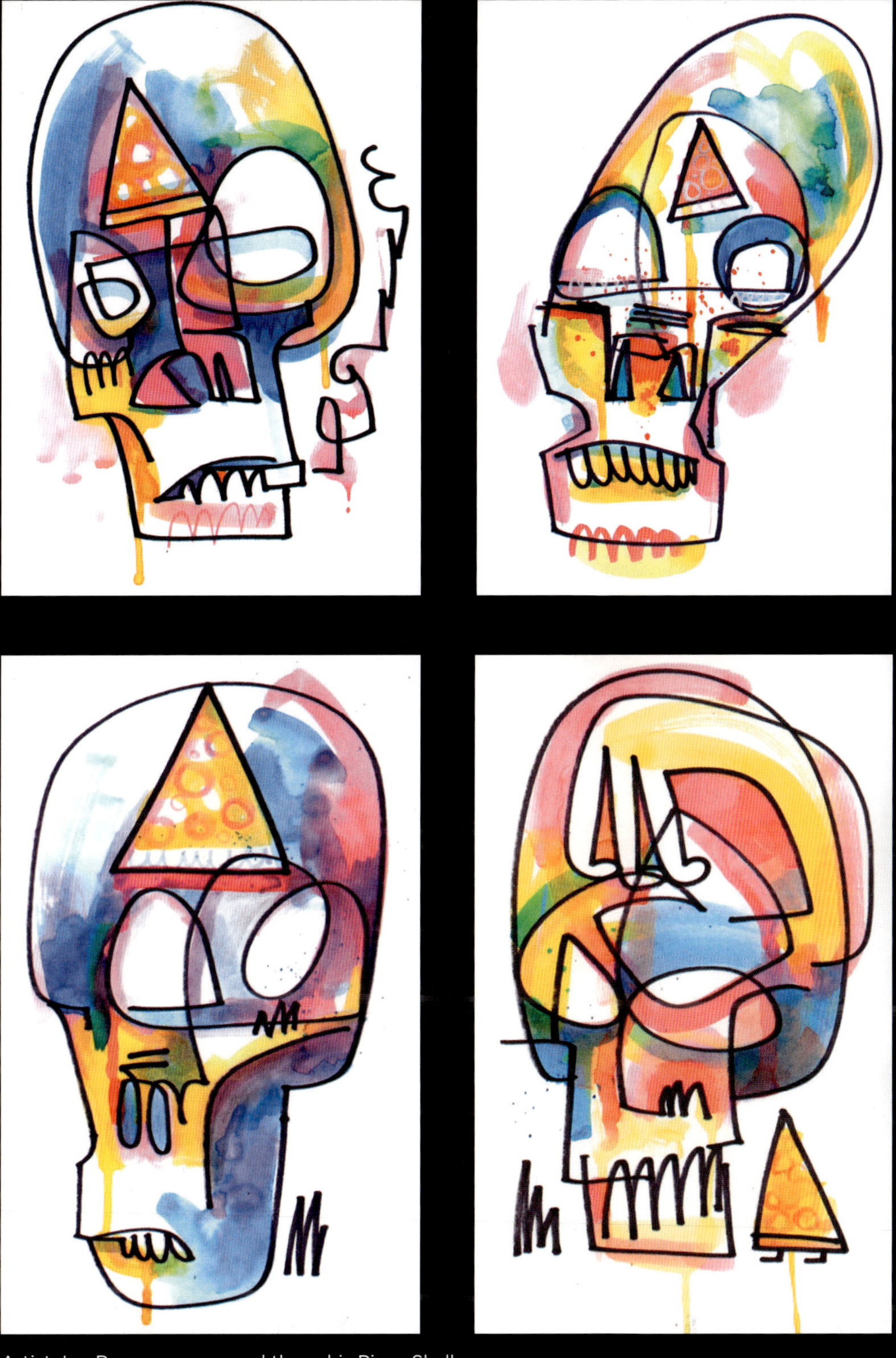

Artist Jon Burgerman named these his Pizza Skulls.

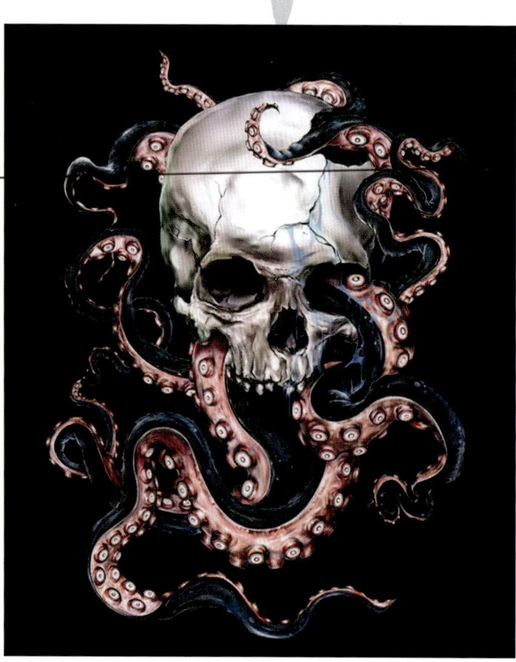

Skulltopus by Eric Pineda combines a skull with octopus arms.

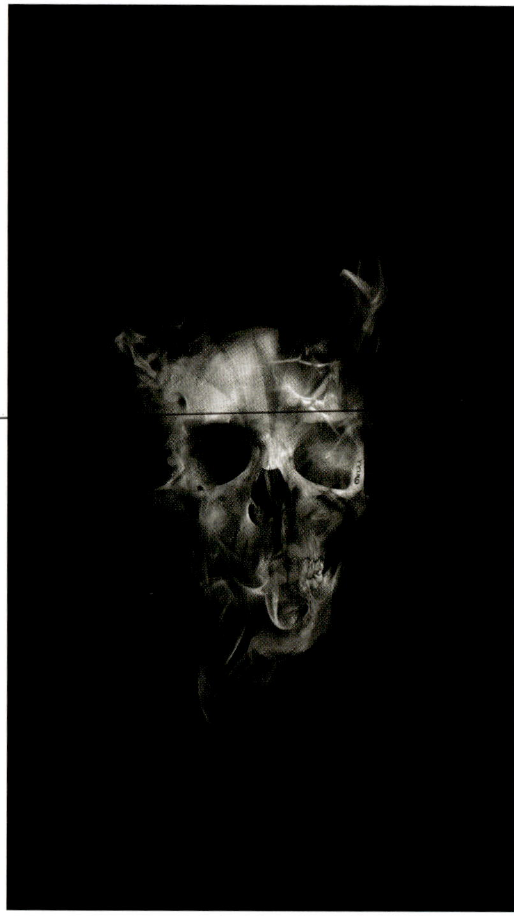

This ghostly skull was designed by artist Gabor Nagy.

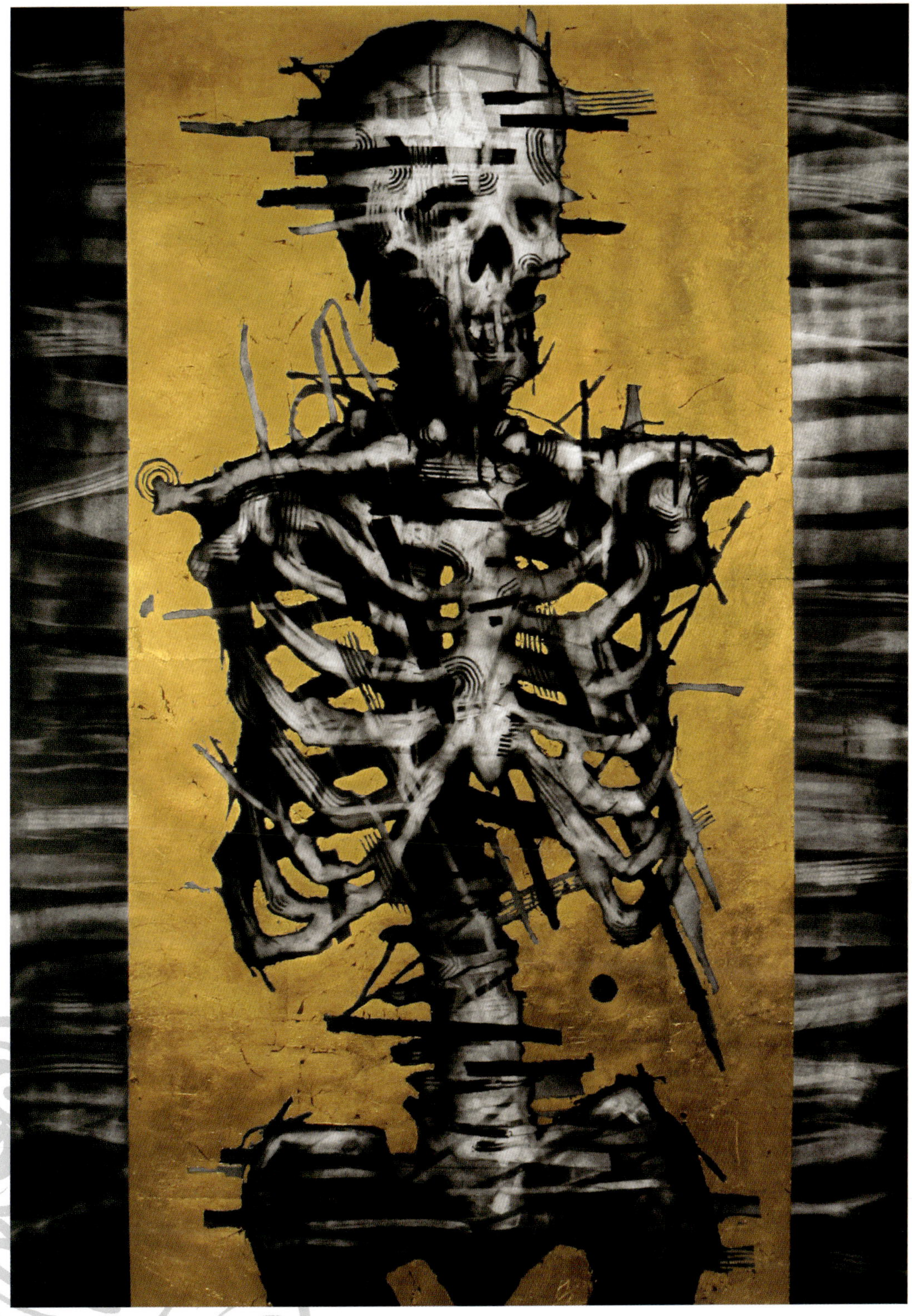

Joseph Loughborough designed this piece, called *Flowers of Alcatraz*.

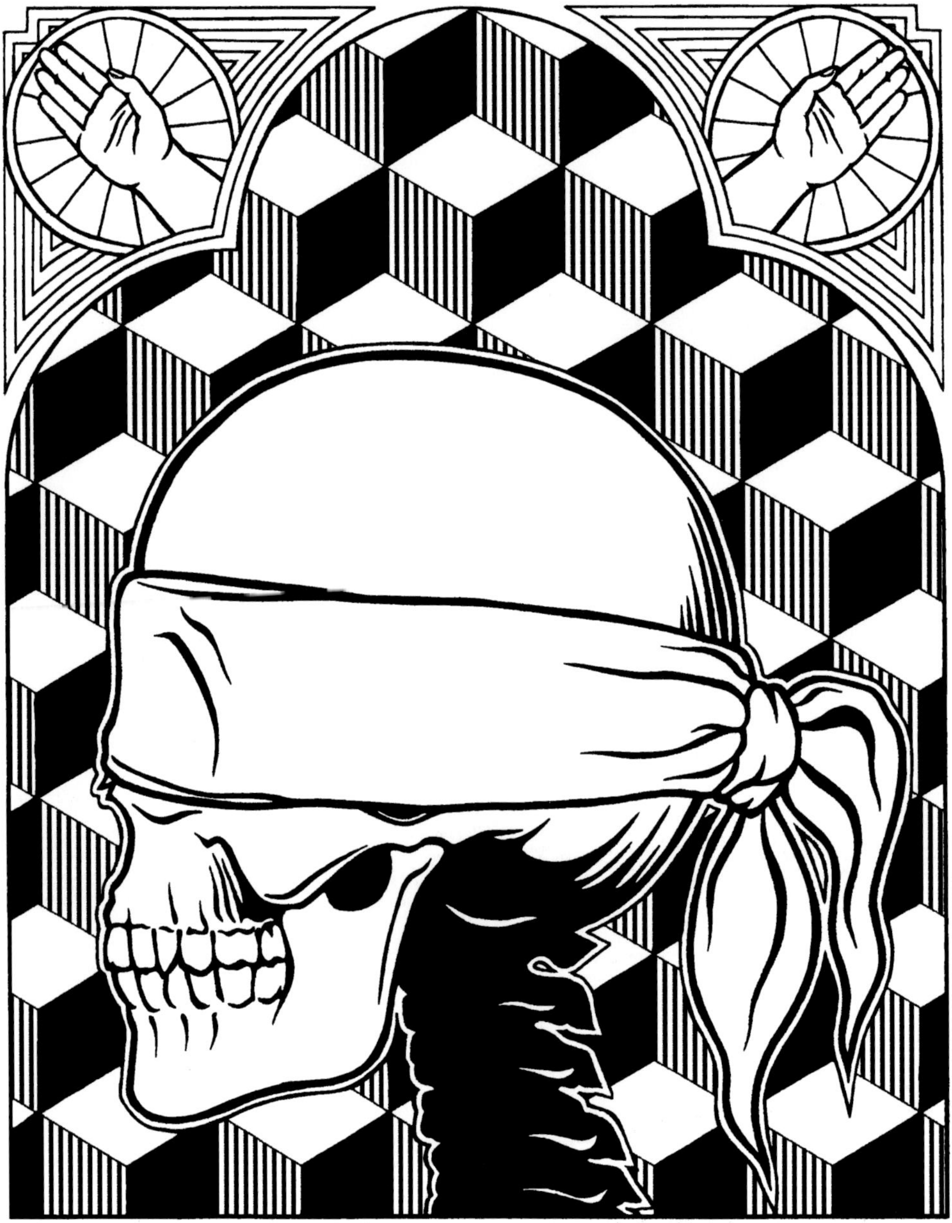

This spread: Work by artist Jeremy Elder.

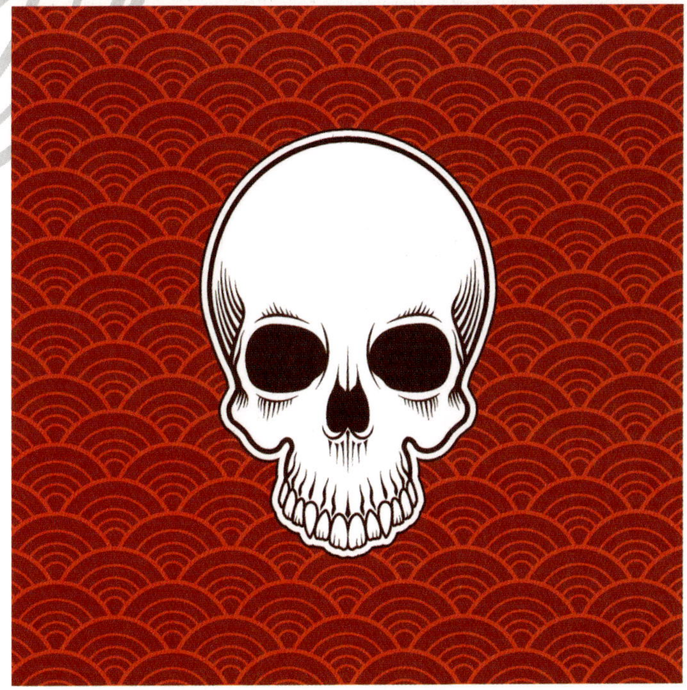

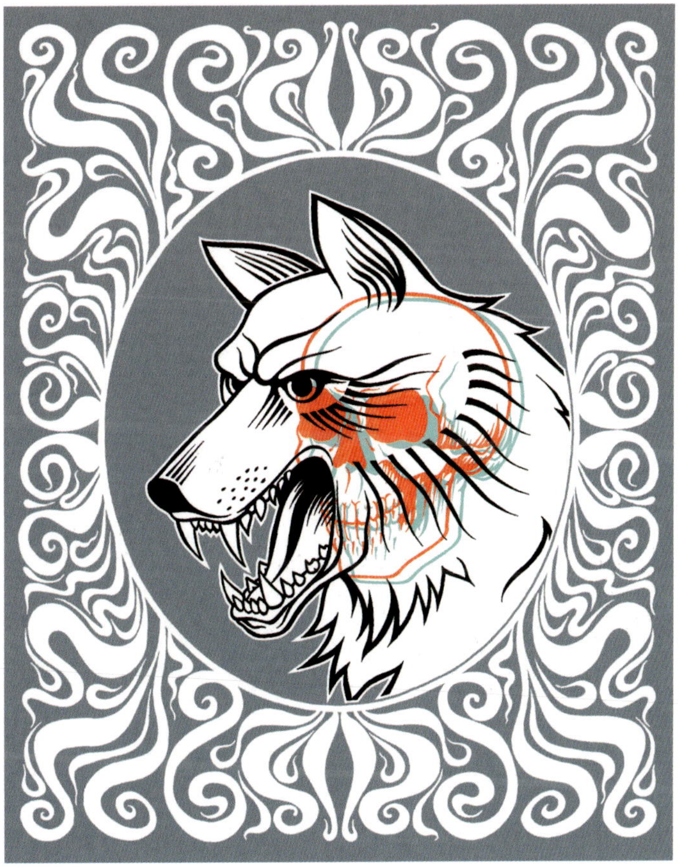

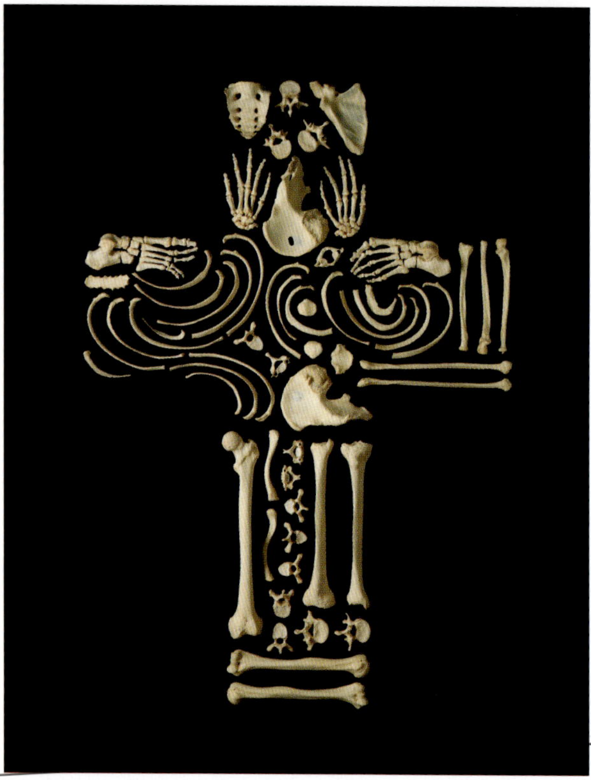

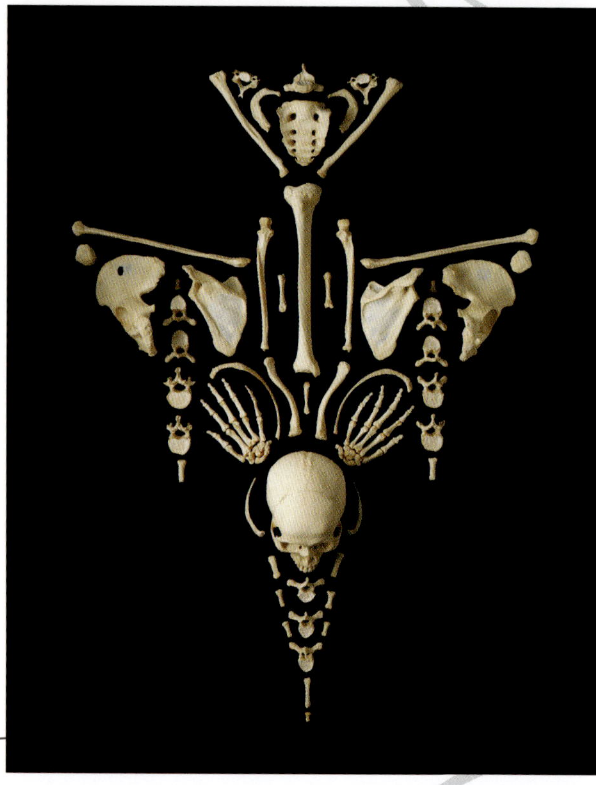

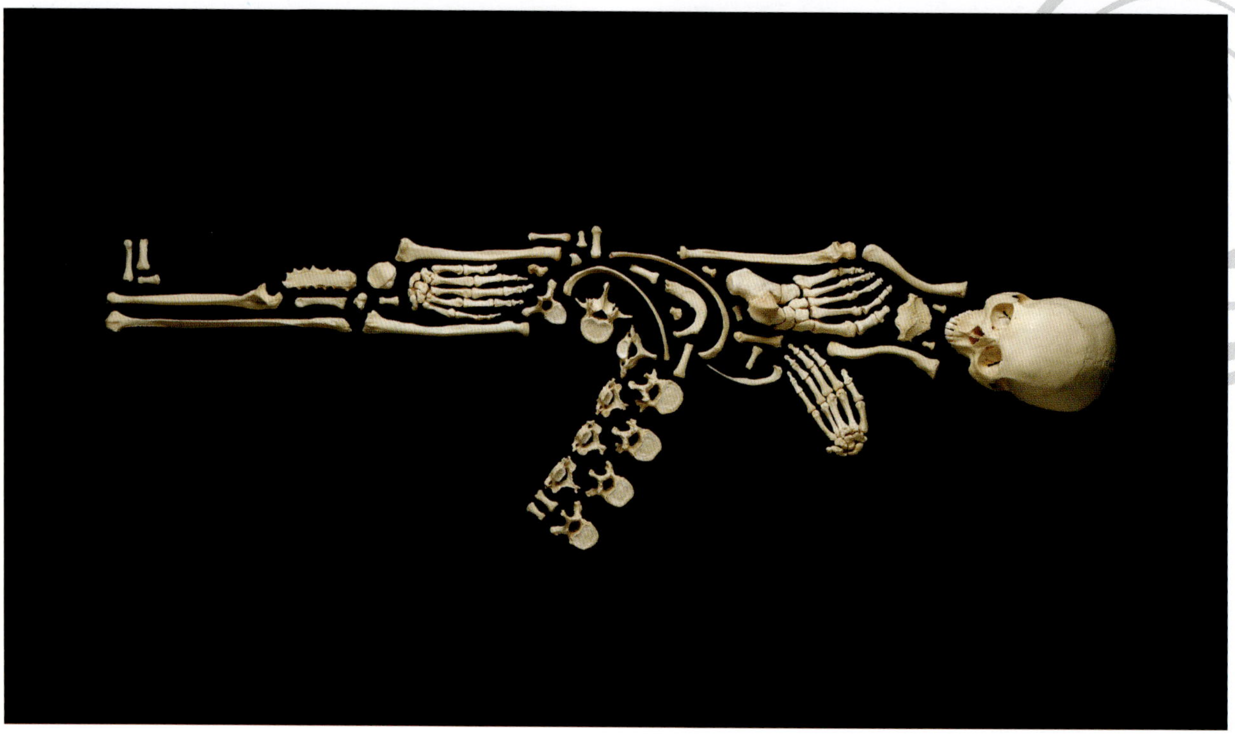

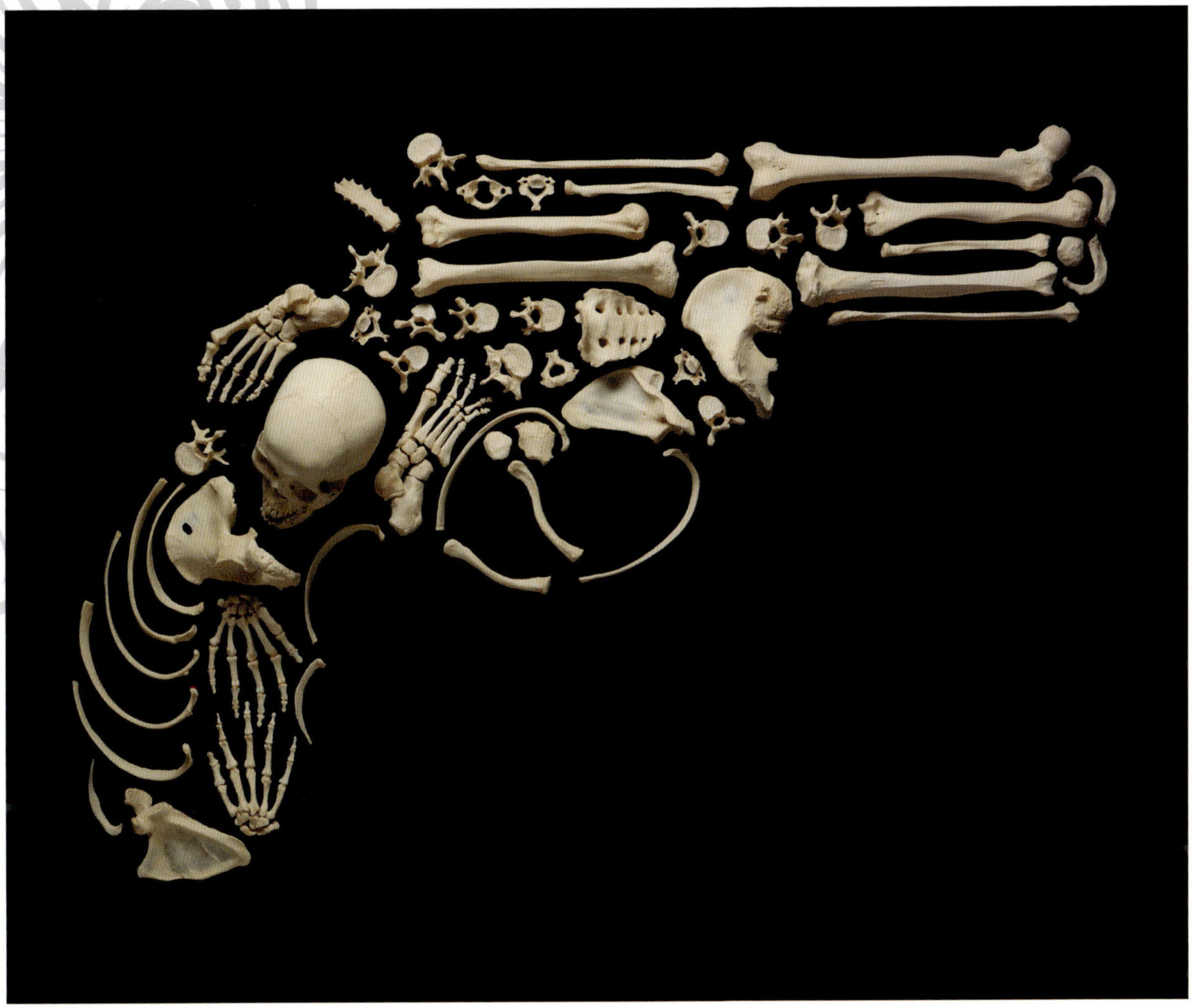

This spread: Selections from Francois Robert's photography series, *Stop the Violence,* including *Cross, Fighter Jet, Kalashnikov,* and *Gun.*

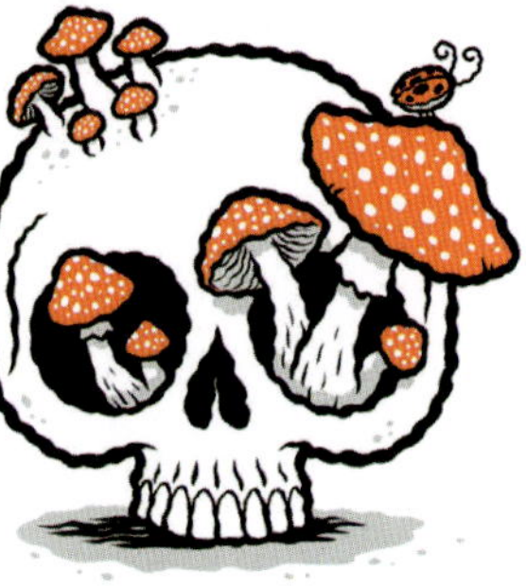

LIFE AFTER DEATH

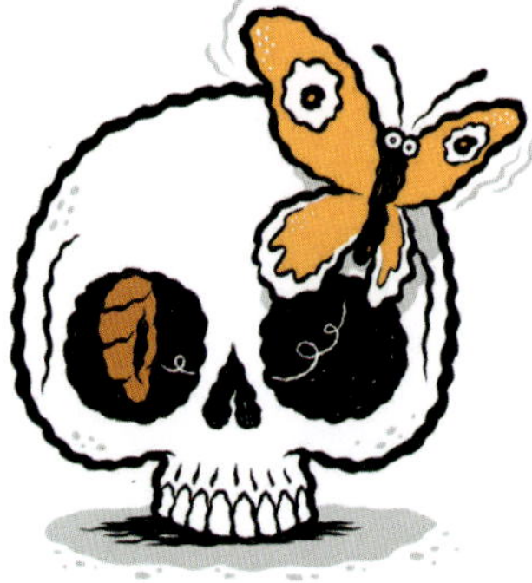

LIFE AFTER DEATH

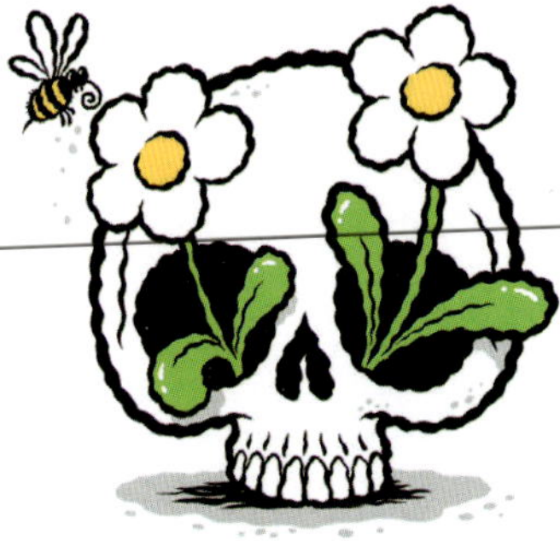

LIFE AFTER DEATH

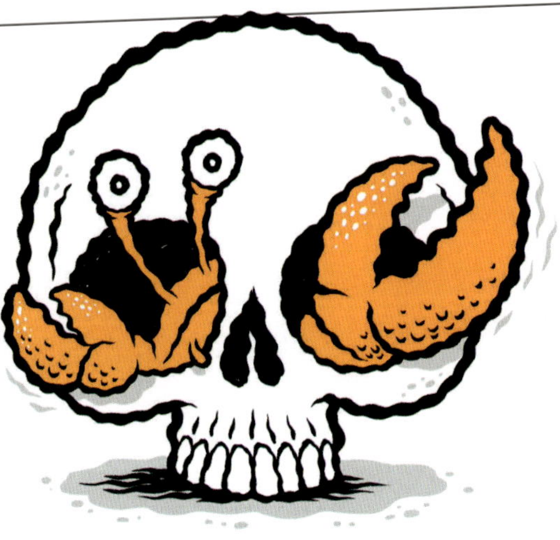

LIFE AFTER DEATH

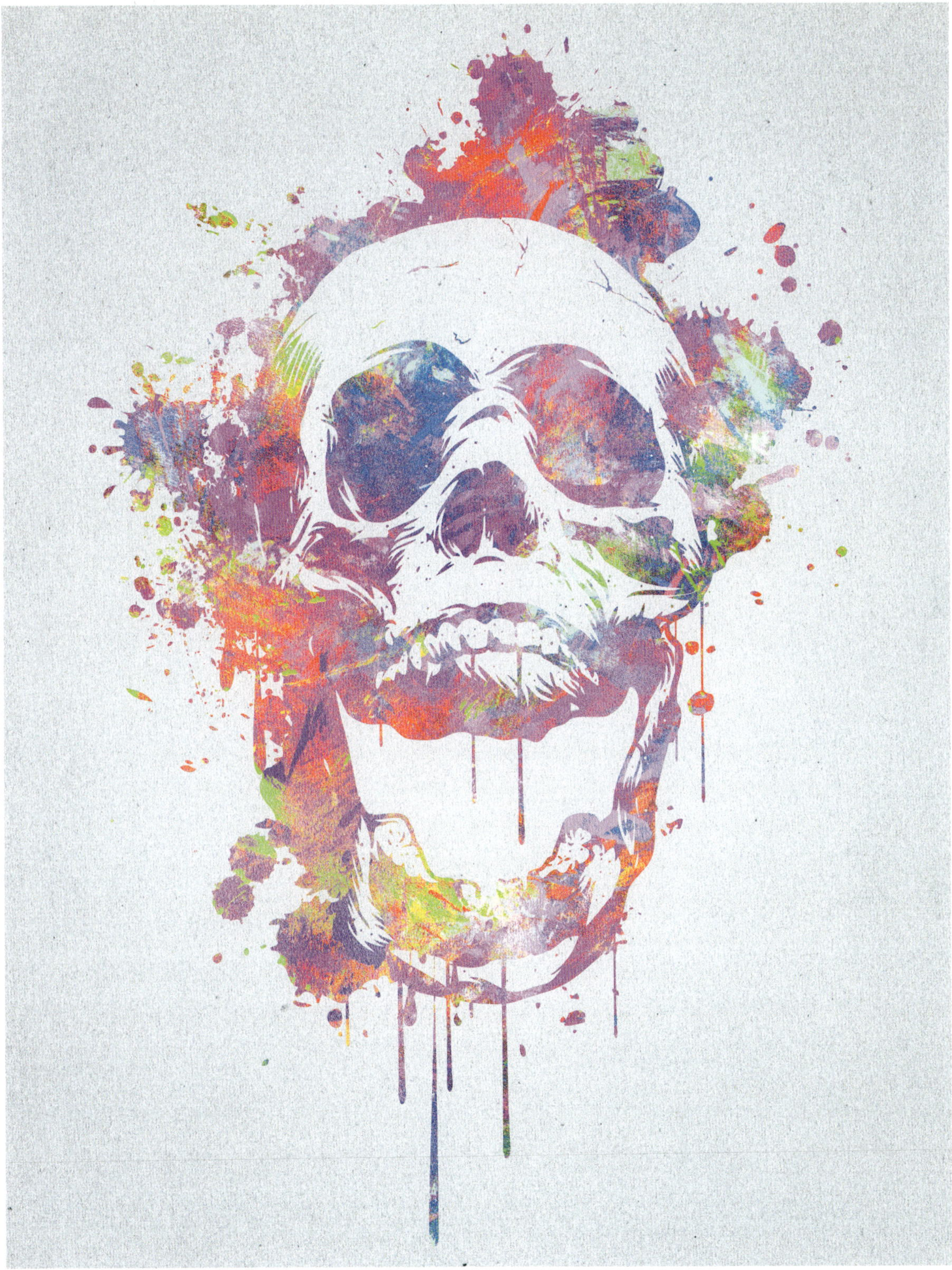

Artwork by Philipp Rietz.

This spread: Drawings by Motohiro Nezu showing celebrities with skull faces; celebrities shown include Biggie Smalls, Bob Marley, Kurt Cobain, Elvis Presley. An "Indian Head" is opposite.

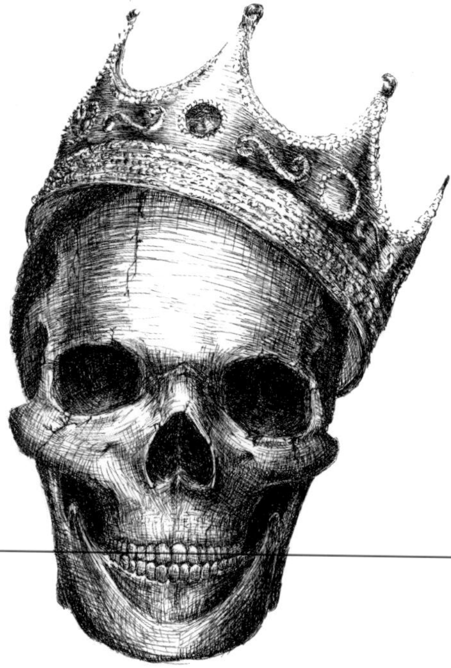

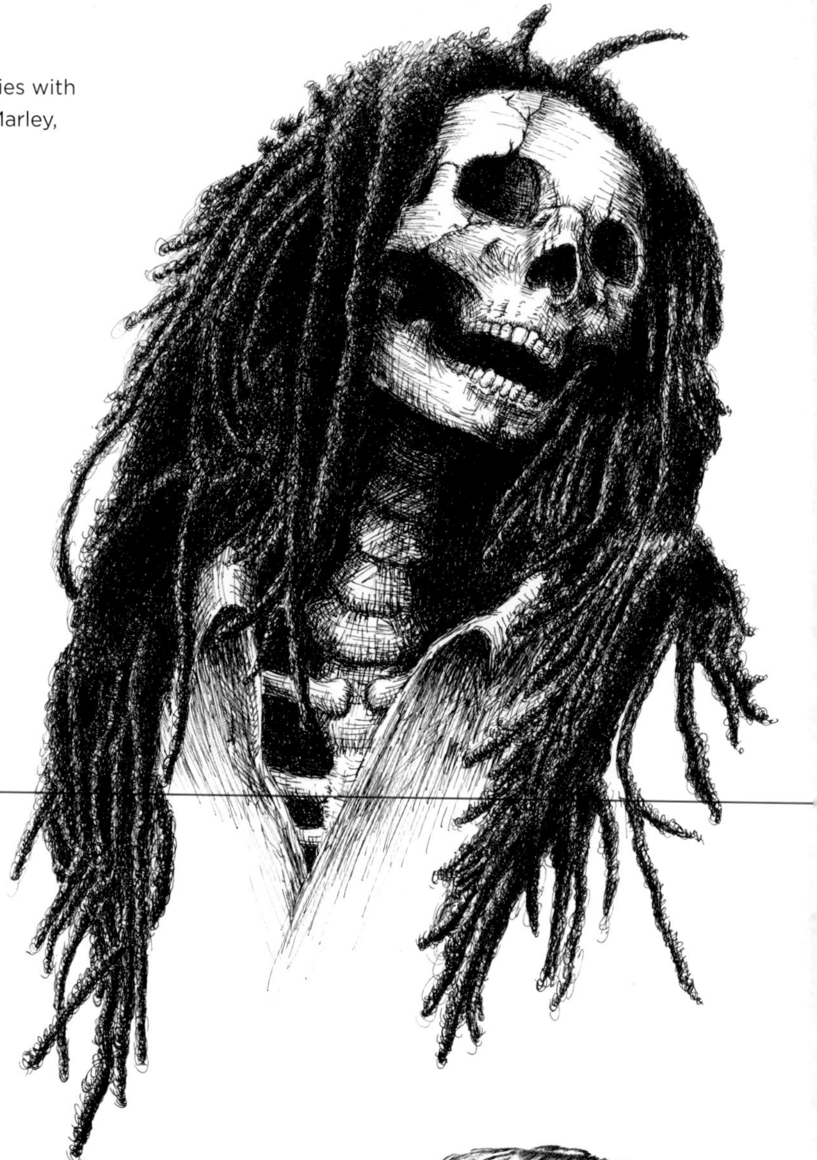

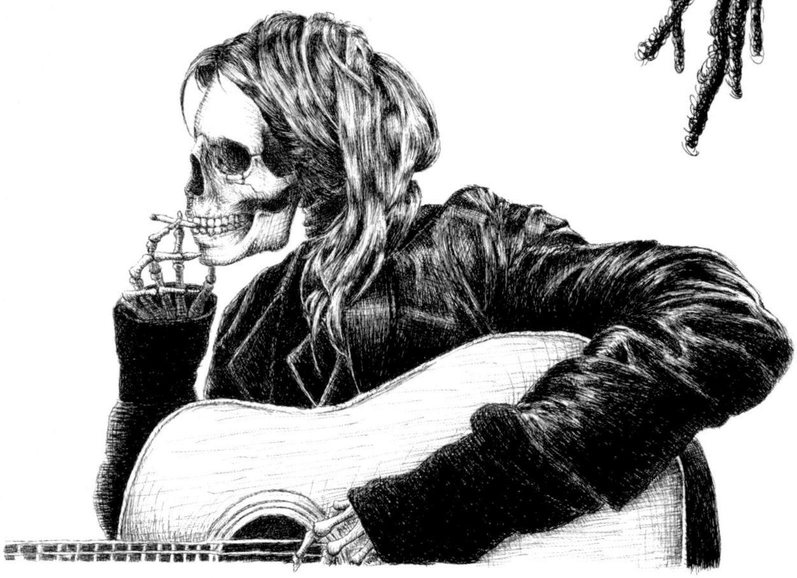

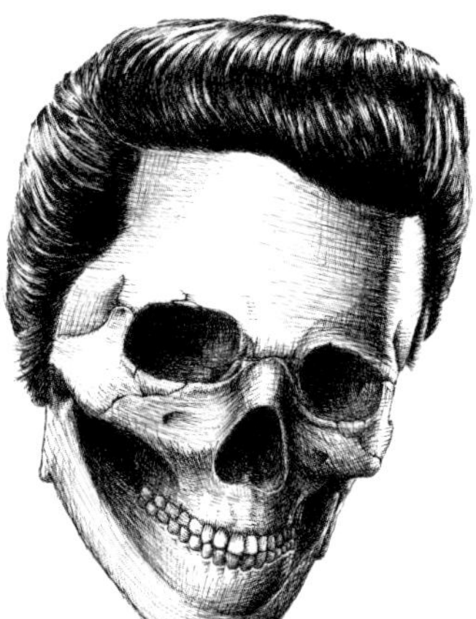

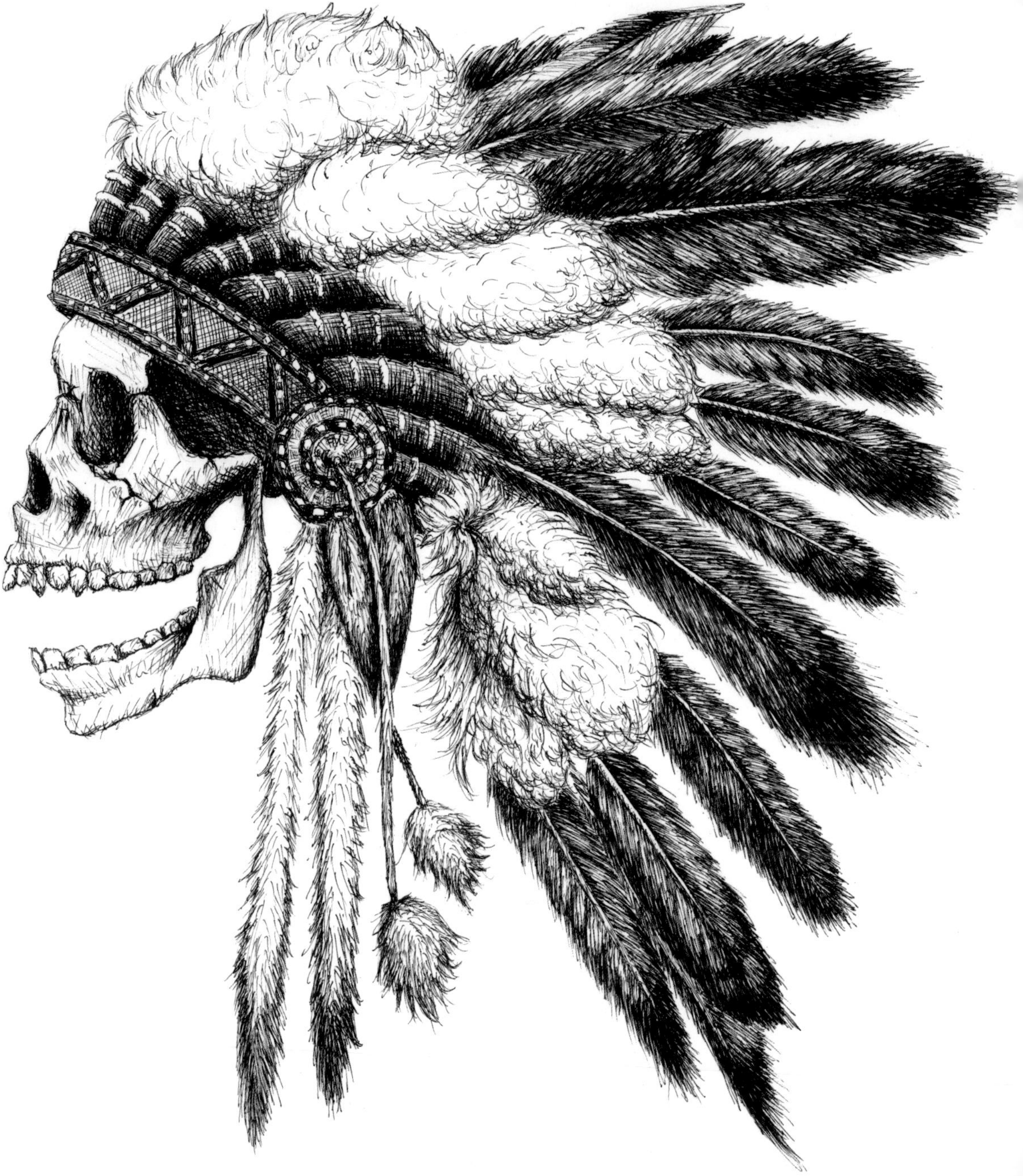

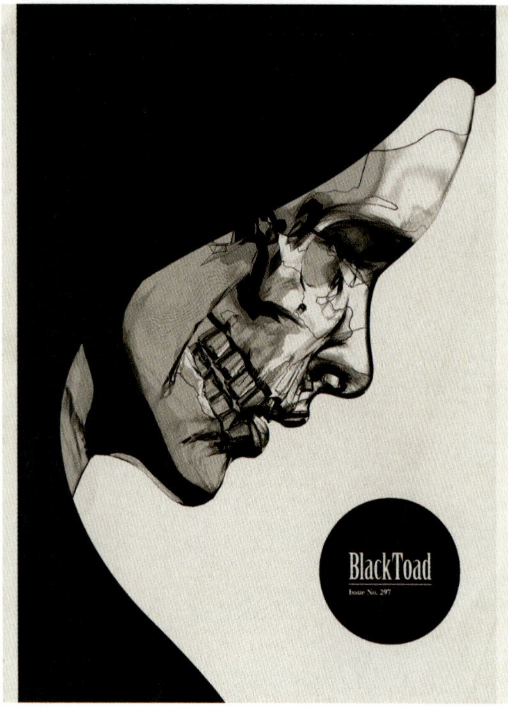

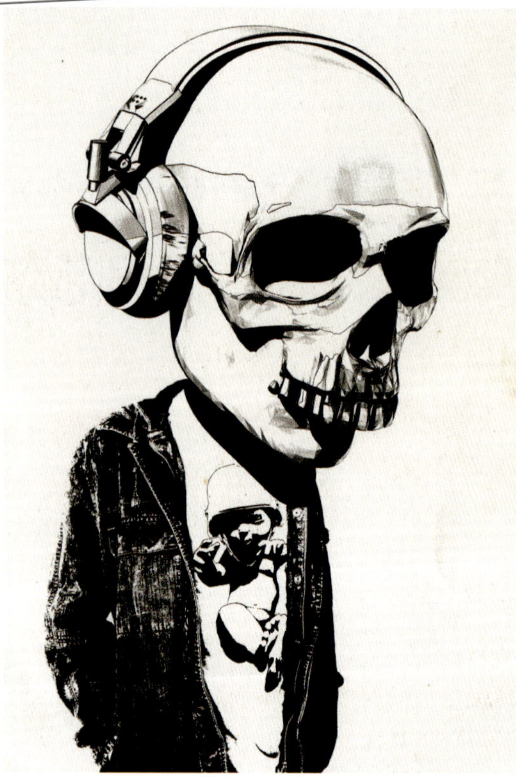

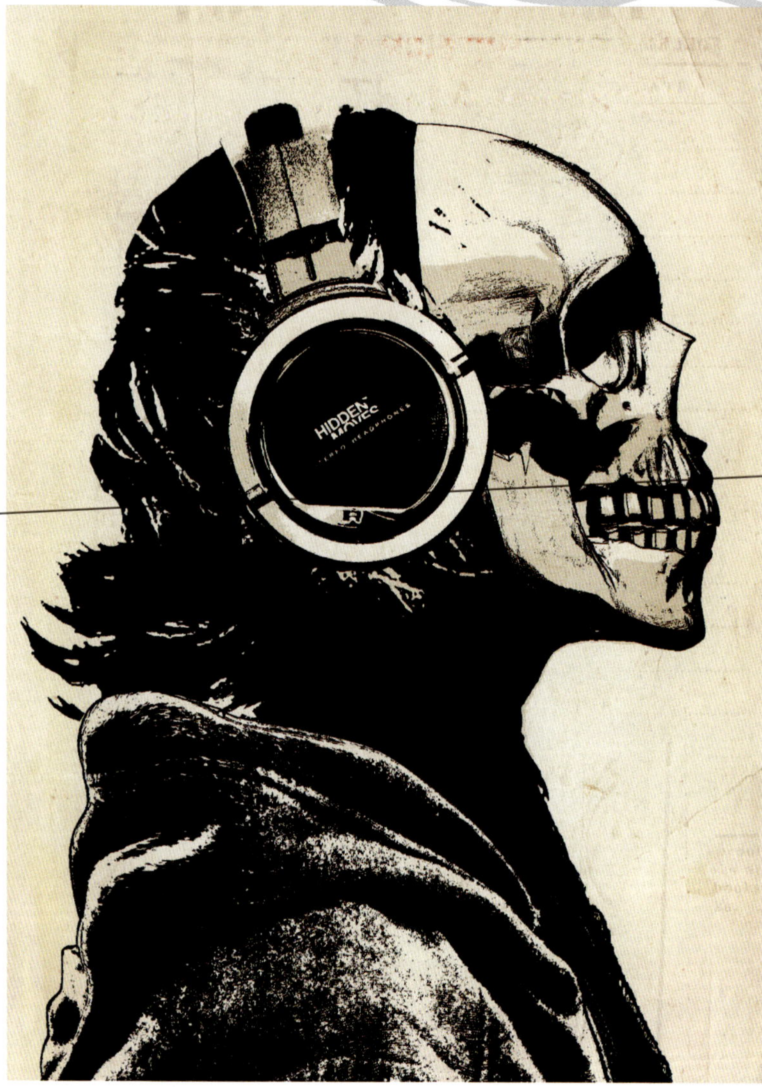

Artwork by Welsh artist Rhys Owens (www.HiddenMoves.com) displaying post-apocalyptic characters.

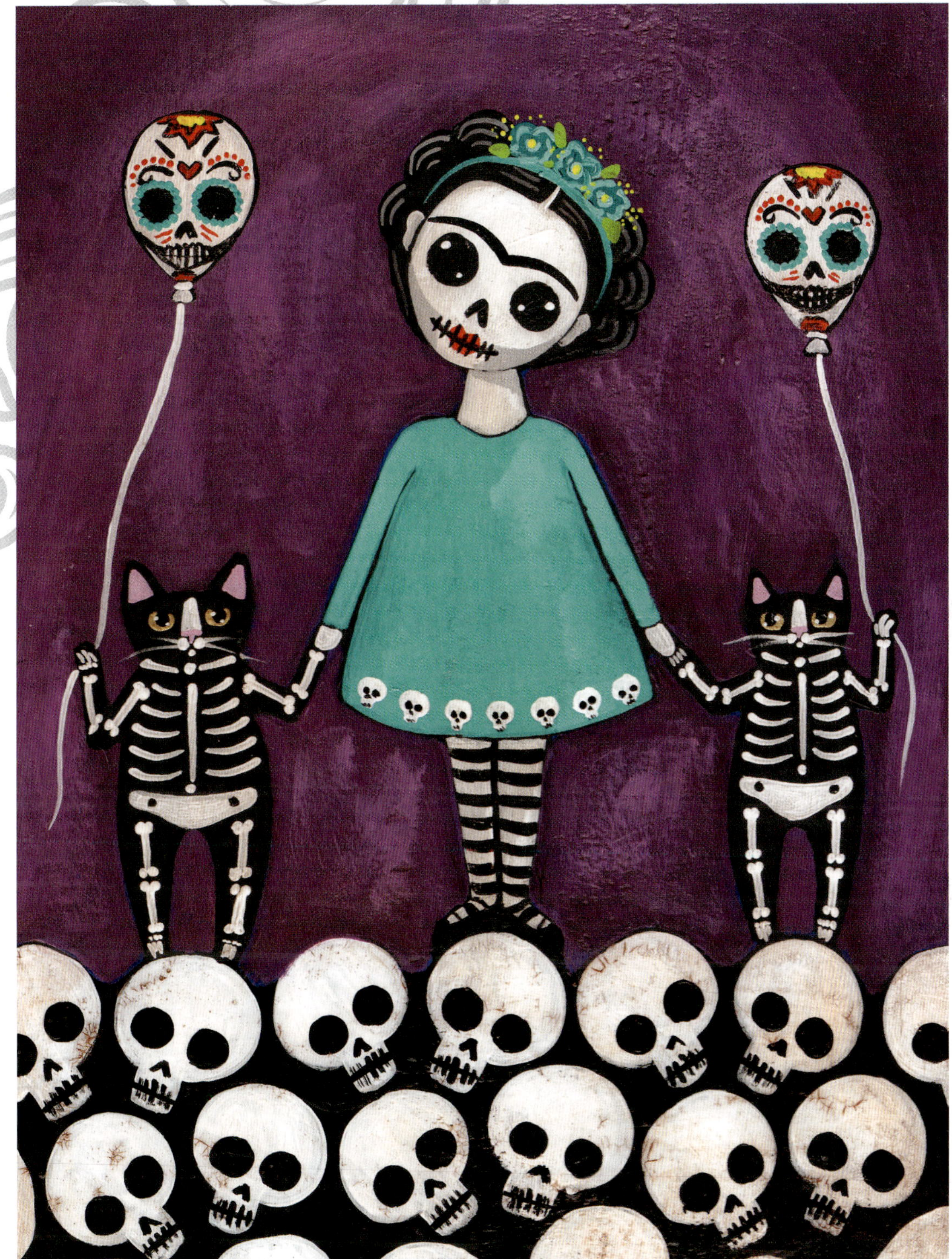

Artwork by Ryan Conners showing Frida Kahlo in a Day of the Dead–style.

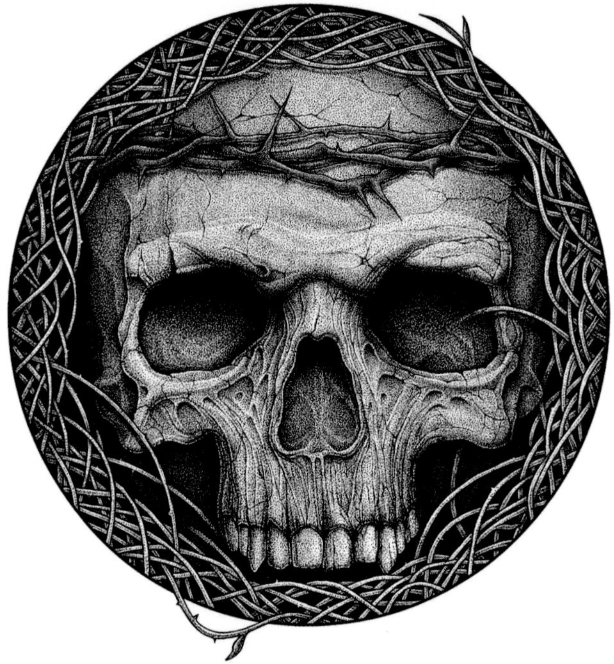

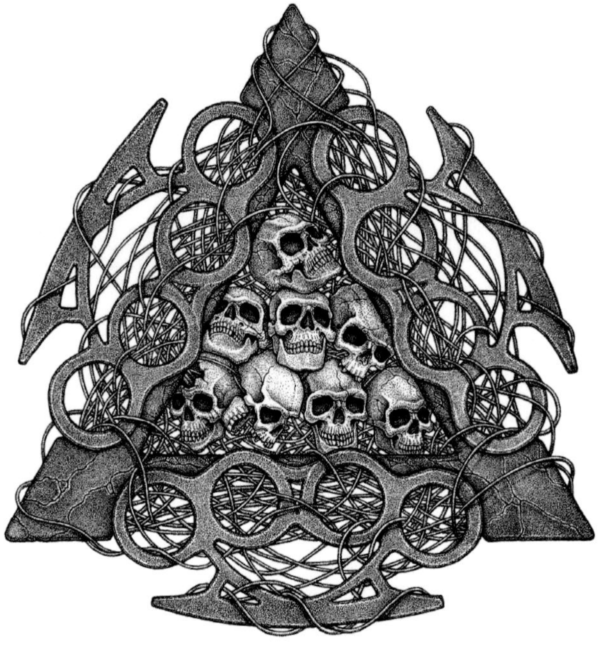

This spread: Work by artist Yuyiy Skorohod.

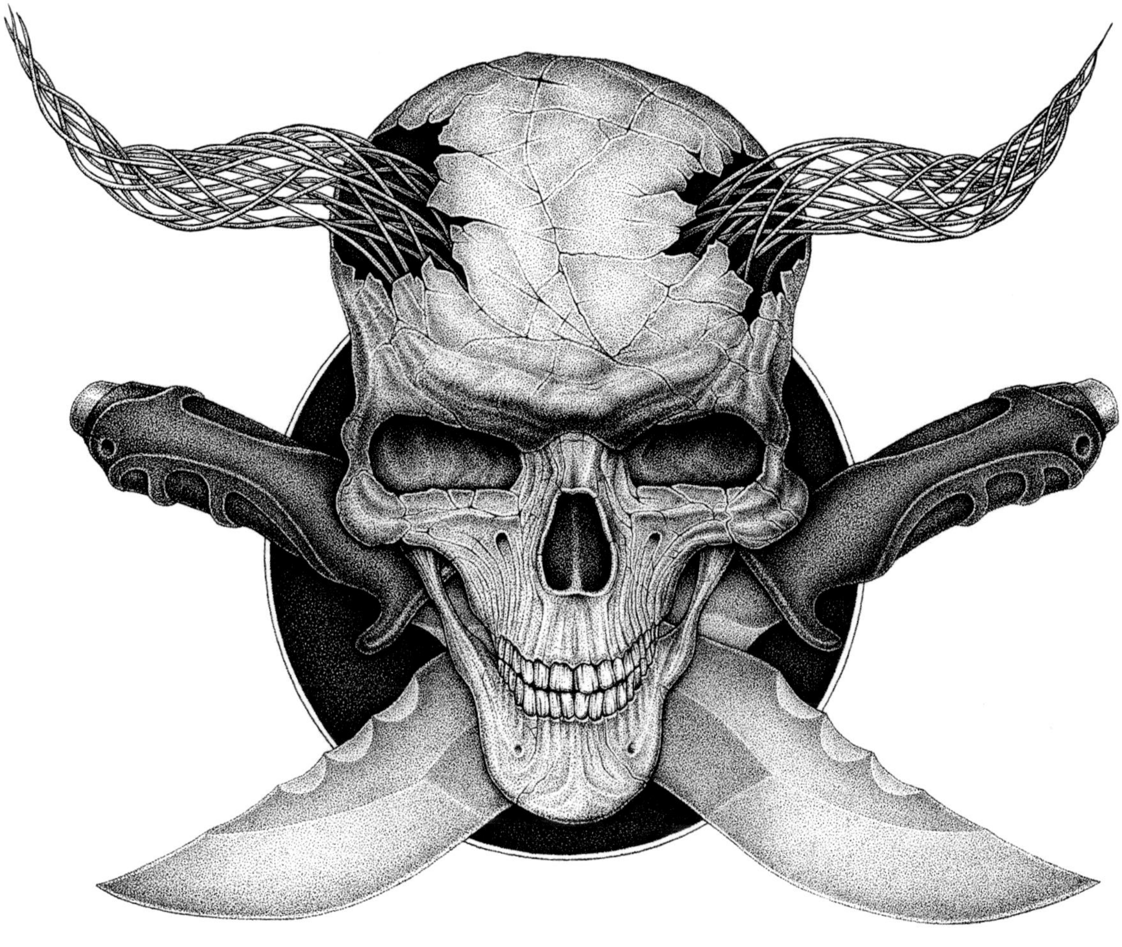

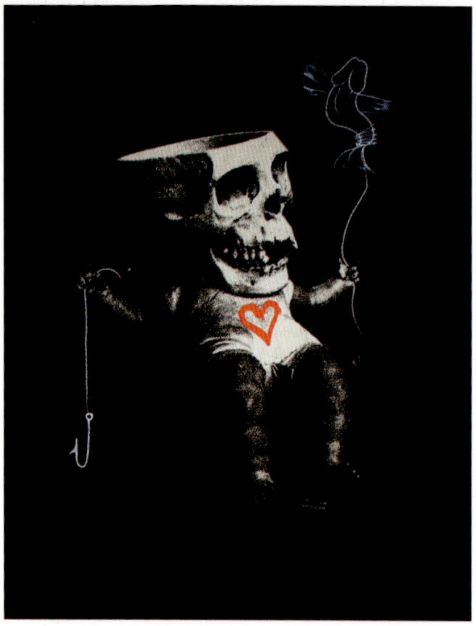

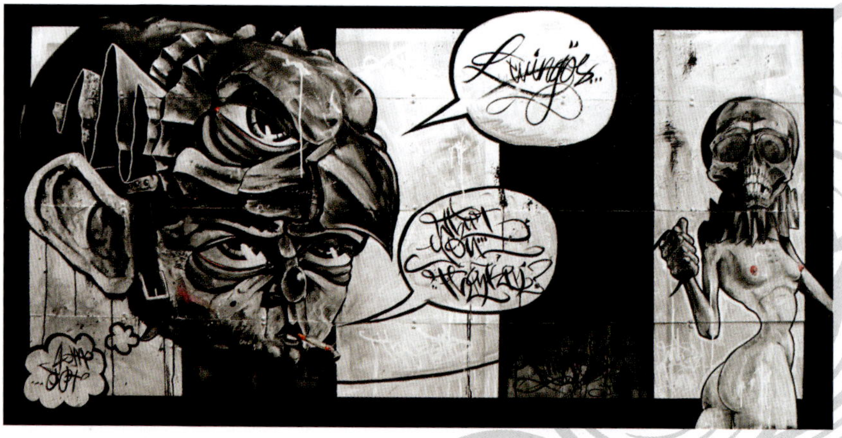

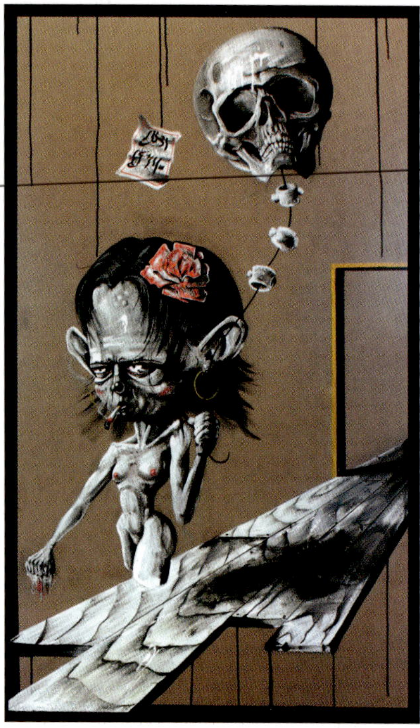

This spread: Art by Tommy Fiendish.

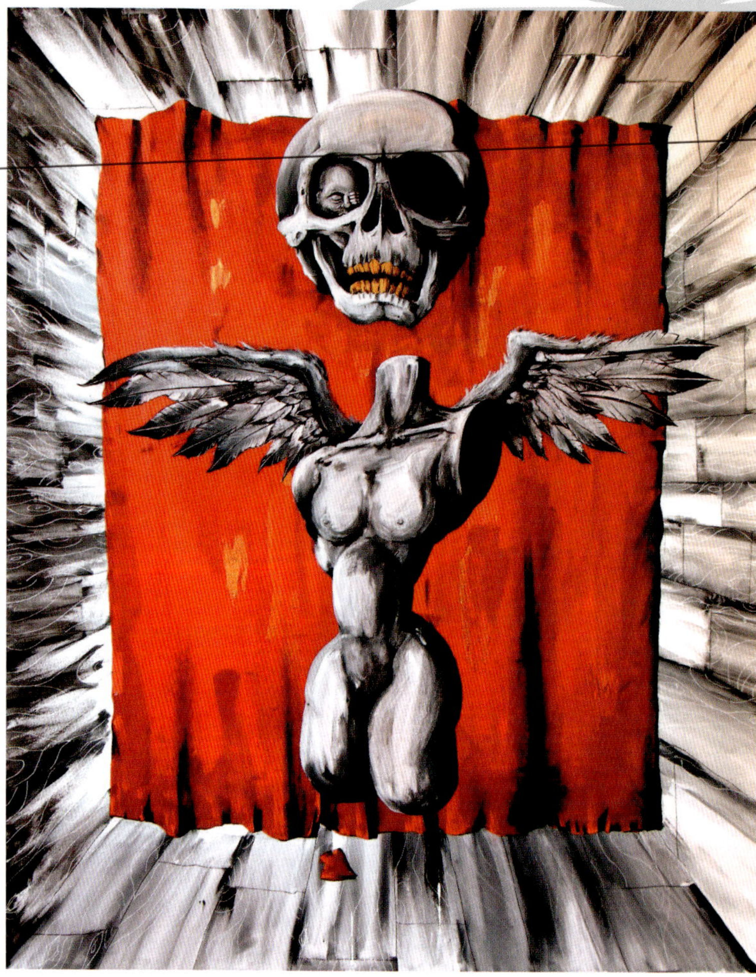

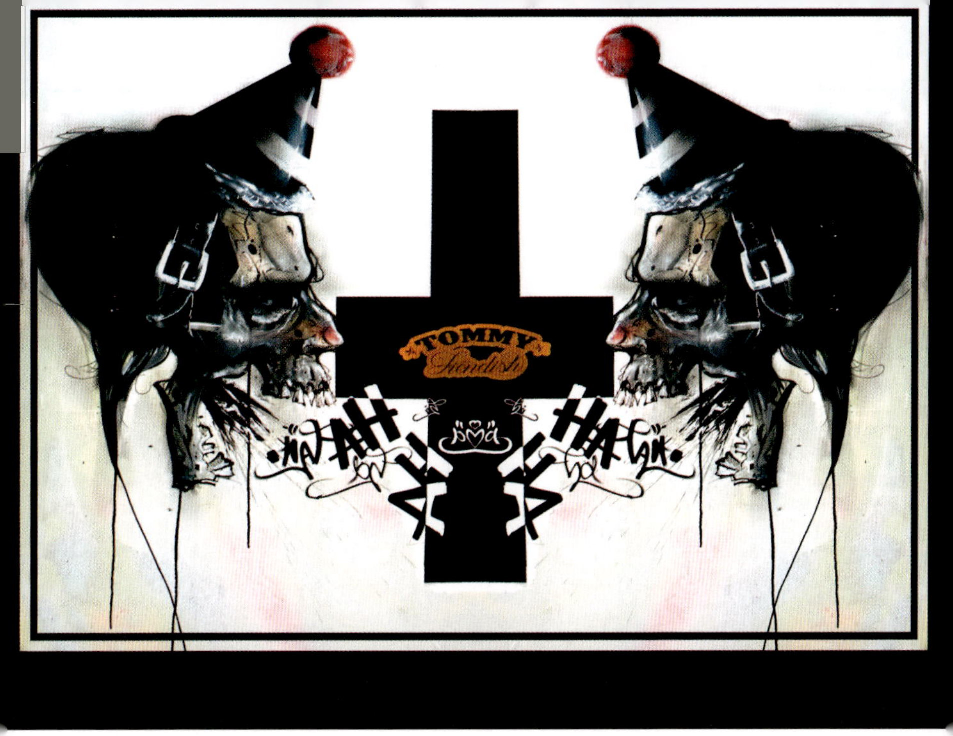
TOMMY
fiendish

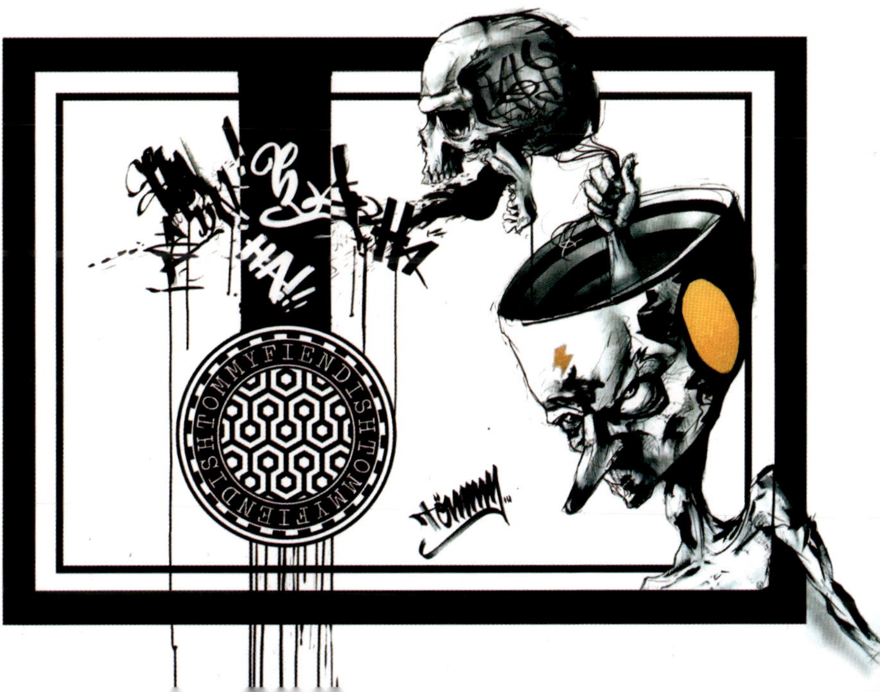
TOMMYFIENDISHTOMMYFIENDISH

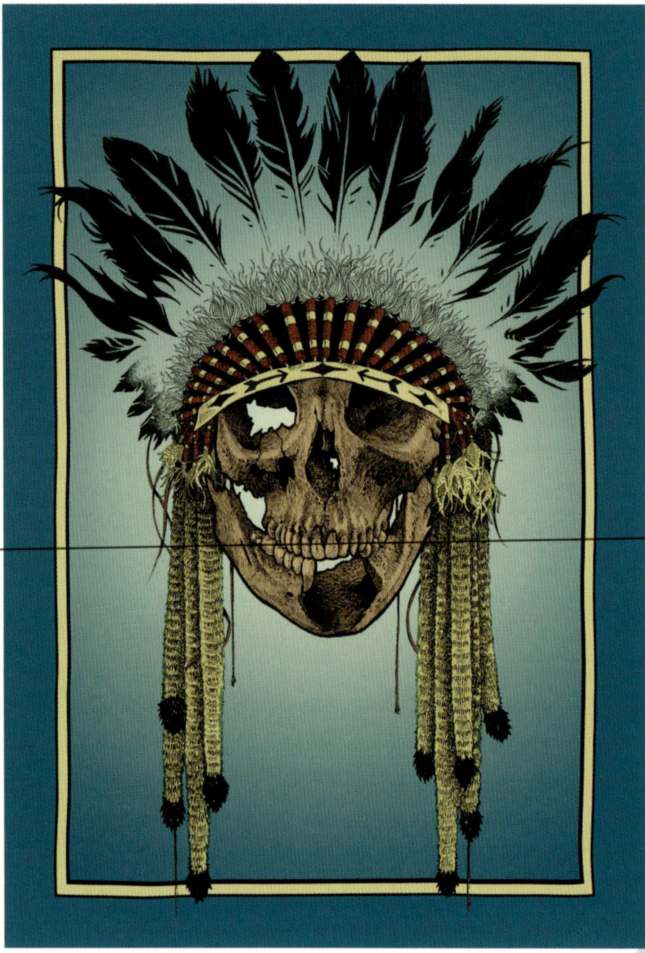

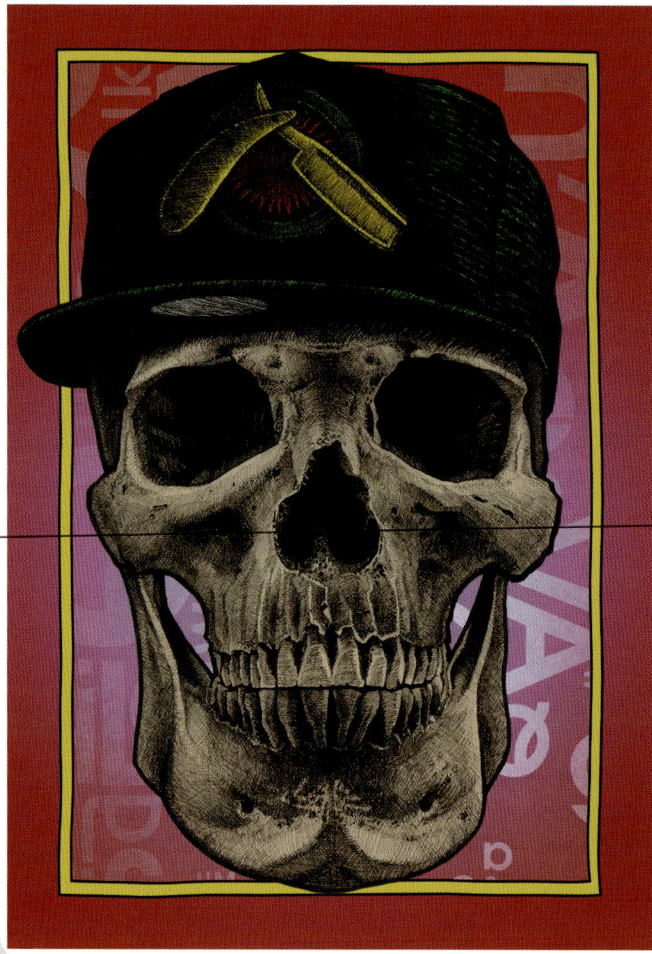

This spread: Art by Vadim Zhulanov. Above are *Redskin* and *Skull*, while a skull-collage is opposite.

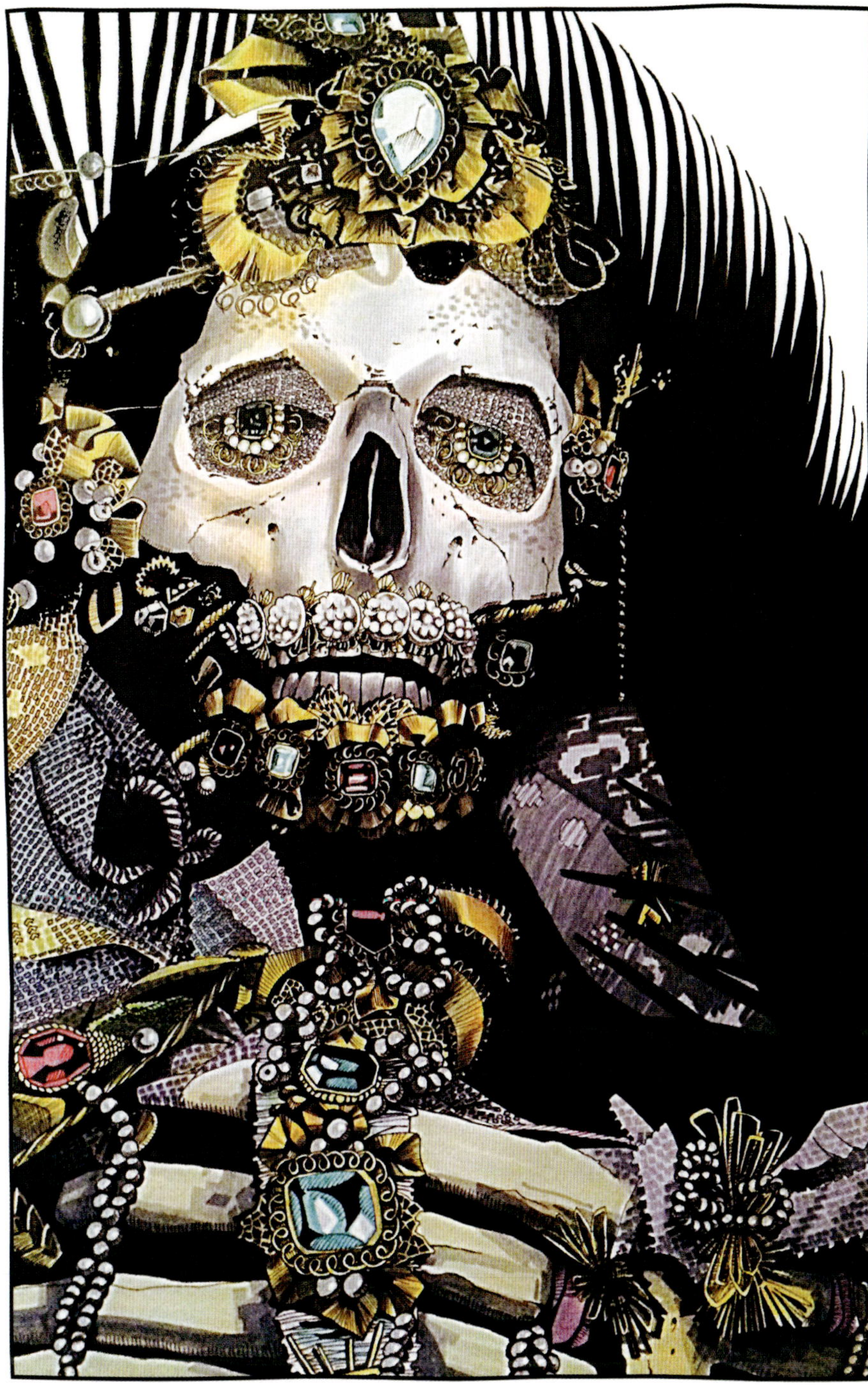

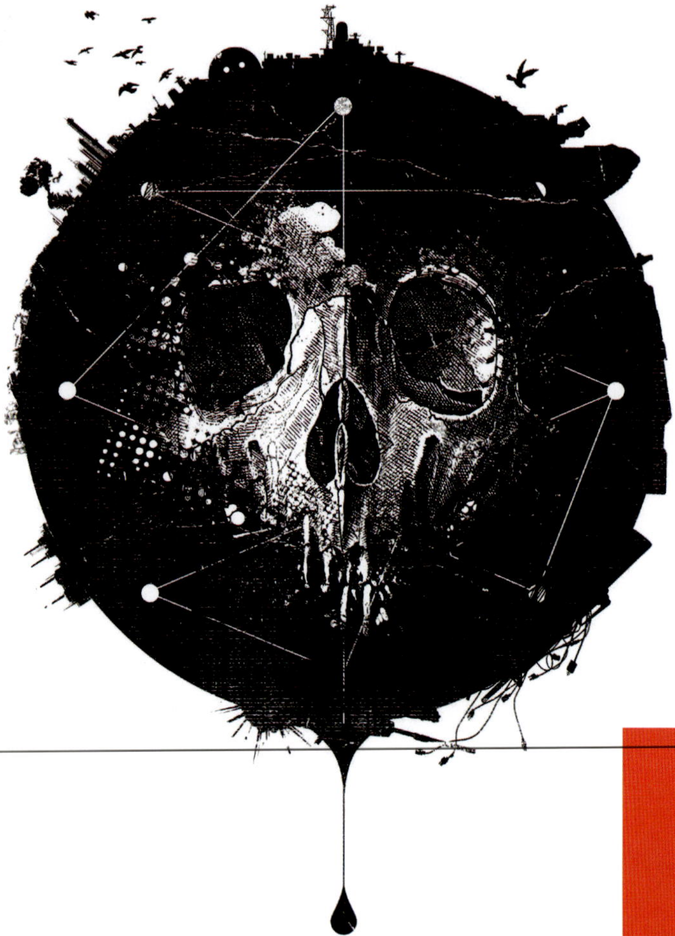

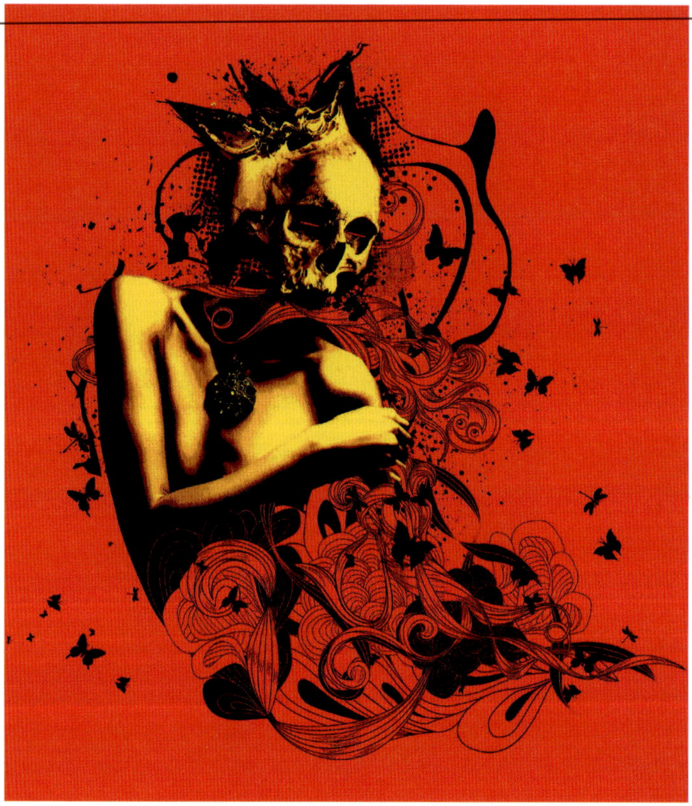

This spread: Skull art by Wilhelm Steiner. *Black Earth* is above; *In Bloom* is at right; *Danse Macabre* is opposite.

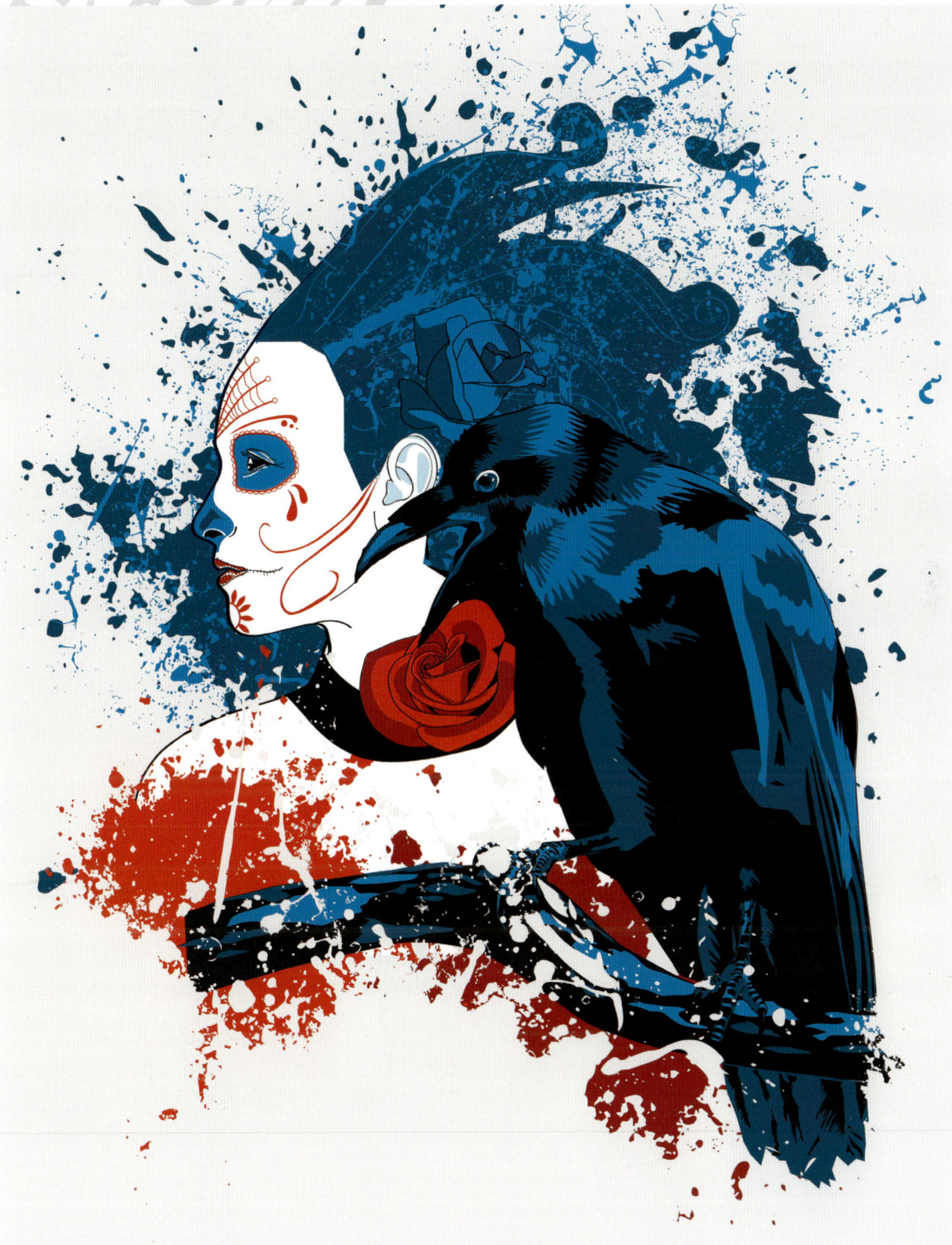

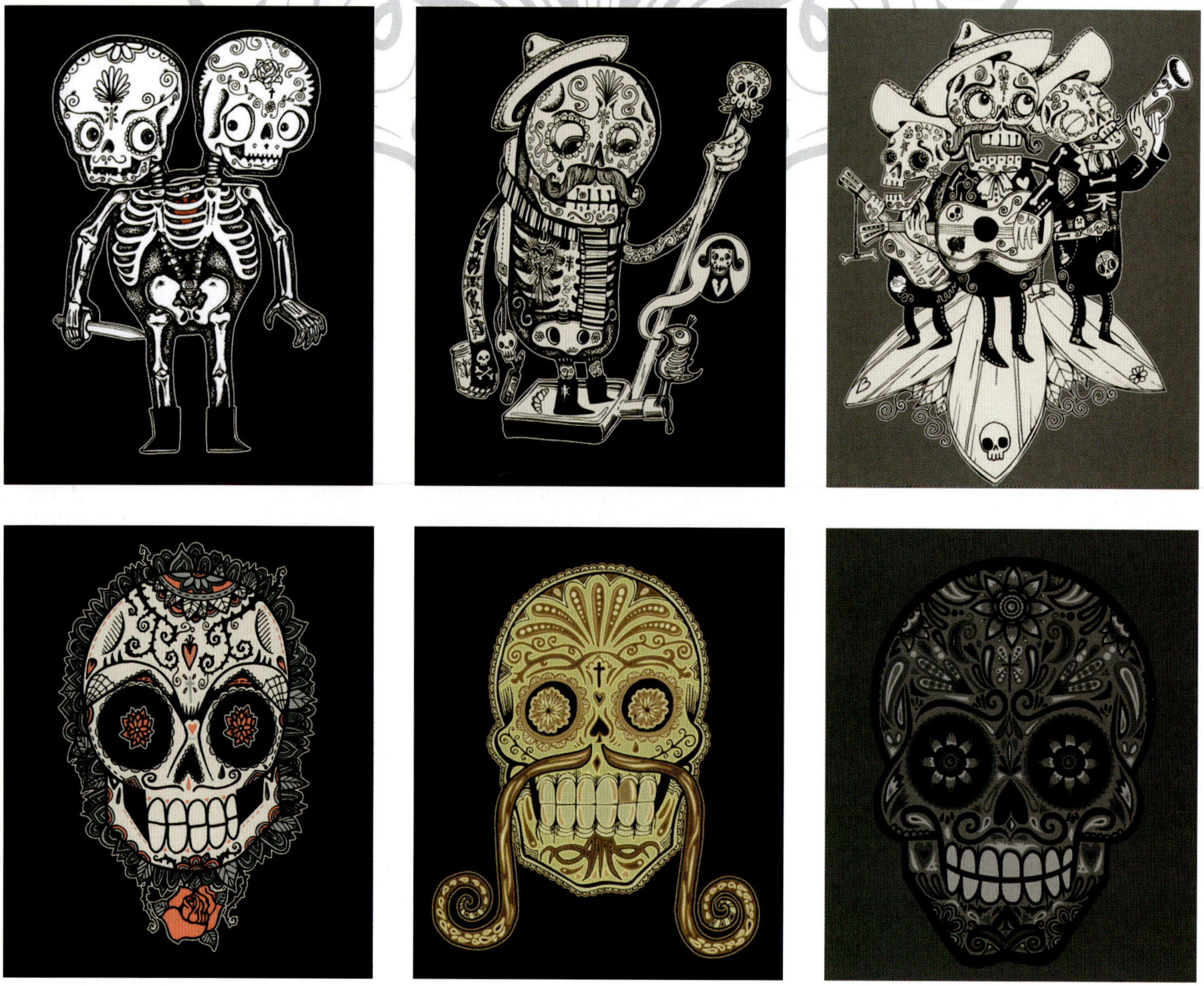

Skull-related artwork by Craig Watkins.

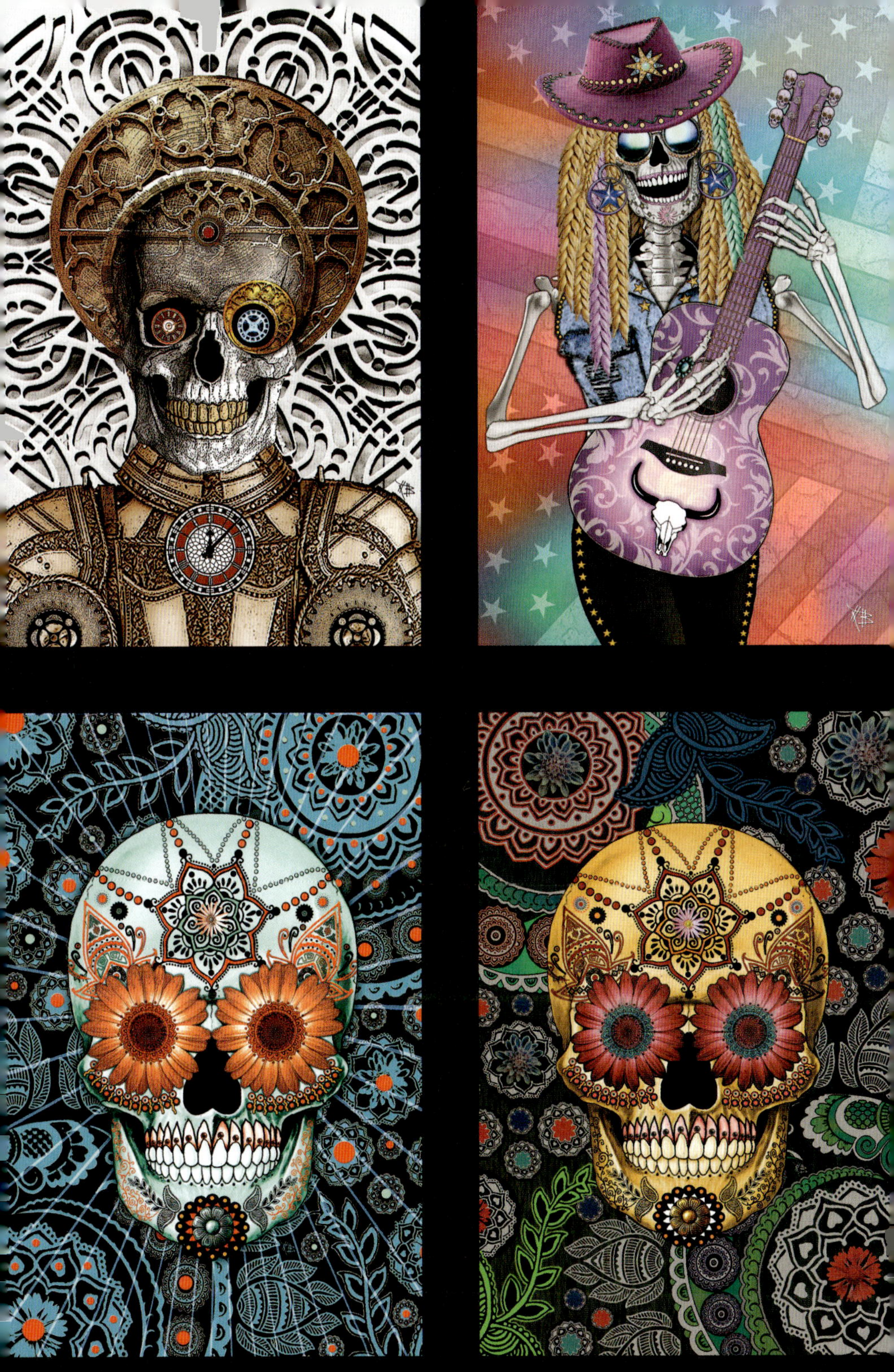

Sugar skull images by artist Christopher Reikmann

SKULLS IN HISTORY

PERHAPS ONE OF THE MORE MEMORABLE CHARACTERS IN SHAKESPEARE'S PLAYS IS ONE WHO NEVER UTTERED A WORD. THAT CHARACTER IS YORICK, THE DECEASED COURT JESTER WHOSE SKULL APPEARS IN THE OPENING SCENES OF *Hamlet*. Hamlet, the price of Denmark, holds the exhumed skull in one hand and bemoans the fate of the dead man to his friend, Horatio, "Alas, poor Yorick. I knew him, Horatio; a fellow of infinite jest...." The speech is a reminder that death awaits us all.

The Aztecs revered the symbolism of the human skull—their artwork is brimming with them. Examples of this fascination include carved stone works, such as the tonalamatl calendars, dating back to the 16th century. The skull occupies the center of the Calendar, with the eighteen months of the Aztec year surrounding it in a series of concentric circles. The design can be interpreted as a reminder that death forms the center of the universe. The Aztecs are also known for their exquisite mosaic skull masks, rendered in precious stones, turquoise, and gold. The Aztec Goddess of Death, Coatlicue, wears a skull pendant in a style similar to some of the Hindu deities of India, Nepal, and Tibet. In particular this is reminiscent of Kali, the goddess of time and death, with her necklace of human skulls. However, while Kali and her skulls were, and remain, a part of the Hindu culture for hundreds of years, Coatlicue wan't so lucky.

The Spanish conquistadors, who arrived in the 1500s, were keen to see Catholicism gain supremacy, so Aztec art and design was suppressed. The skull disappeared from the culture of the Americas for several centuries, returning only when Mexico gained its independence in 1821. Then, the skull started to re-appear, with artists such as José Guadalupe Posada using the symbol as a statement of freedom and a return to the old culture. Perhaps the most common example today is the Mexican Day of the Day, which is known the world over. Its origins are believed to go back over 3,000 years. Today, Dia de los Muertos is celebrated across Mexico and in parts of the United States with people of Mexican origin. Colorful skeletons rendered in sugar, some with the name of a dead person written on its forehead, are depicted in various light-hearted ways. It is a holiday that celebrates an odd combination of cartoonish and ghoulish. People, too, wear wooden masks, called "*calacas*," which are in the shape of a skull. More wooden skulls are placed on graves and altars, as a way of honoring the dead.

Ancient elongated skulls—or skulls which otherwise veer from the norm—have been found in various places around the world, including Peru, Malta, and the Crimea. Some people believe that the skulls belong to alien beings; others cite the ancient practice of "molding" the skull of an infant, or artificial cranial deformation, a practice best known amongst the Germanic tribes of the Huns. This practice dates back to prehistoric times, and is still practiced today in Vanuatu. To accomplish the deformation, the skull of a child is bound with cloth from the age of one month to six months, while the skull is still relatively flexible. Sometimes, wooden reinforcement molds help achieve the desired shape. Artificial cranial deformation may have been practiced as part of a tribal tradition or to give a ferocious or frightening appearance. Whatever the reason, it left strange skulls!

One of the most famous skull images is that of the crossed skull and bones, commonly used as a symbol by pirates in movies. The skull and crossbones, however, could well be explained by the Freemasons and the Knights Templars, secretive and exclusive organizations. Also called the "Jolly Roger," the skull and crossbones started life as the "Jolie Rouge," the name given to a particular French branch of the Knights Templar. The flag was flown by the Templars' ships as a warning to any enemies; subsequently the design was adopted by pirates. The ancient organization known as the Freemasons made use of a myriad of symbols, including the skull and crossbones. The ritual "tracing board," an essential part of rituals, features six skull and crossbones images. In this usage, the symbol represents the irrefutable connection between life and death.

Crystal skulls are perhaps the most modern of the historically used skull images here. These "artifacts" were objects of Victorian fascination because their creators sought to portray them as pre-Columbian artifacts. They were at one time thought to be somewhere between 5,000 and 36,000 years old, and were believed to have healing powers! They became so entrenched in societal lore that the British Museum purchased a crystal skull in 1897 from Tiffany and Co. in its collection. The crystal skull is presented beside an informational card, stating that the item is "not an authentic pre-Columbian artefact." There is no doubt that crystal skulls are man-made and there is rigorous archaeological analysis showing that none of them are pre-Columbian in origin.

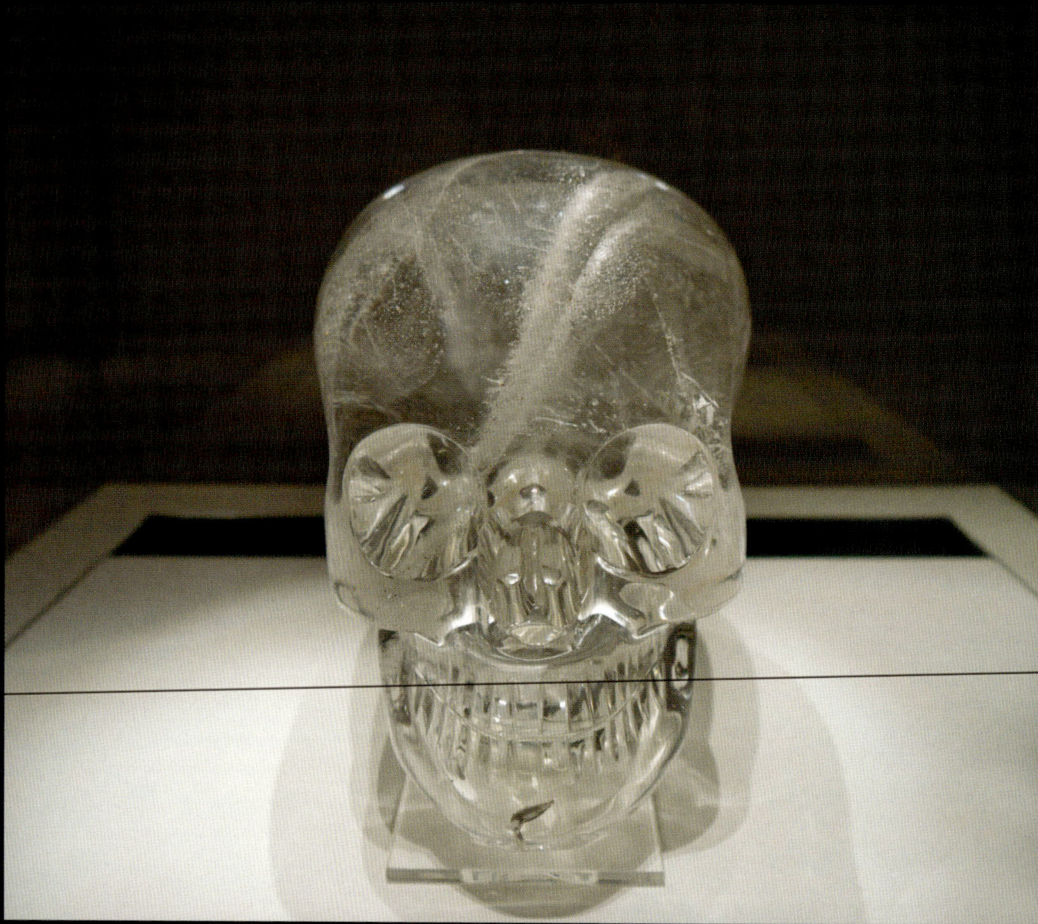

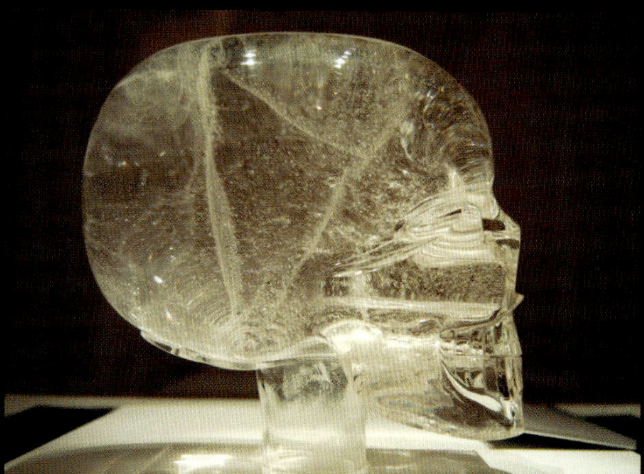

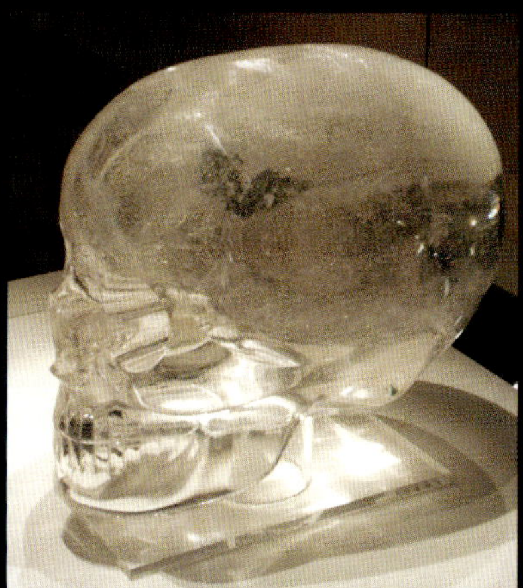

Multiple perspectives of a crystal skull, held in the British Museum's collection.

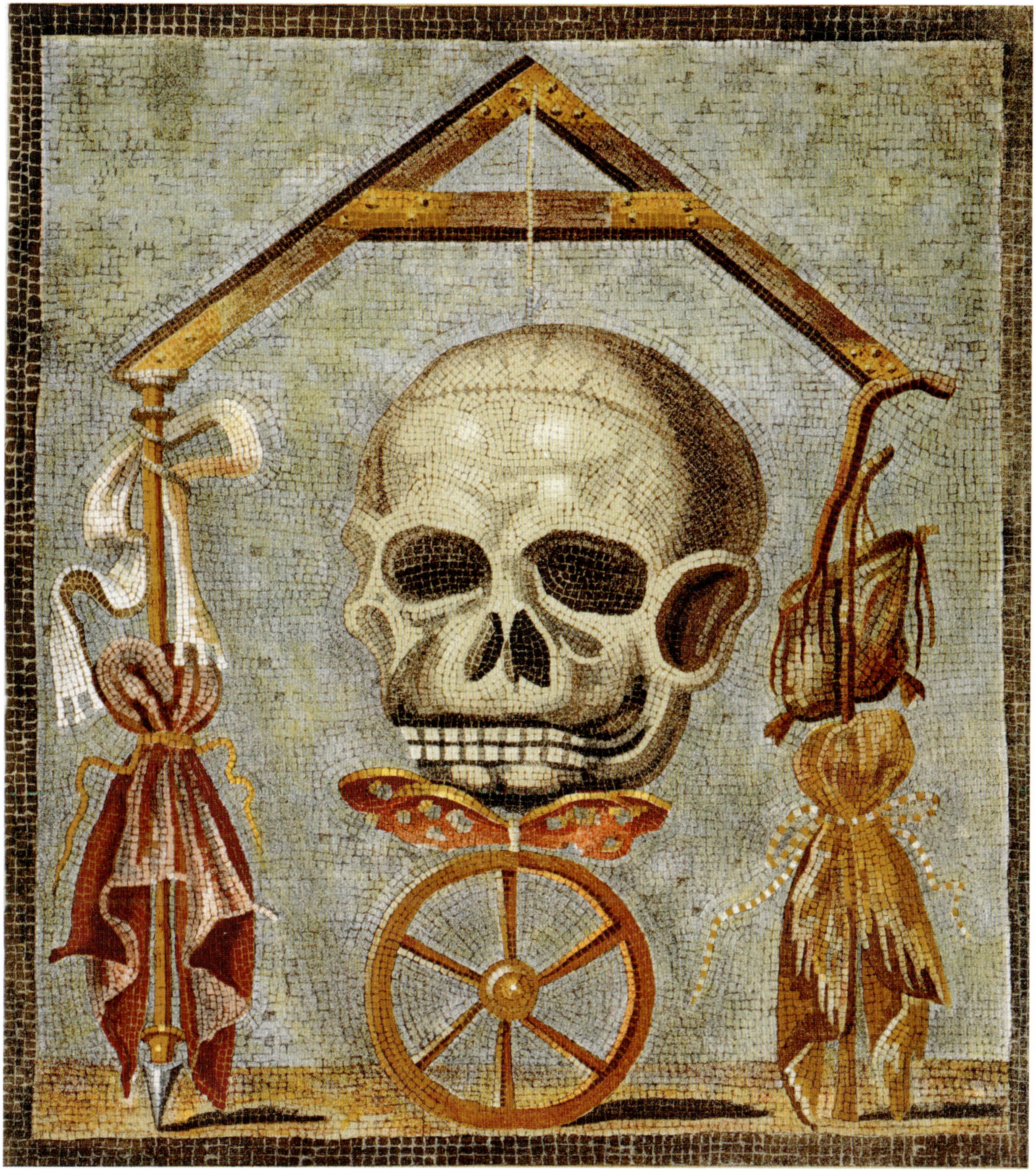

A reproduction of a Masonic mosaic from *The Houses and Monuments of Pompeii*, by Fausto and Felice Niccolini.

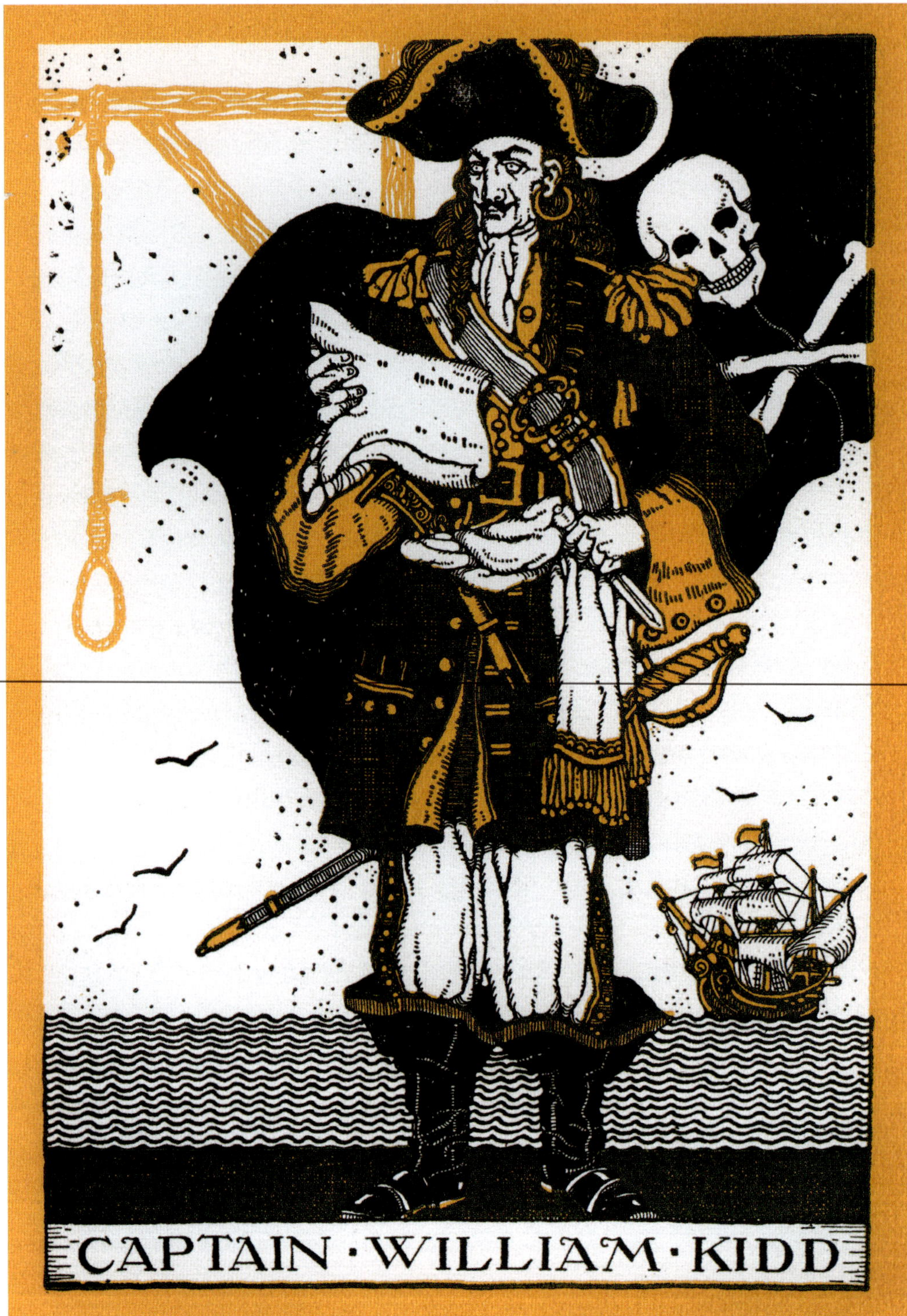

A 1930 lithograph portrait of Captain William Kidd, complete with skull and crossbones flag and hangman's noose.

The traditional skull and crossed bones is known as the Jolly Roger.

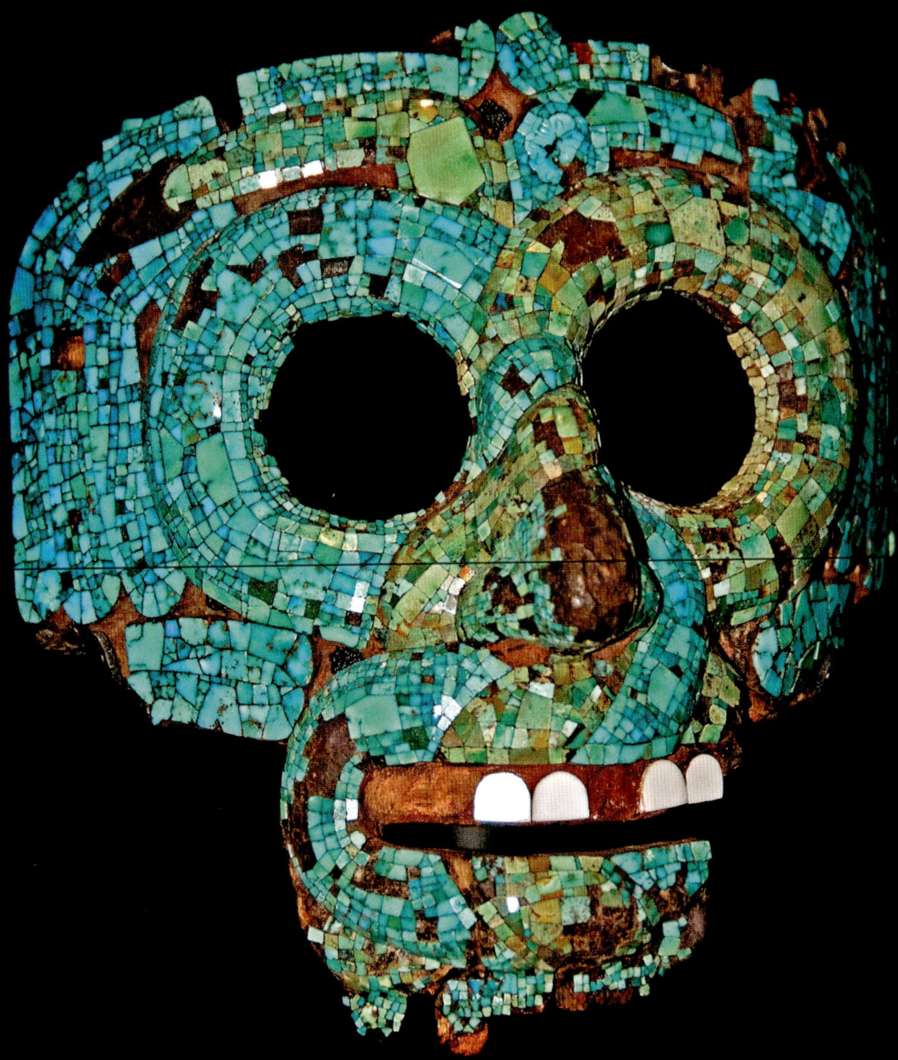

This spread: Aztec turquoise mosaic masks estimated to originate from the fourteenth through the sixteenth centuries.

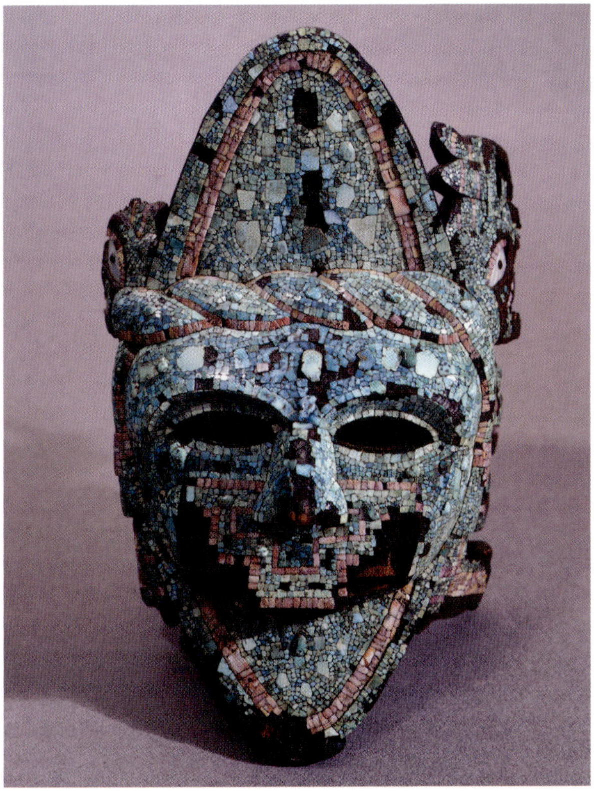

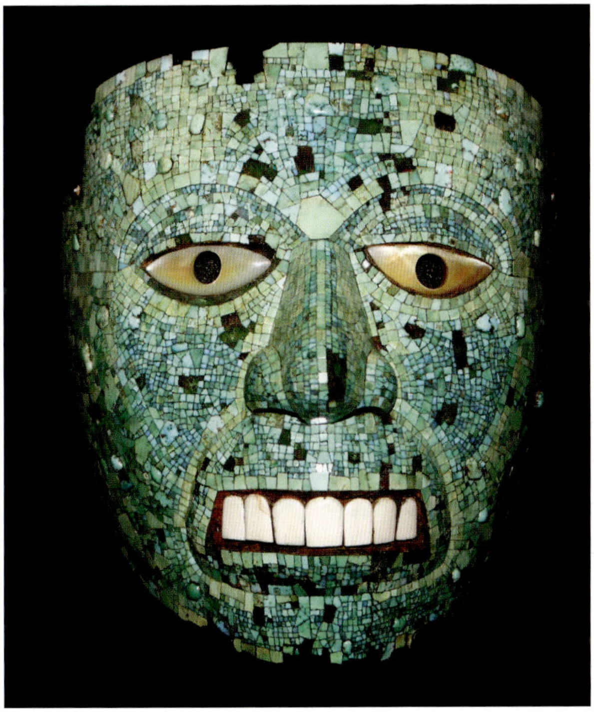

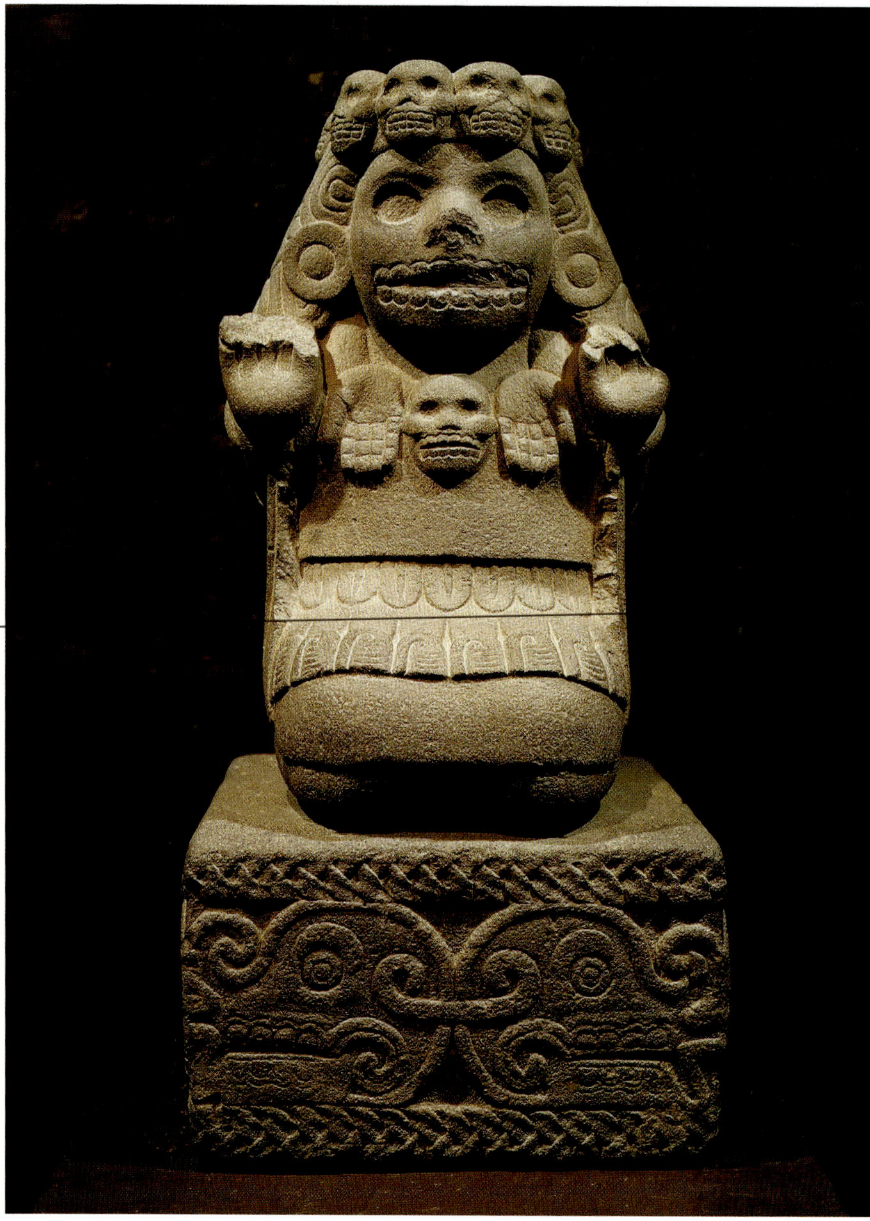

This spread: Statue of Coatlicue, the Aztec earth goddess of life and death.

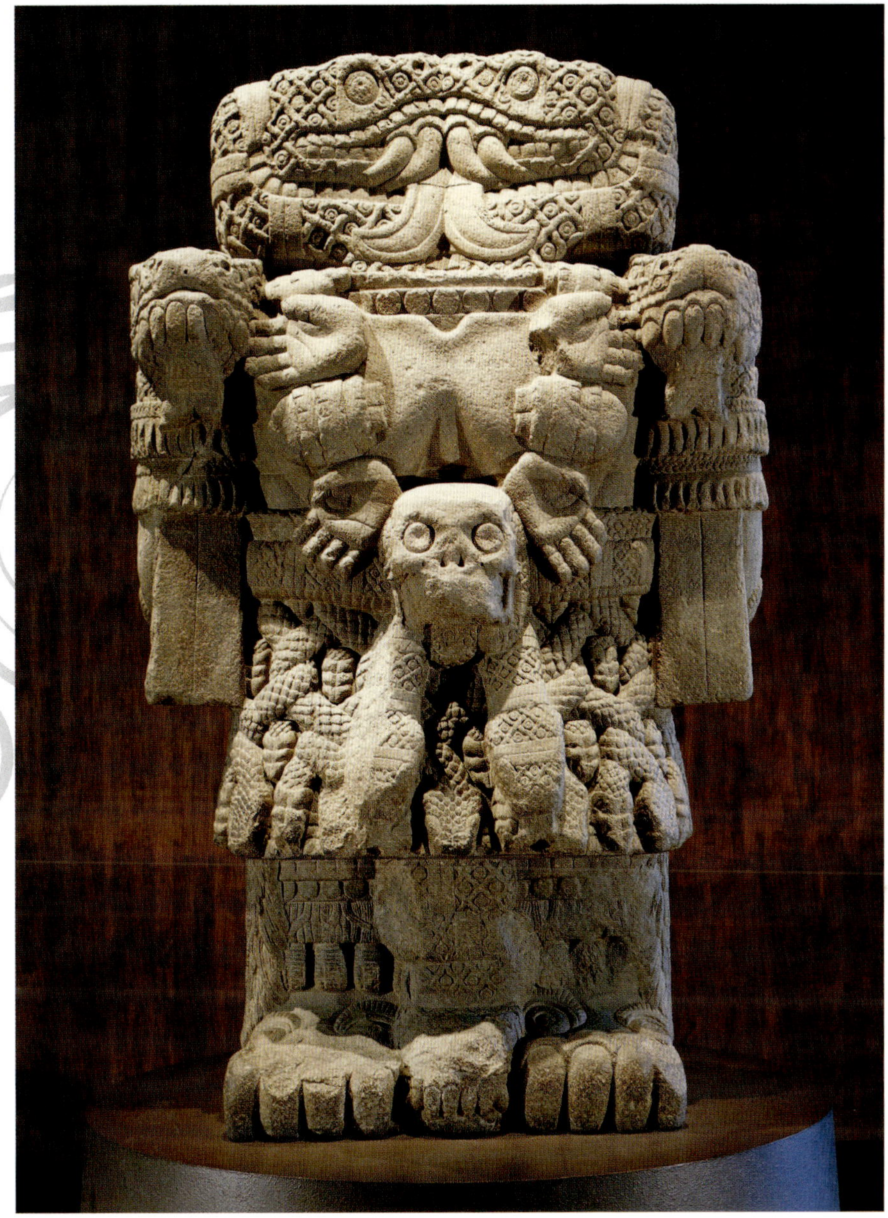

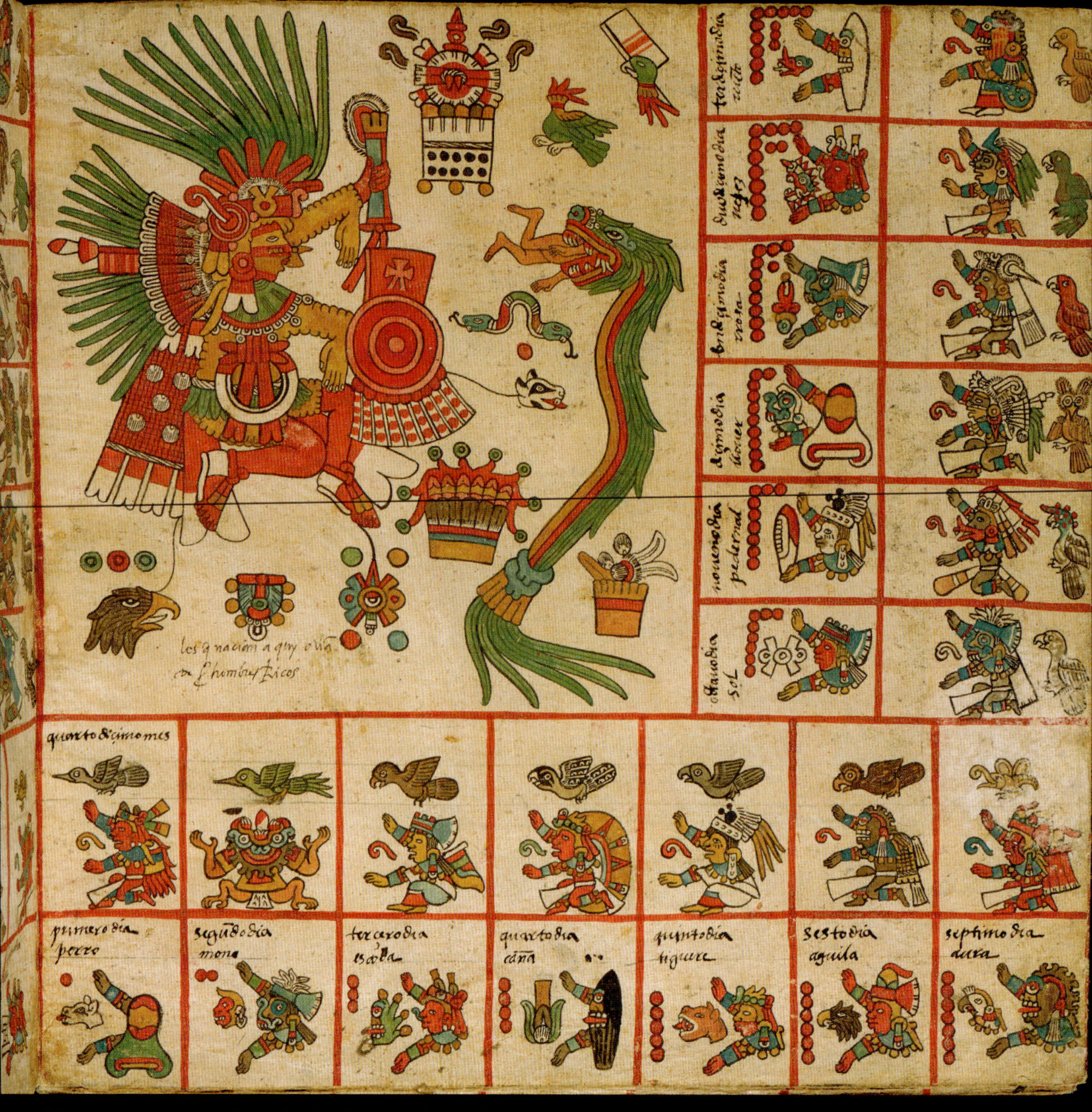

Detail from the Aztec Tonalamatl calendar from the *Codex Borbonicus*.

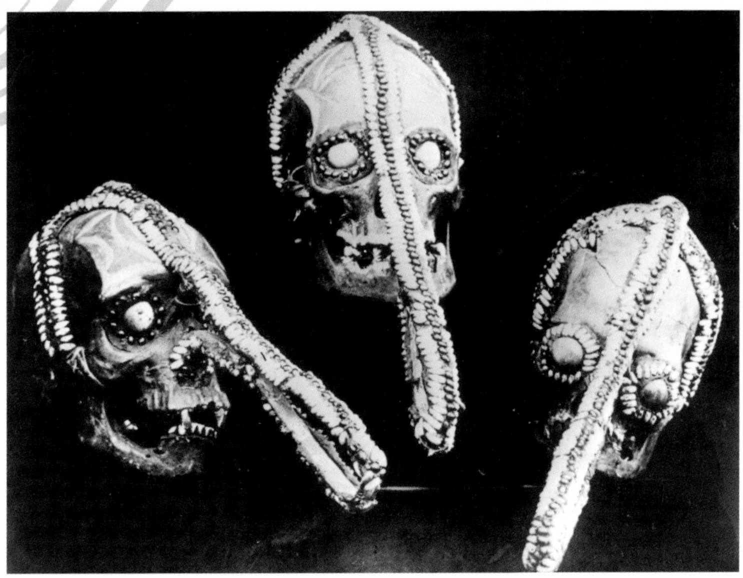

Decorated human skulls used by witch doctors in Ghana.

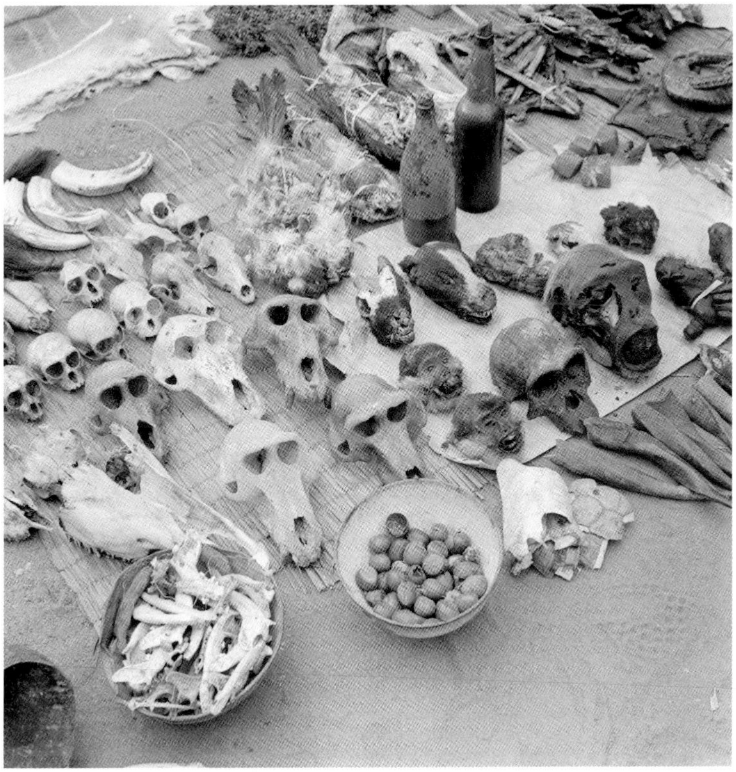

A collection of animal skulls at a witch doctor's market stall in Mali, around 1956.

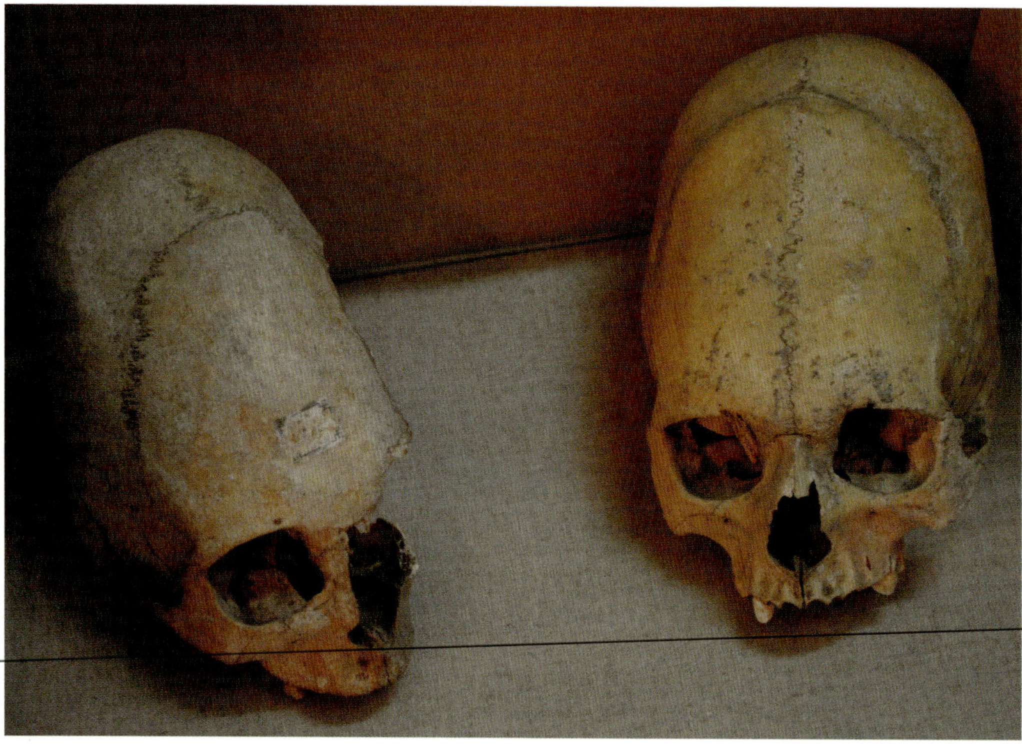

Modified skulls believed to be from the Sarmatians, a group that inhabited Eastern Europe and Central Asia during the time of the Roman Empire.

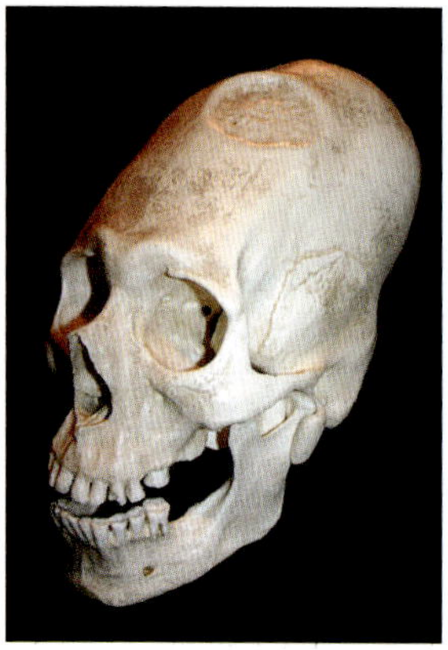

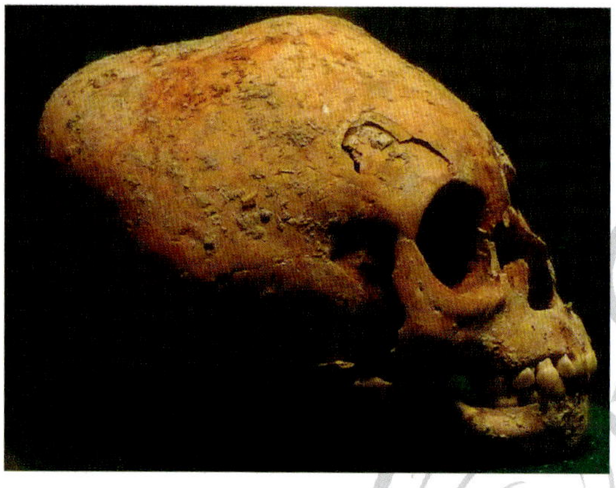

Two modified skulls: above left was discovered in Peru; above right is of Mayan origin.

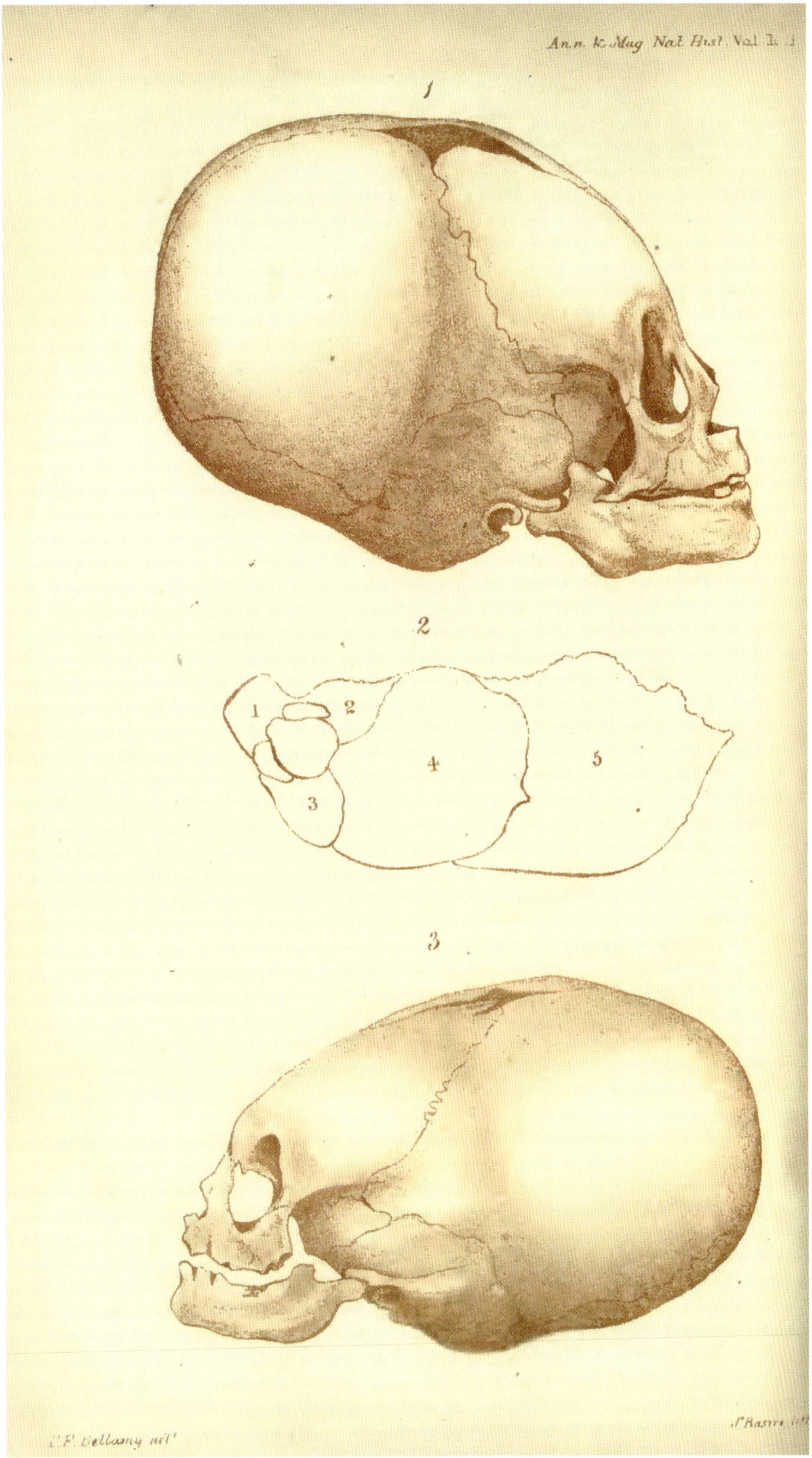

Lithographs of two deformed Peruvian skulls, drawn by Dr. P.F. Bellamy in 1842.

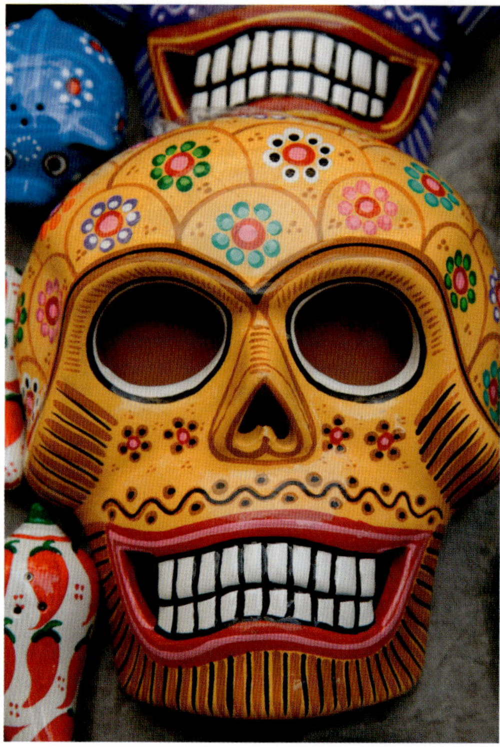

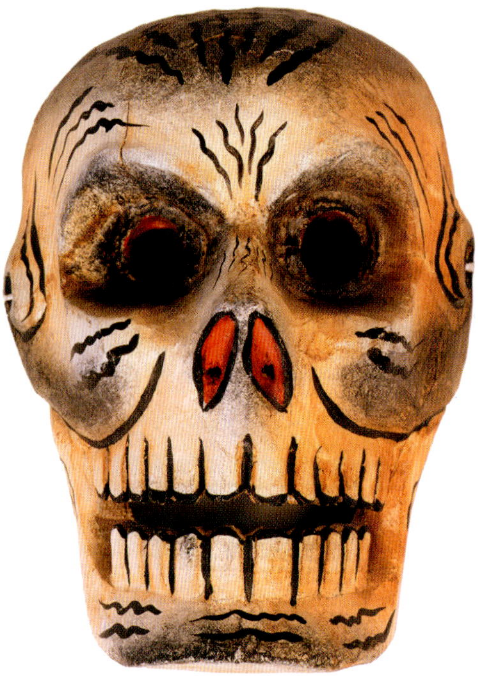

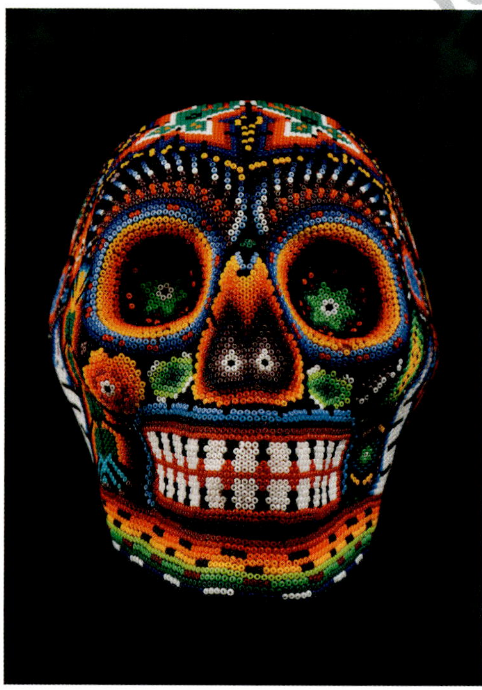

Skull masks celebrating the Mexican Day of the Dead.

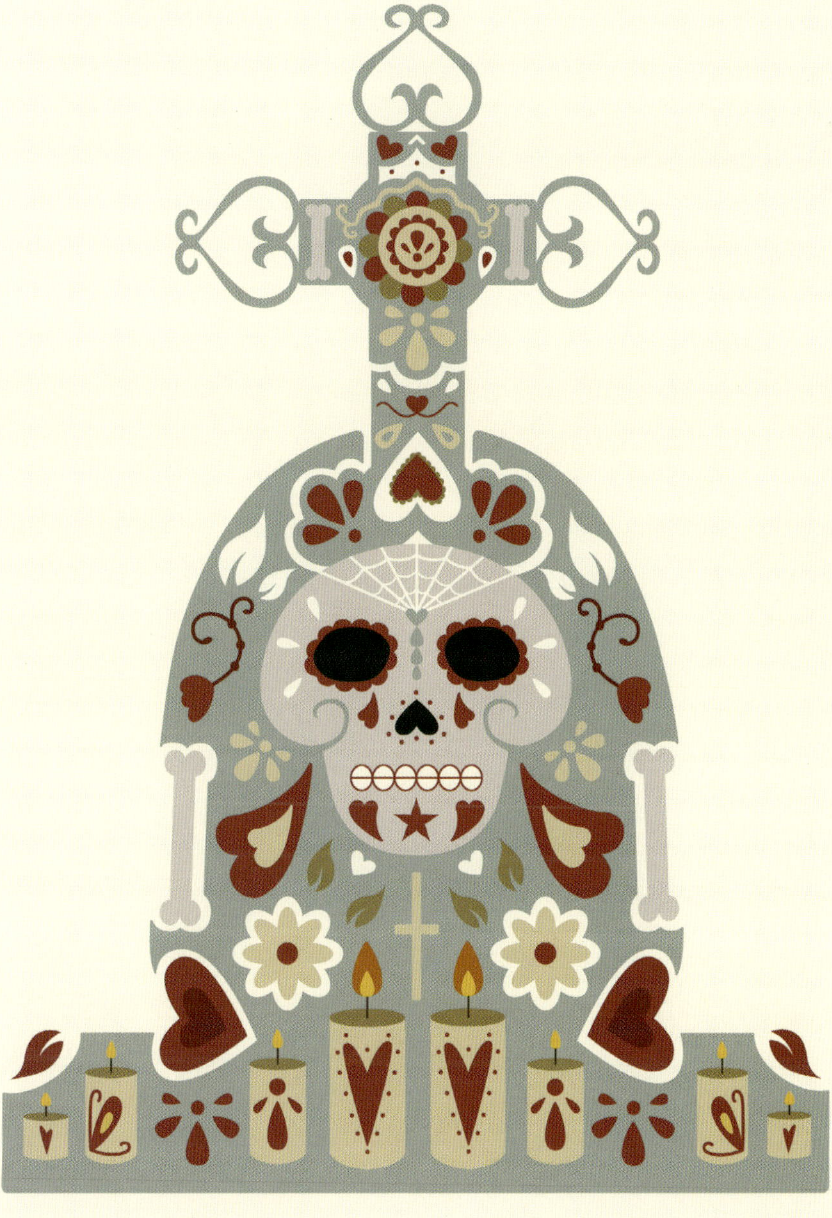

A Day of the Dead illustration in the Horror Vacui style.

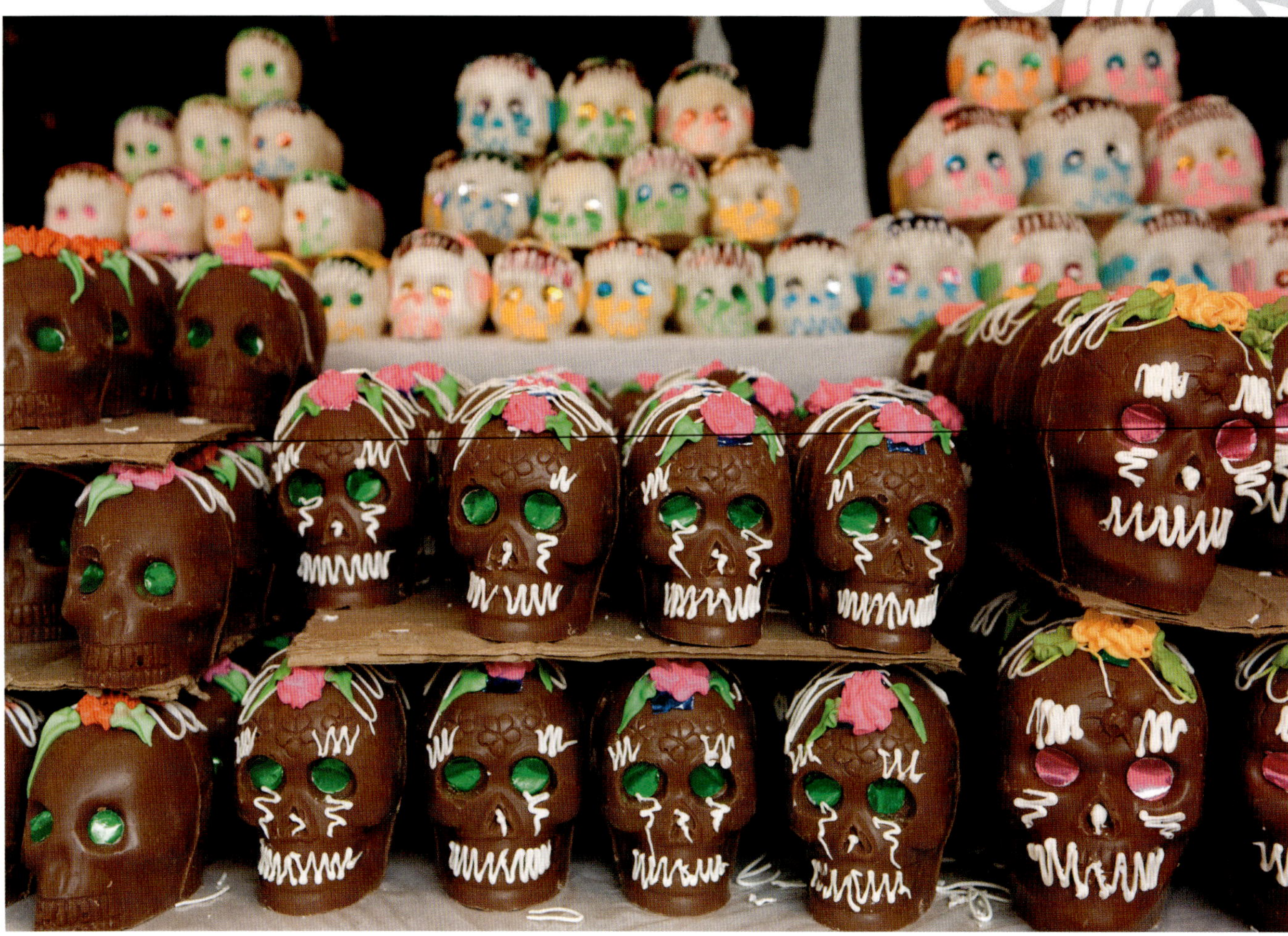

Skull-shaped candy for the Mexican Day of the Dead.

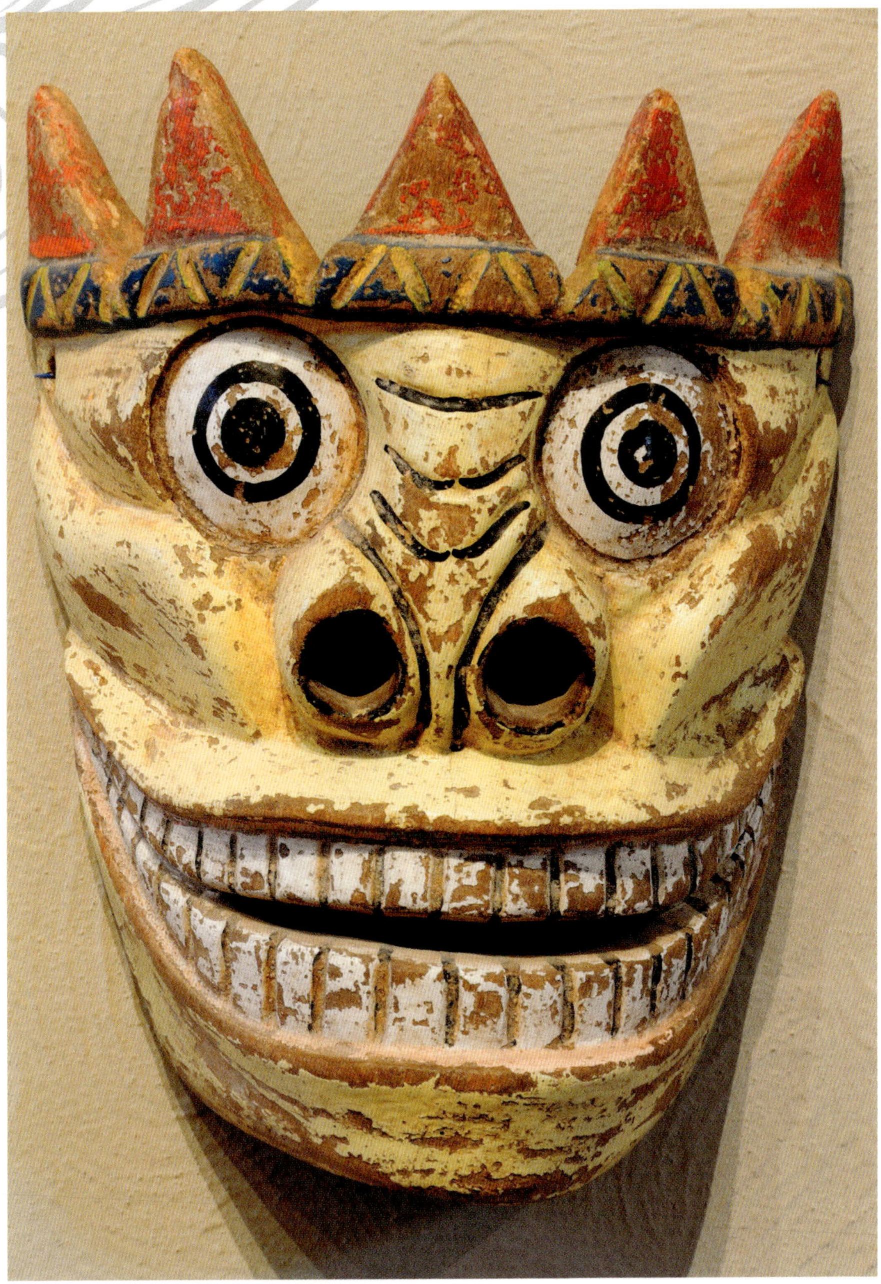

A Mexican Day of the Dead mask from the collection of Museo Rafael Coronel.

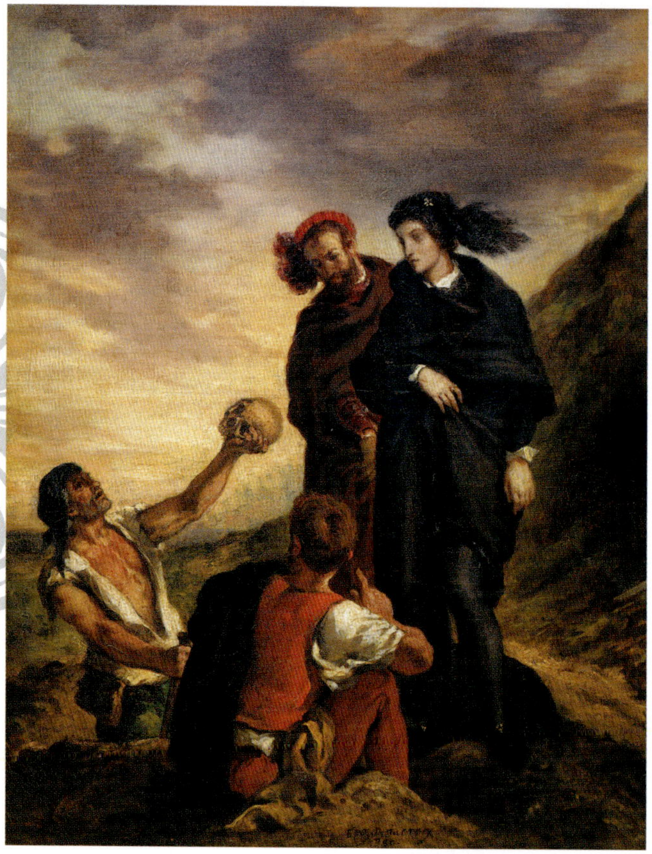

Art inspired by Shakespeare's *Hamlet*, showing Hamlet and Horatio with Yorick's skull.

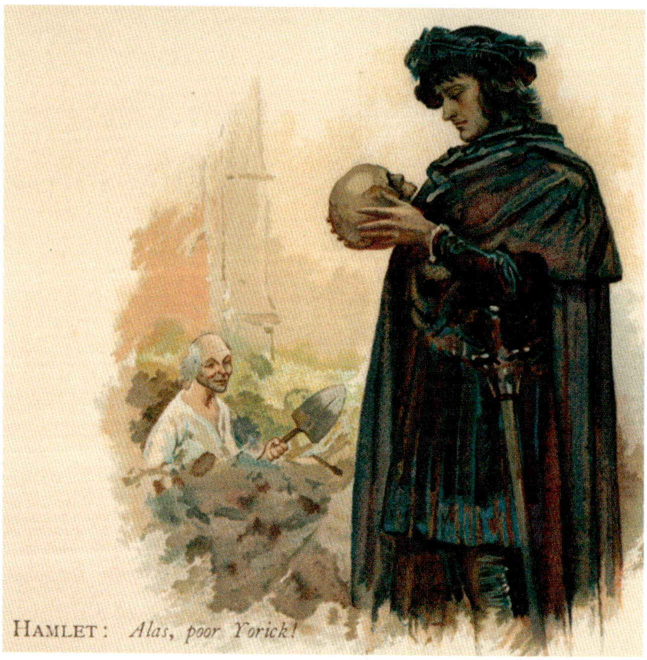

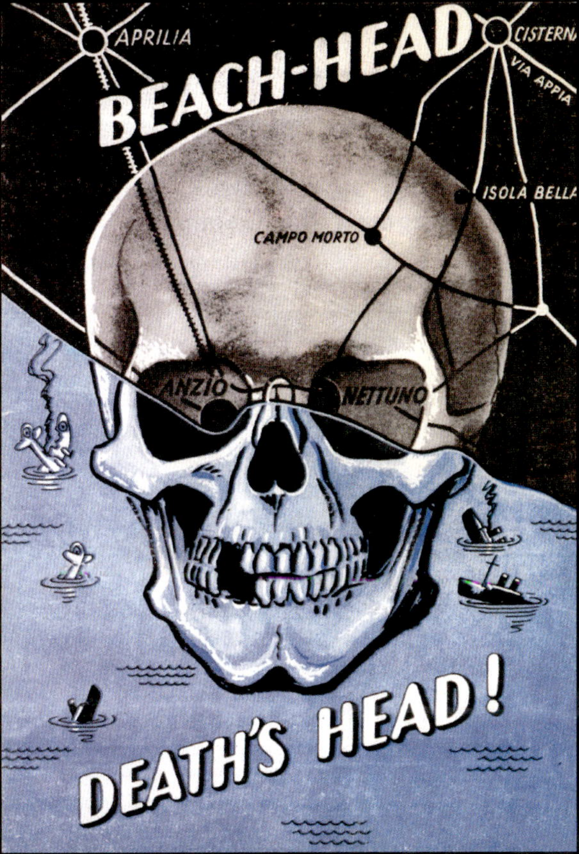

A 1943 German World War II propaganda poster showing Allied landing sites on the Italian coast.

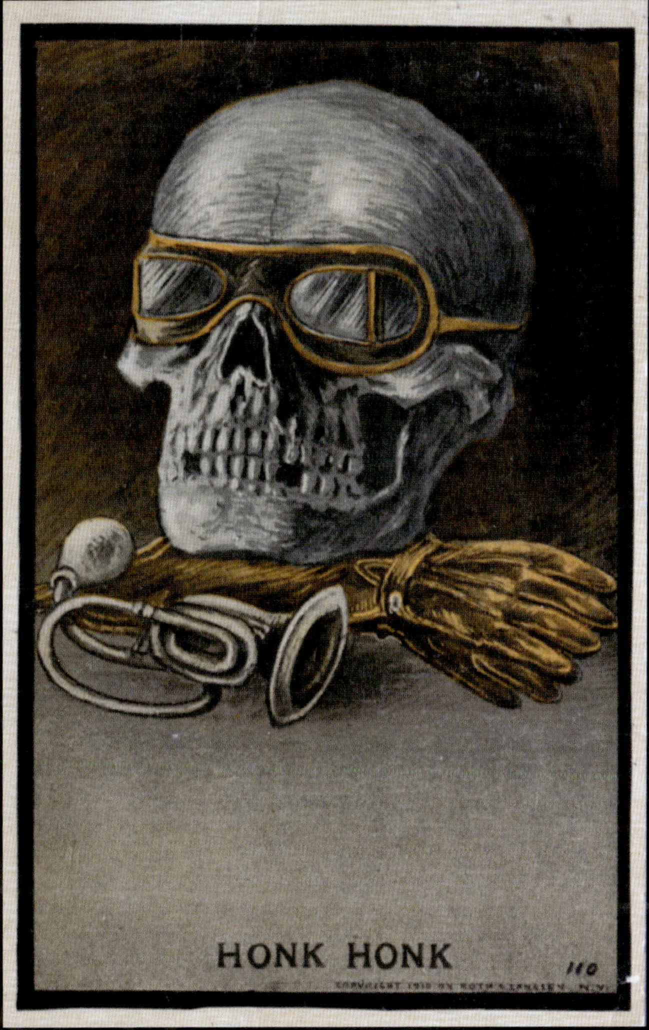

A 1910 postcard entitled "Honk Honk" containing a skull with driving goggles, leather gloves, and a horn.

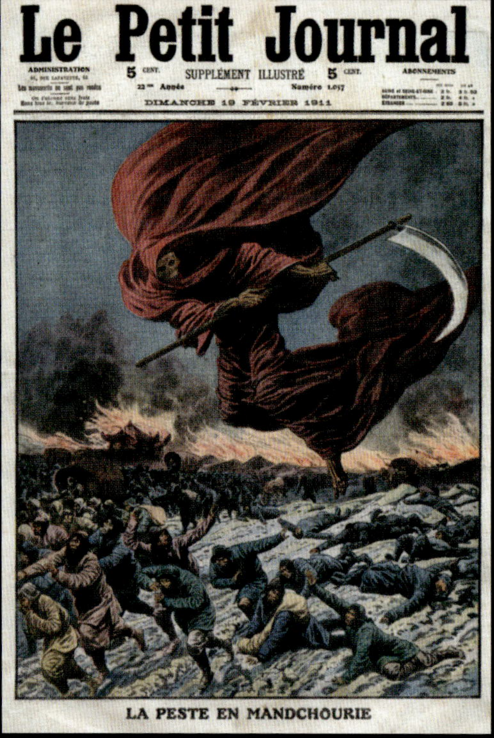

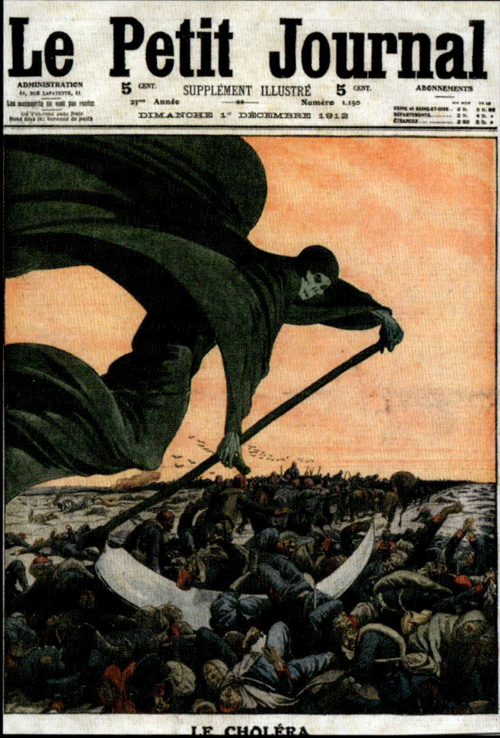

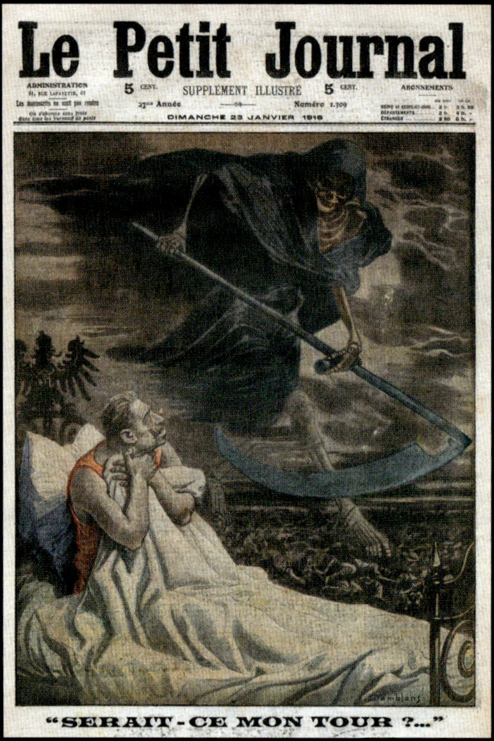

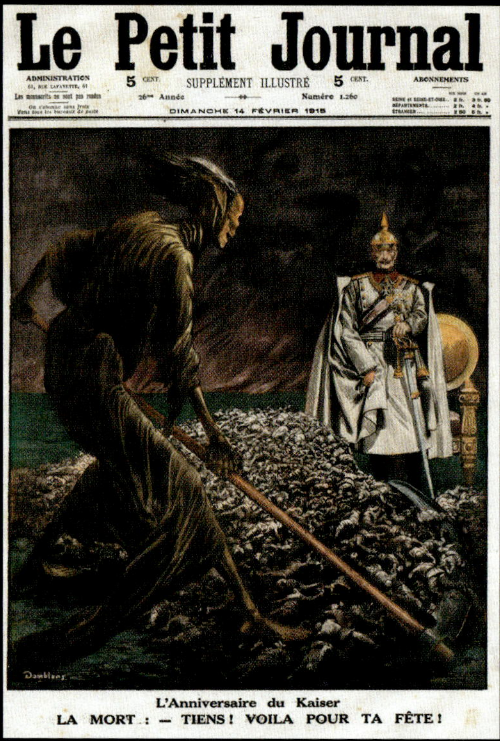

Front pages of *Le Petit Journal*, a French language newspaper (clockwise from top left): The plague in Manchuria (February 19, 1911); Cholera epidemic during the First Balkan War (December 12, 1912); Death comes for the Kaiser's birthday (February 14, 1915); and Now it's the Kaiser's turn (January 23, 1916).

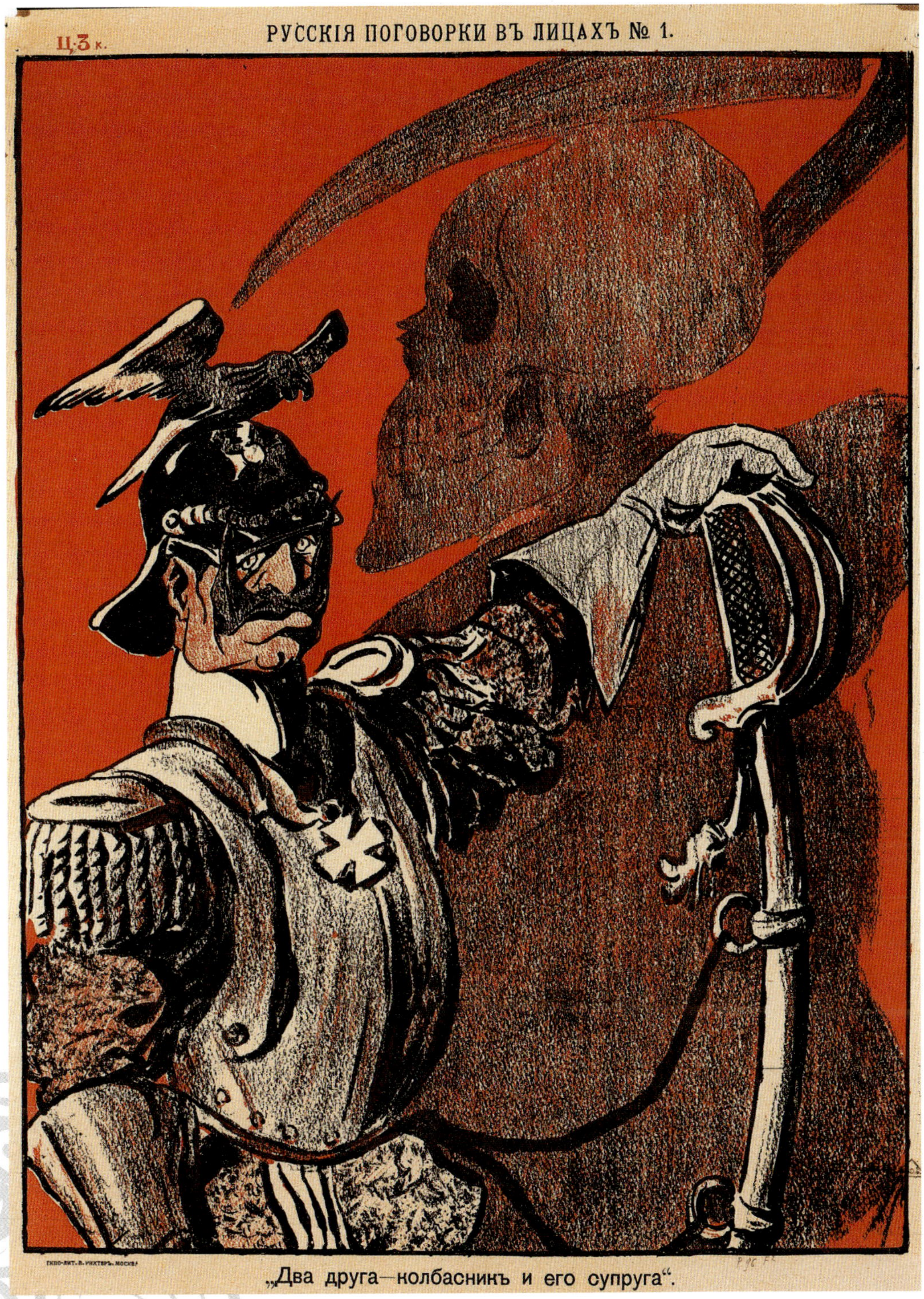

A Russian caricature of King William II of Prussia, the Kaiser of Germany, during the First World War.

SKULLS IN SCI-FI, FANTASY, AND HORROR

THINK OF AN ICONIC SCIENCE FICTION CHARACTER WITH THE HEAD OF A SKULL, WHO HAS BEEN FEATURED IN SEVEN MOVIES. TO ALL INTENTS AND PURPOSES, C-3PO, THE TALL, GANGLY, EAGER-TO-PLEASE ROBOT CHARACTER with a key part to play in George Lucas' *Star Wars* series of films, has the head of a skull. That is, all except for the wide mouth and the grinning teeth. Large, hollow eyes and a truncated nose make him arguably the most famous skull character in science fiction. Unlike most skulls, which promote fear, C-3PO's gangling gait and kindly nature account for his near-universal adoration.

The haunting skull of the Grim Reaper strides through the myths and legends of antiquity and into latter-day sci-fi and fantasy genres. Nearly every society on earth has some equivalent of the Grim Reaper, whether it goes by that name or not. In fact, its appearance—cloaked in black, often with a scythe—also seems to transcend cultures.

Crystal skulls—man-made "artifacts" from Victorian times in reality—are also infamous in the realms of fantasy and science fiction. *Indian Jones and the Kingdom of the Crystal Skull* features crystal skulls so prominently that it is in the title. The fourth film in the Indiana Jones franchise, it features the eponymous title character searching for a telepathic crystal skull.

Hundreds of thousands of other movies have featured skulls, which may be one of Hollywood's favorite motifs. Horror films of the 1970s and 1980s, especially liked to use skulls in their advertising.

However, the trend toward skulls was by no means started in the 1970s. Alfred Hitchcock's 1950 film, *Stage Fright*, did not use skulls in their American advertising, but did overseas. The Polish poster, for example, features a design reminiscent of a crossword puzzle, with the lines of the puzzle forming a skull around the text. Another example, though not nearly as subtle as Hitchcock, was *The Screaming Skull*, a 1958 horror movie which was advertised with movie posters which screamed the promise: "FREE! We guarantee to bury you without charge if you die of fright during *SCREAMING SKULL!*" A tempting offer. The movie is based on a British folk tale regarding the skull of an ancestor, kept, in a safe place, at the ancestral home. Should the skull be removed, so the story goes, then a series of poltergeist activity, including intermittent screaming, will begin. One such tale belongs to Burton Agnes Hall, in Yorkshire. Here, the skull of Anne Griffiths, daughter of the 17th century Baron, Sir Henry Griffiths, is reputed to be hidden inside a bricked-up alcove, yelling occasionally.

The stylized design of Hitchcock's 1950s skull fell out of favor by the 1960s, as can be seen in the almost cartoonish drawings of movie posters of that period. *Macumba Love* has several voodoo skulls on the advertising poster that are laughable to us today. Other posters of the era are similar in their depictions of skulls, such as that of *Dead Eyes of London*, and *Astro-Zombies*. *The Skull*—a 1965 movie based on the Robert Bloch book *The Skull of the Marquis de Sade*—actually starred a skull as a main character. In this movie, a man who buys the skull of the Marquis learns, too late, that it is inhabited by a malevolent spirit which entices him to commit murder.

By the 1970s, society had progressed to the point where gruesome advertising was acceptable. The skulls on movie posters of this era aren't subtle or cartoonish and often drop bits of blood and gore or spew fire. This can be seen on the gruesome advertising of movies such as *The Legend of Hell House*, *From Beyond the Grave*, and *The House on Skull Mountain*. They exemplify a society becoming comfortable with the macabre.

More recent movies employ a variety of skull art techniques, including both overt and hidden skulls. The posters advertising the original *Final Destination* in 2000 use a combination of techniques, either imposing skulls over the characters' faces or using a reflection to accomplish a similarly-sinister goal. 2011's *Fertile Ground* has a poster showing a bloodstained skull lying on the ground. The poster advertising for 2010's *The Dead*, however, is much more subtle. At first glance it is simply a map of Africa with standard advertising copy around it. Upon closer examination, however, a skull is revealed.

Alchemists, occult pseudoscientists known for their quest to transform base metals into gold, use the skull in combination with symbols of the raven and the grave to represent a phase of their transformational practice known as "mortification," a throwing-off of earthly delights and material goods in exchange for the eternal life of the spirit as well as the hopeful belief in rebirth.

The skull of a horned goat is frequently seen in connection with popular horror story images of witchcraft, although most self-respecting witches would probably laugh at the crude use of such imagery. Here, the horns of the goat represent the horned God of the Witches, based on the Greek's pre-Christian image of the god Pan, who also has the cloven hooves of a goat in place of normal feet.

Steampunk is a relatively new fantasy genre that features steam-powered machinery rather than advanced technology. It combines Victorian fashions with modern ideas, relying on imagery such as gears, clocks, and keys. Steampunk skulls are often decorated with all three of these, representing the connection between life, machinery, and death.

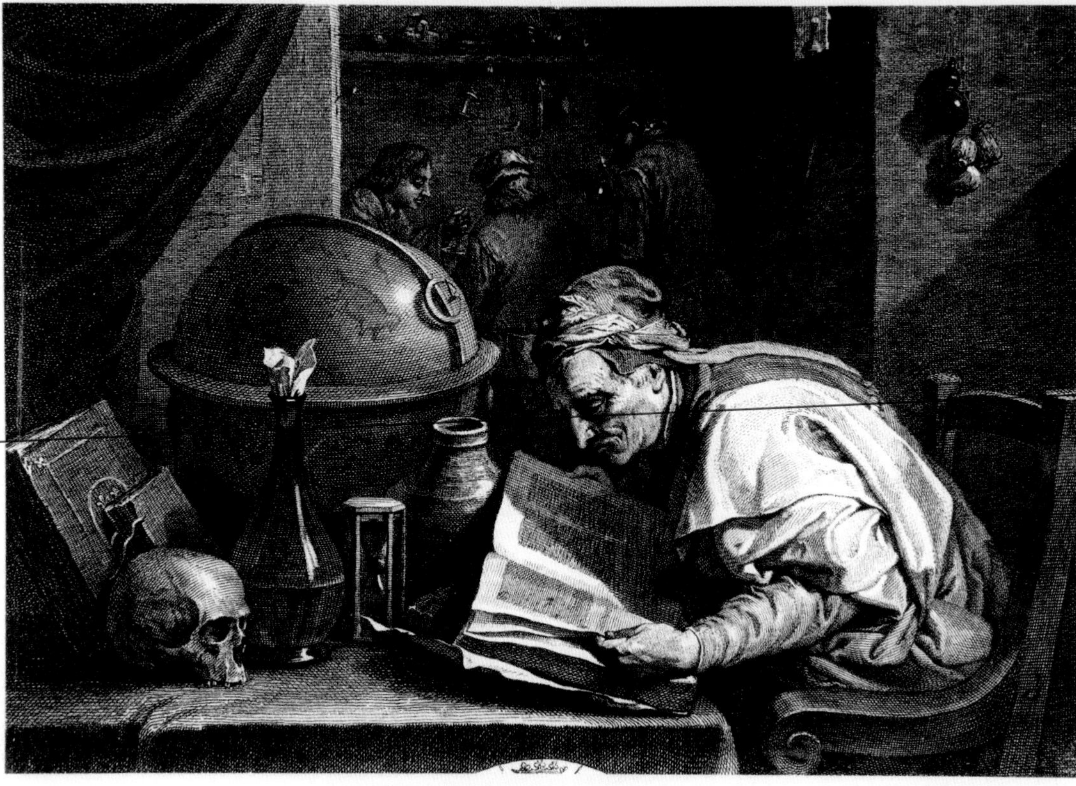

An engraving by Jacques-Nicolas Tardieu entitled *Le Docteur Alchimiste*.

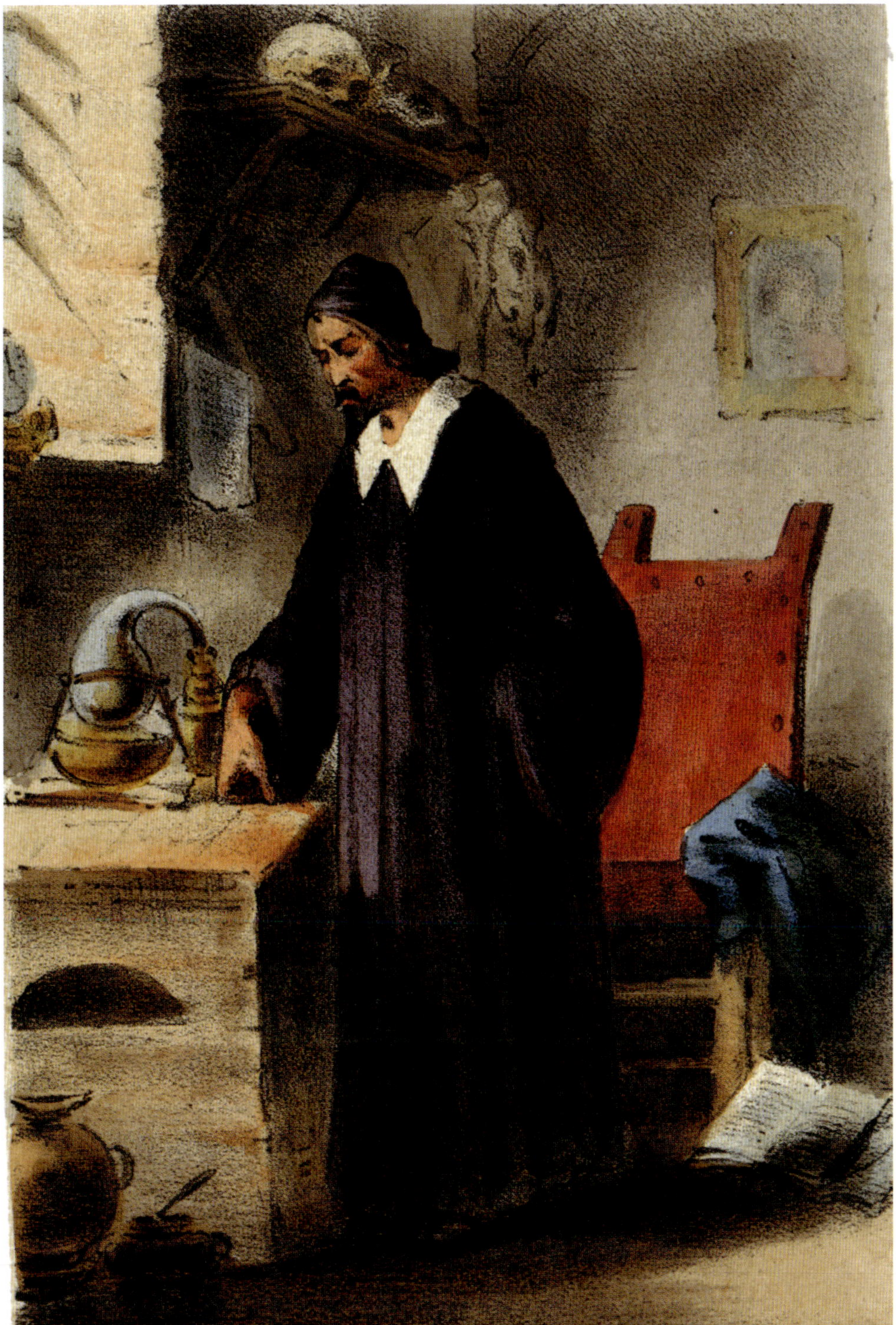

The Alchemist, an engraving by Nicolas Barnaud.

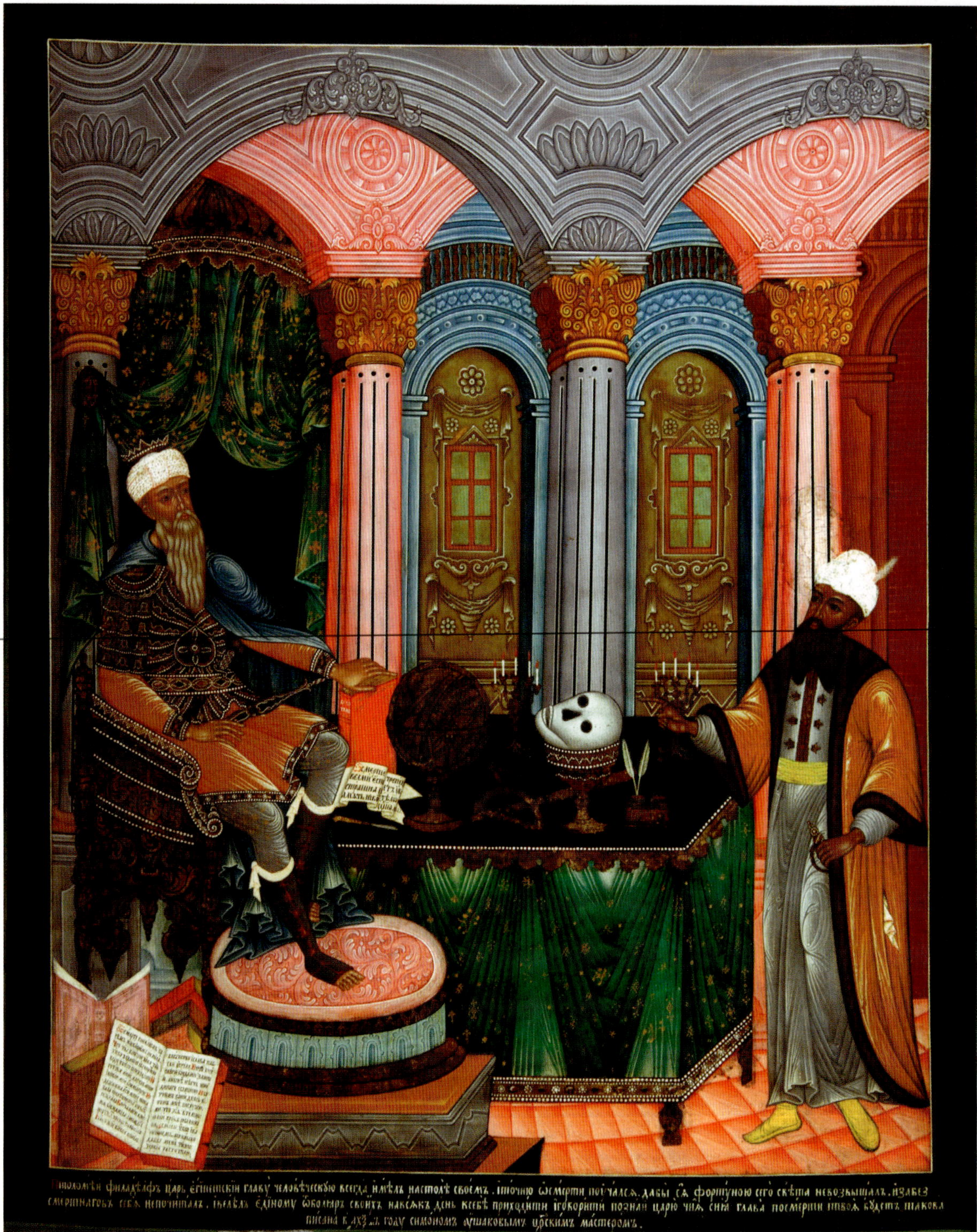

A painting with a skull at the center, titled *Ptolemy II*, by an unknown artist.

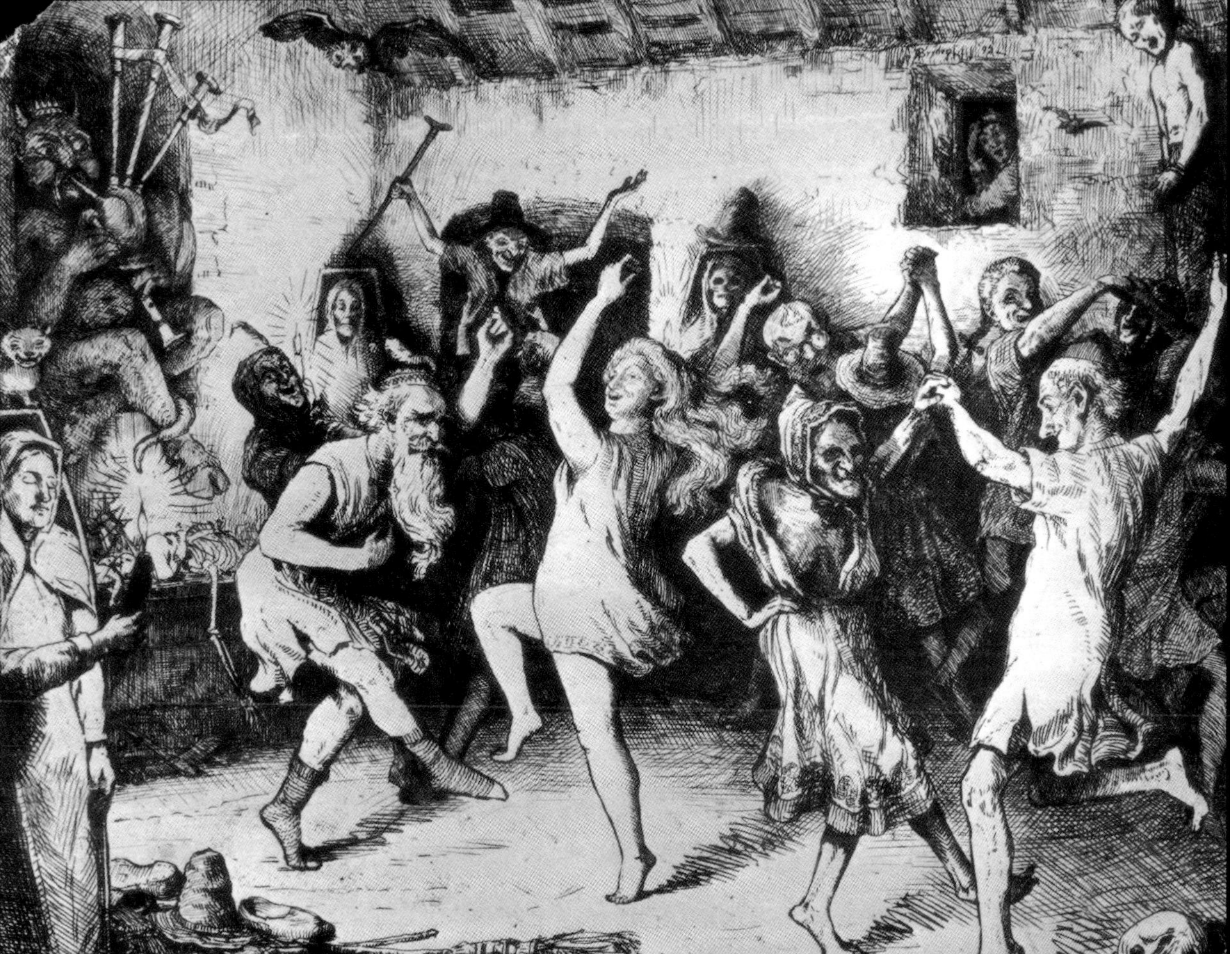

A seventeenth century drawing showing a Sabbat, a ceremony parodying the Christian mass believed to be held by witches.

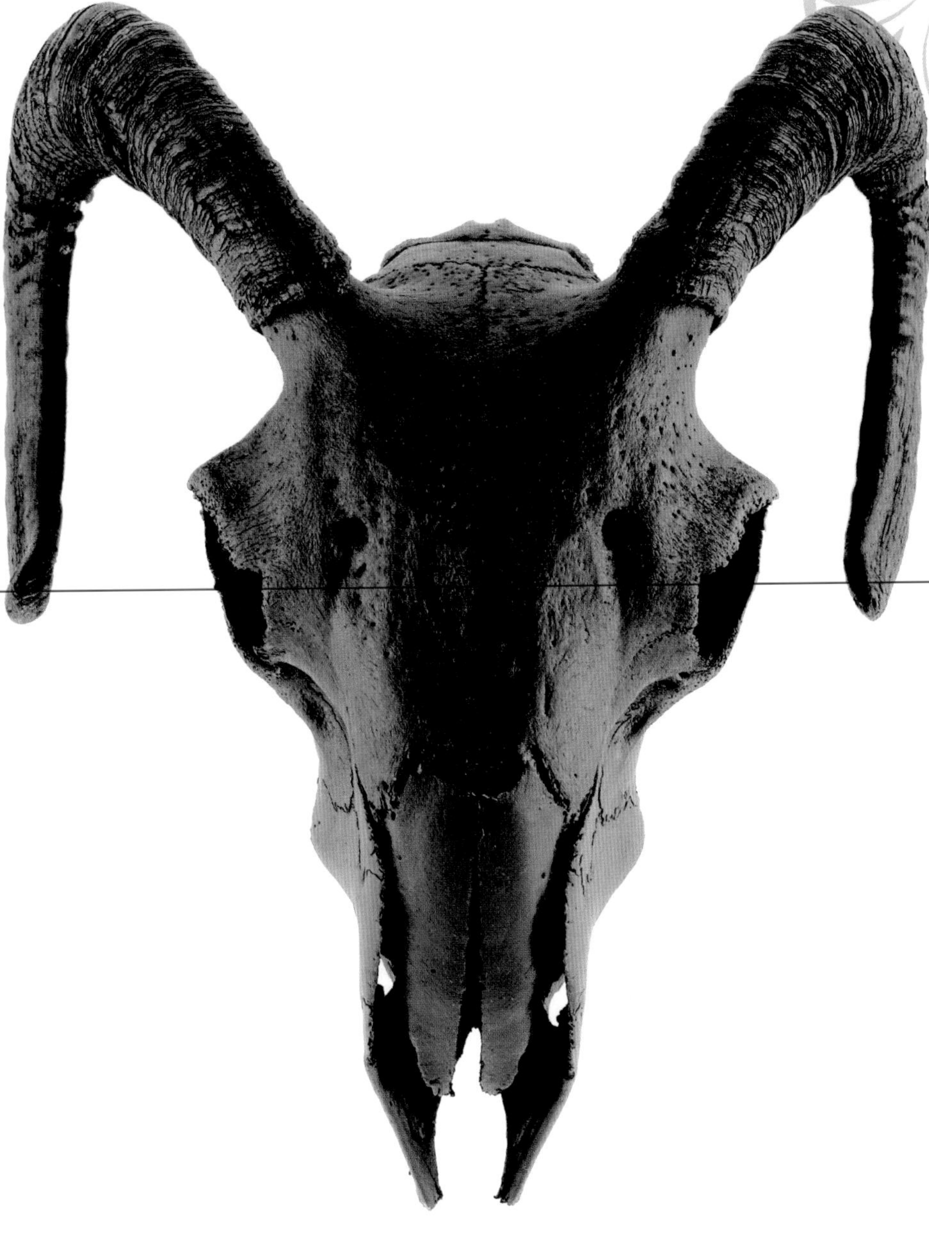

This horned goat skull is painted black.

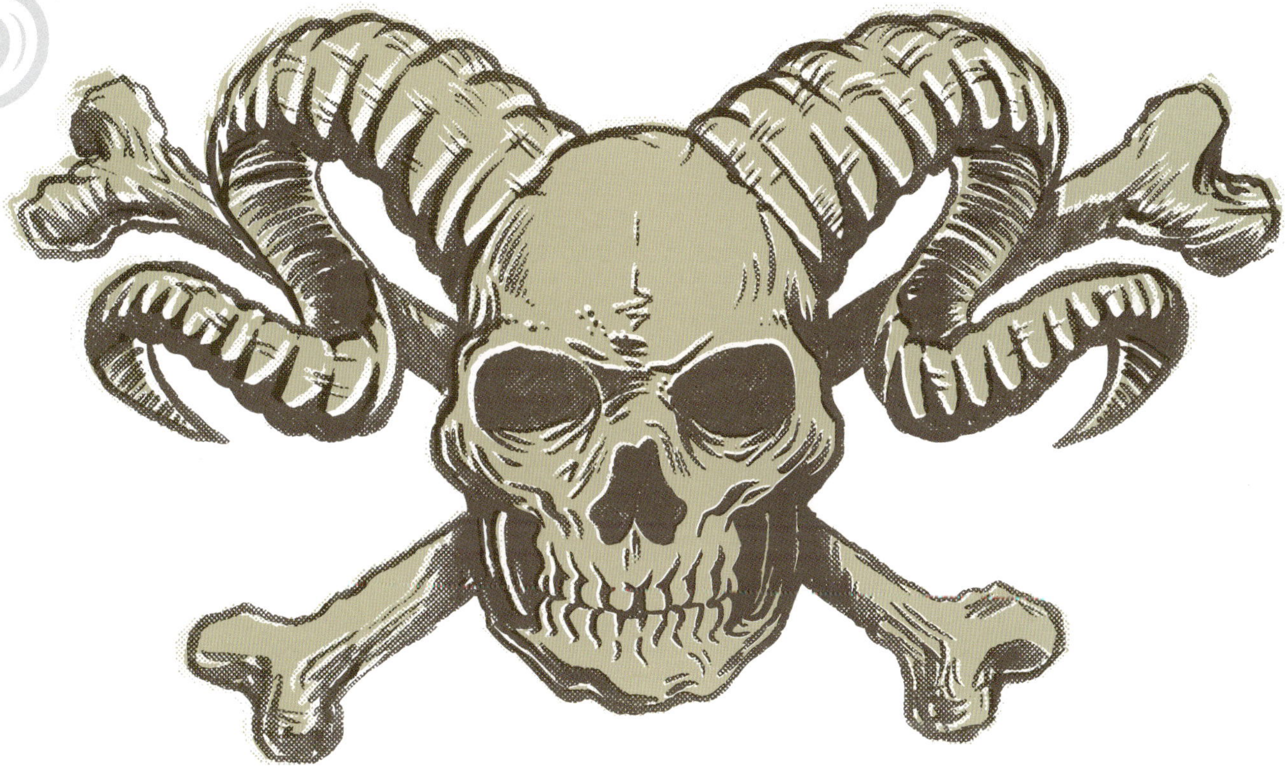

A horned goat skull is a common element of fantasy stories.

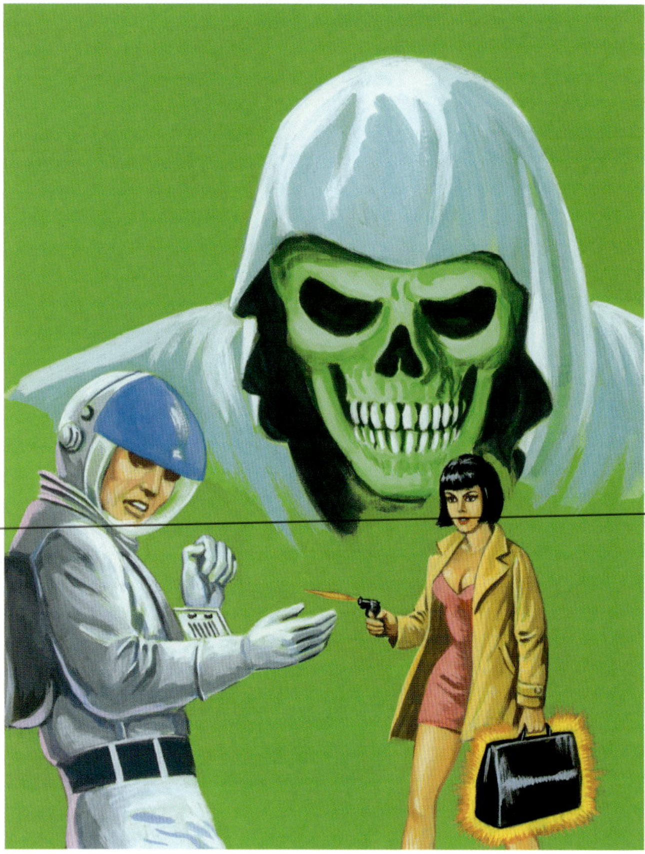

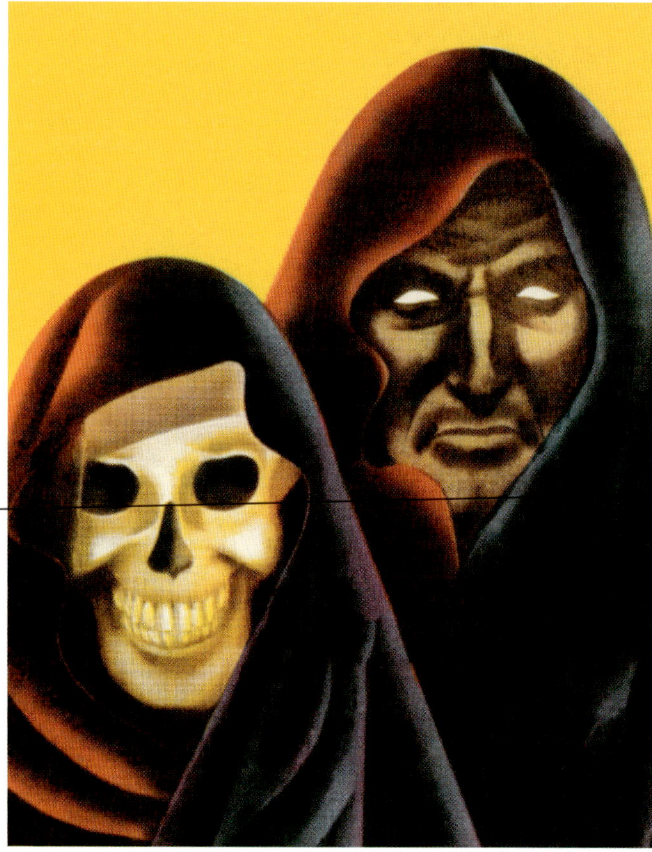

This spread and page 216: The Grim Reaper is a popular character in pulp art.

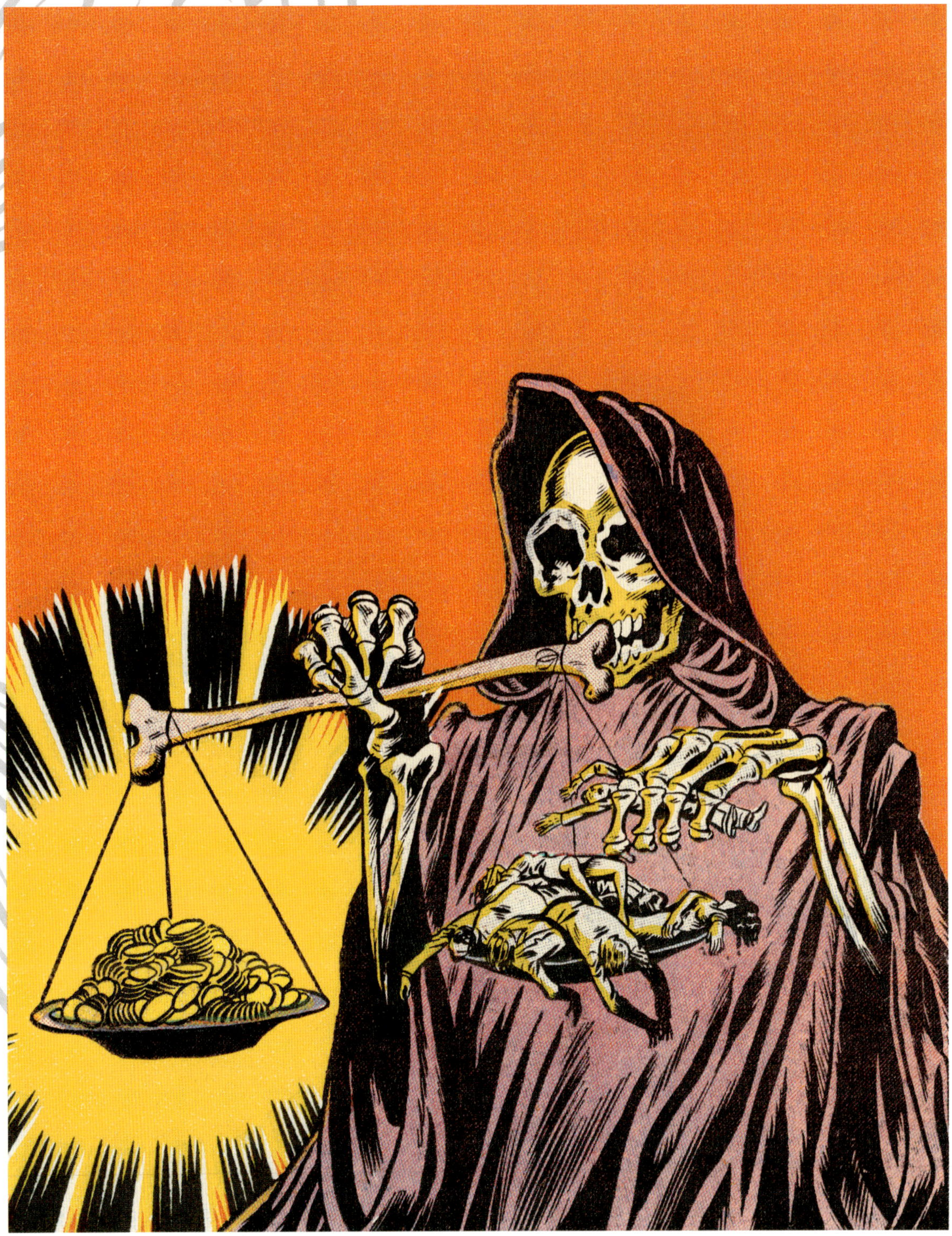

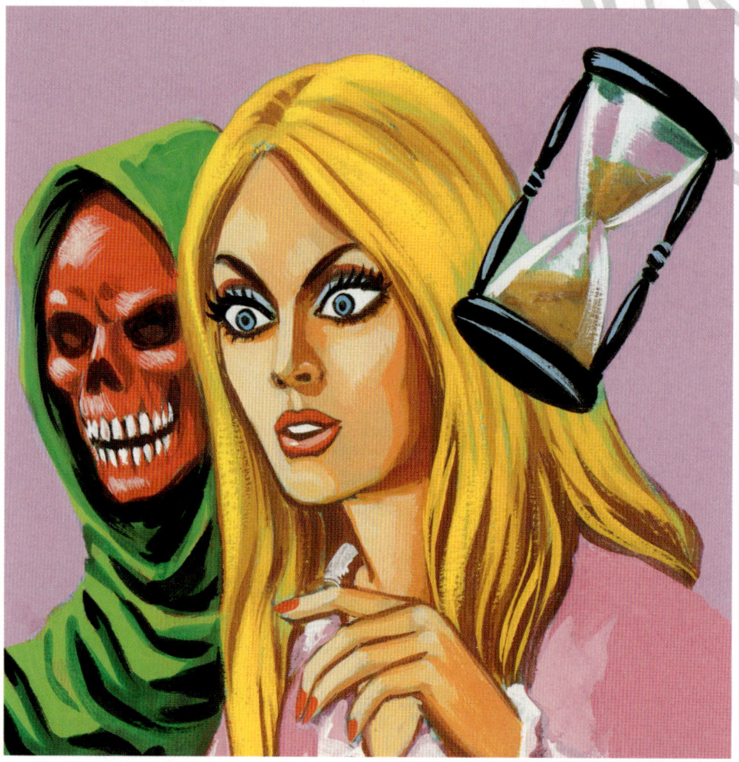

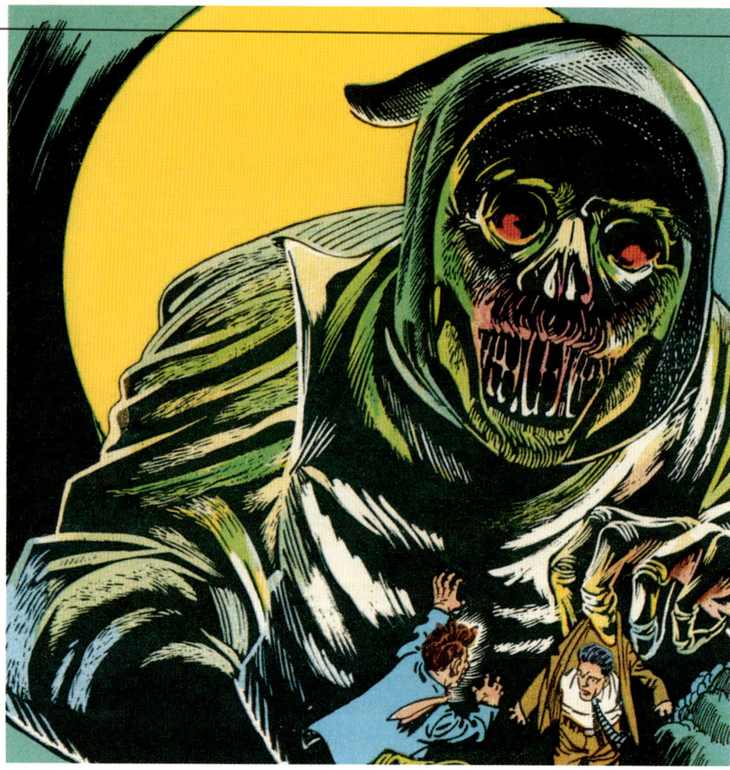

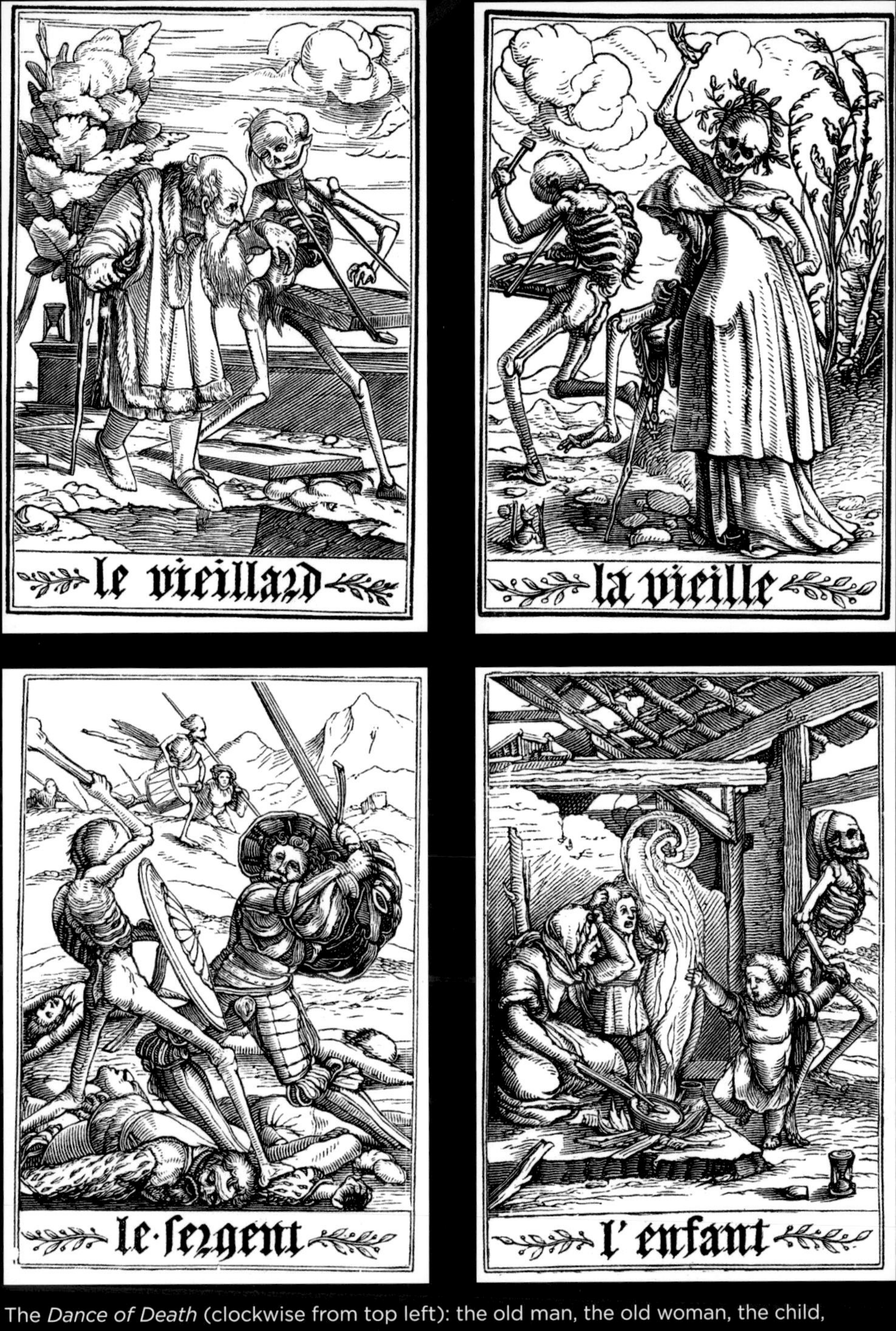

The *Dance of Death* (clockwise from top left): the old man, the old woman, the child, and the soldier.

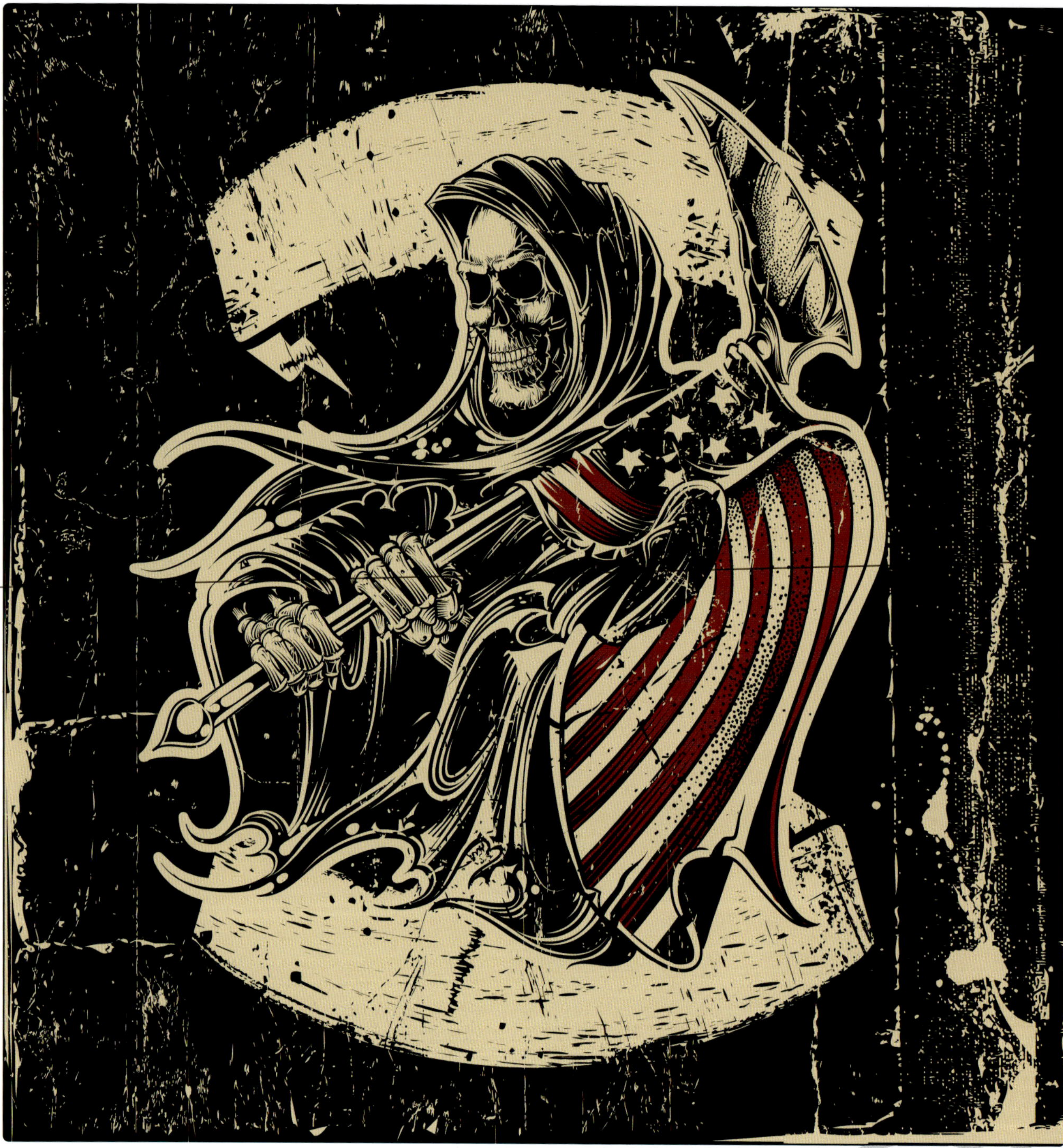

A motorcycle patch featuring the Grim Reaper holding an American flag.

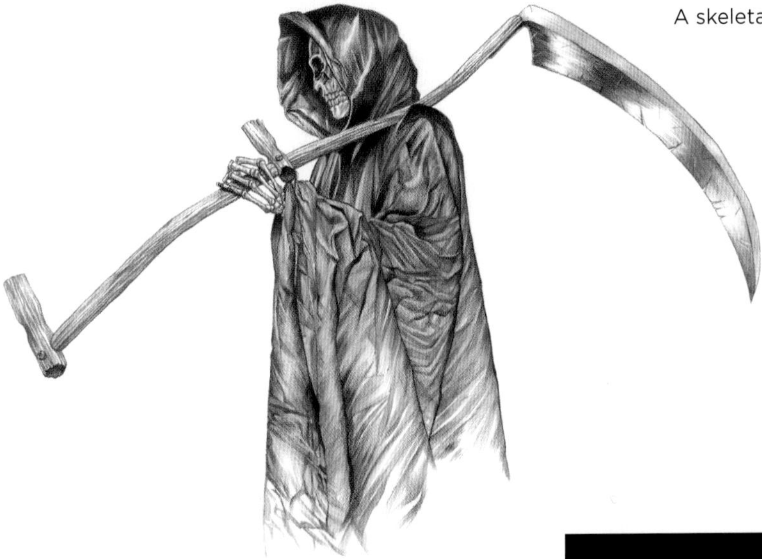

A skeletal Grim Reaper carrying a scythe.

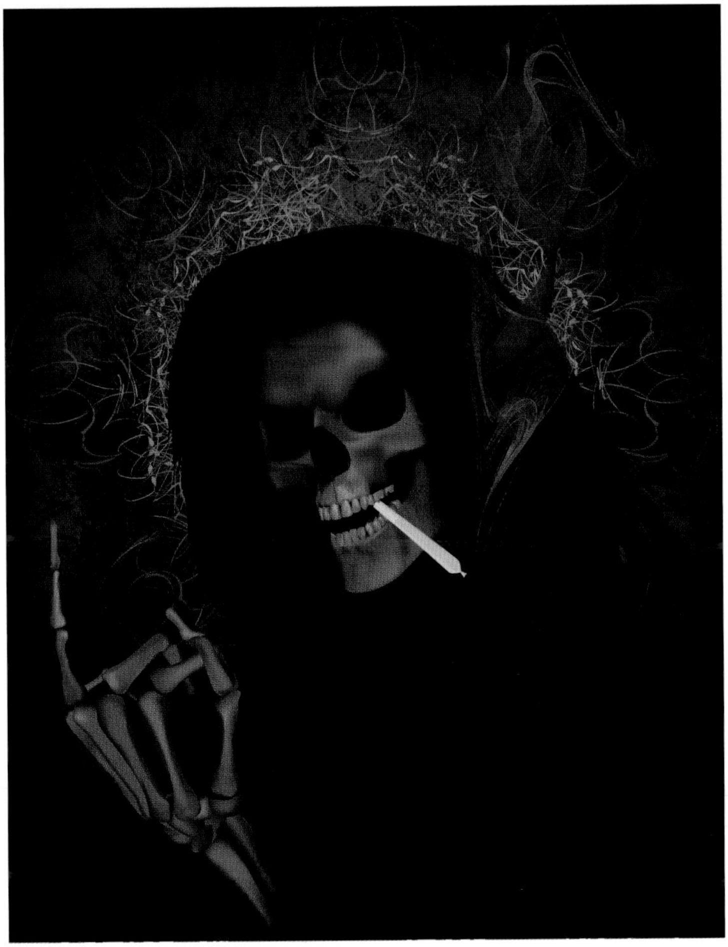

The Grim Reaper smoking a cigarette.

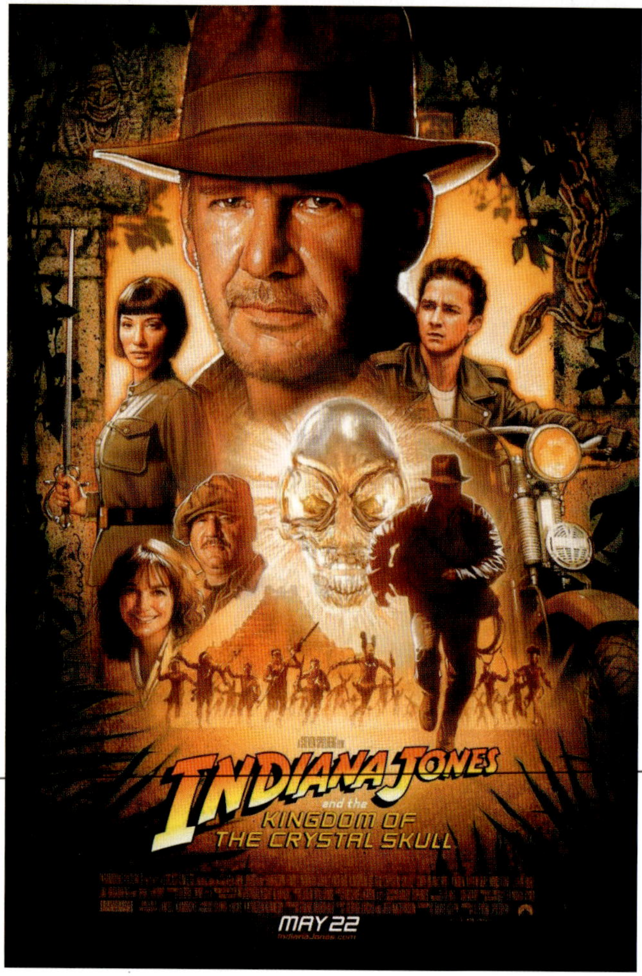

Poster advertising 2008's *Indiana Jones and the Kingdom of the Crystal Skull*.

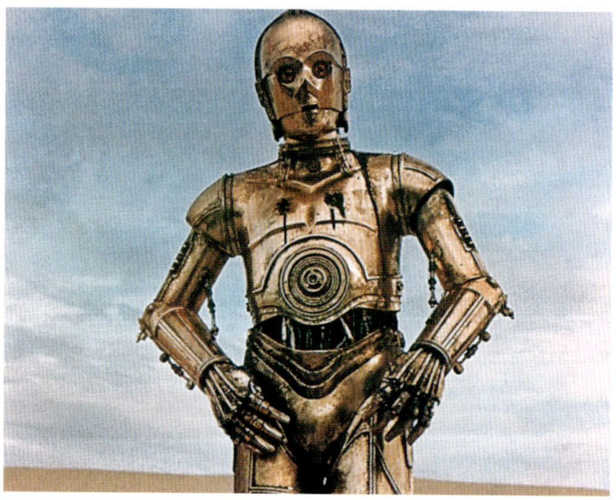

Star Wars character C-3PO's head takes inspiration from a human skull.

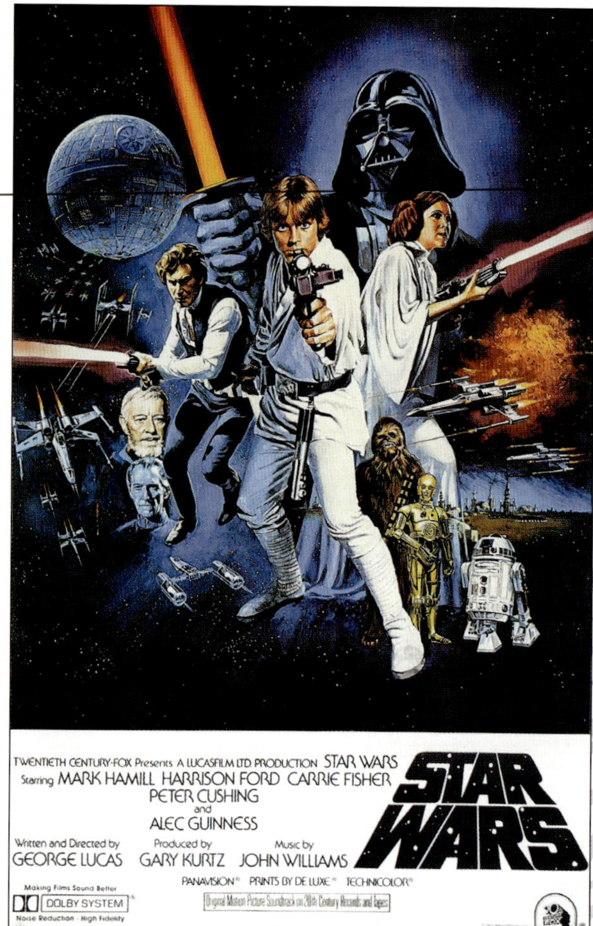

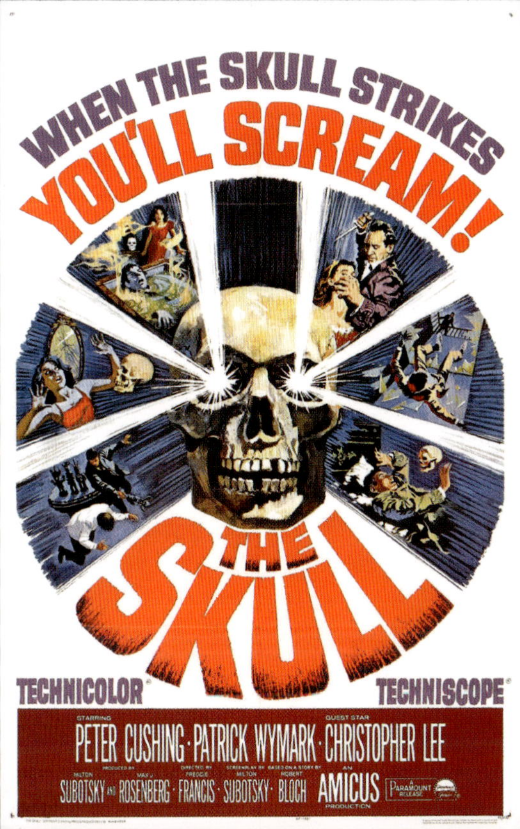

This movie poster advertised 1965's *The Skull*.

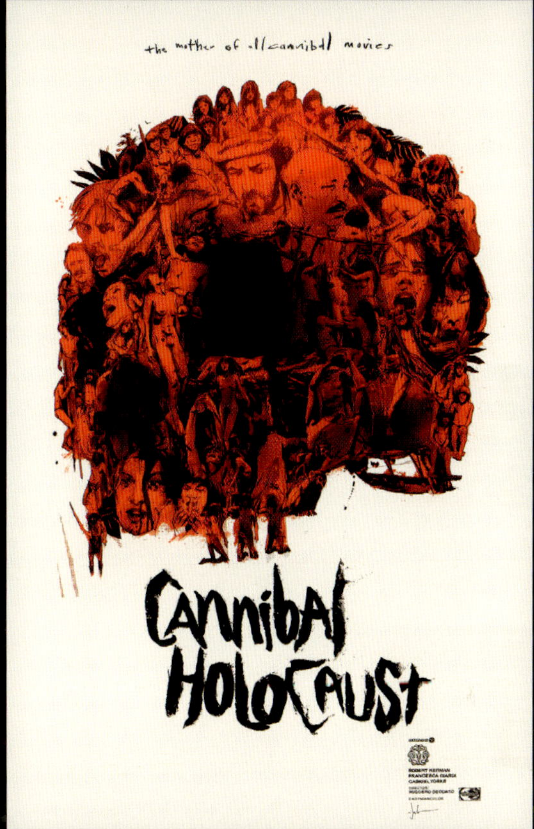

A poster for 1985's *Cannibal Holocaust*.

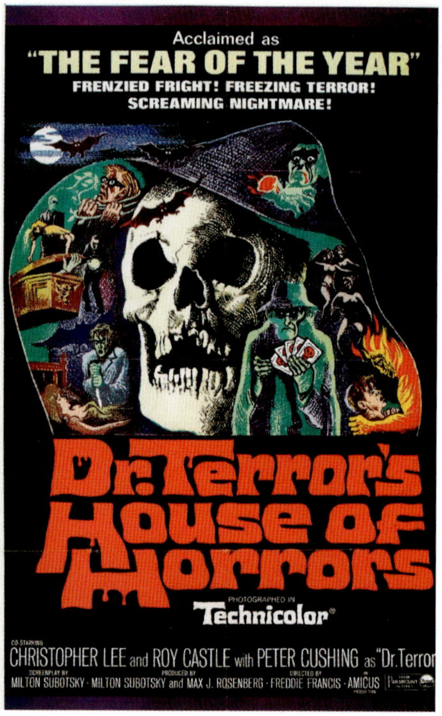

Movie poster from 1965's *Dr. Terror's House of Horrors*.

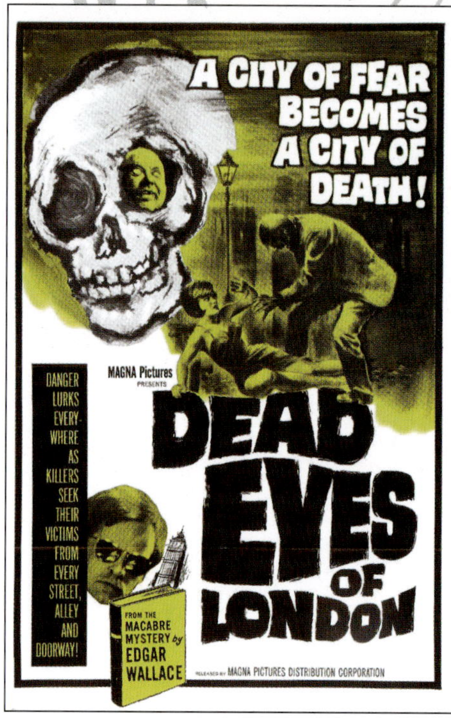

A 1961 advertisement for *Dead Eyes of London.*

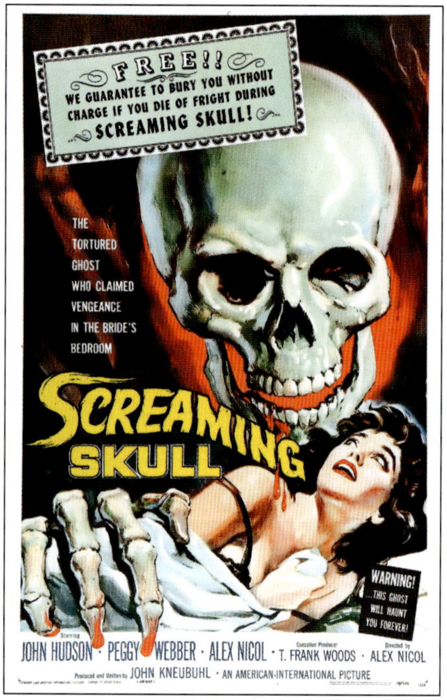

This movie poster advertised 1958's *Screaming Skull.*

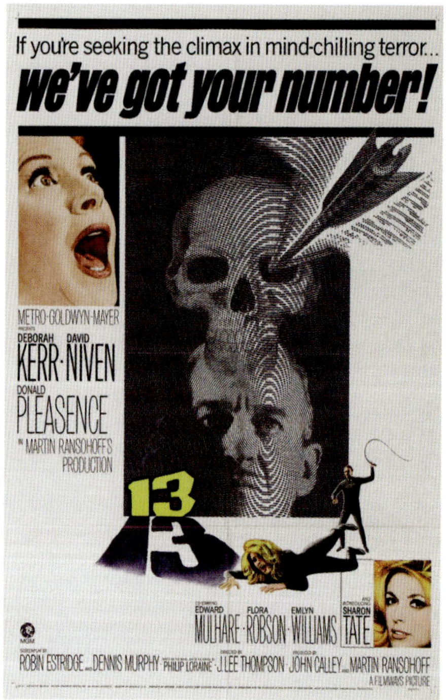

A poster advertising 1966's *13.*

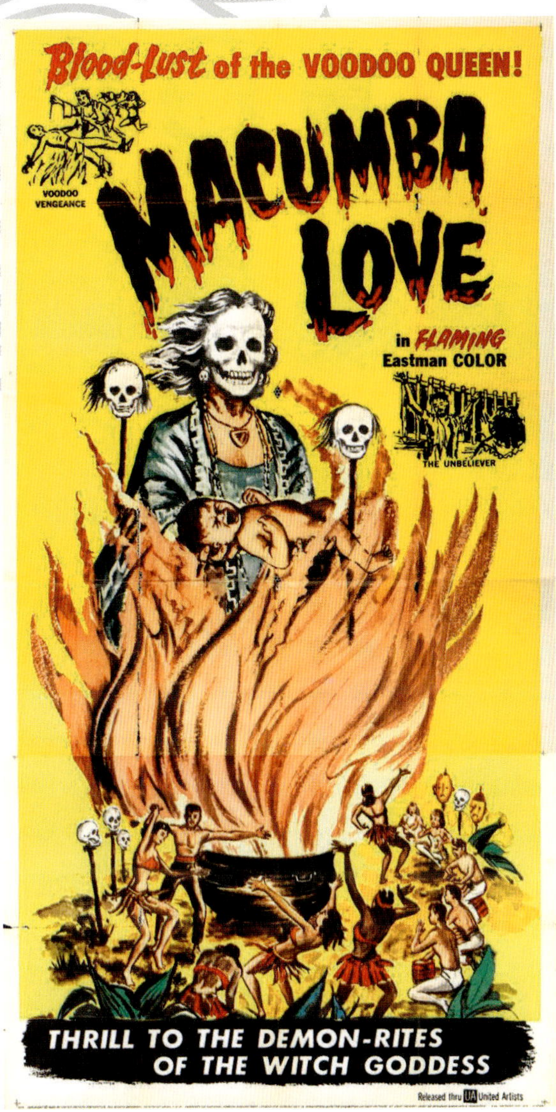

A poster for the 1960 movie *Macumba Love*.

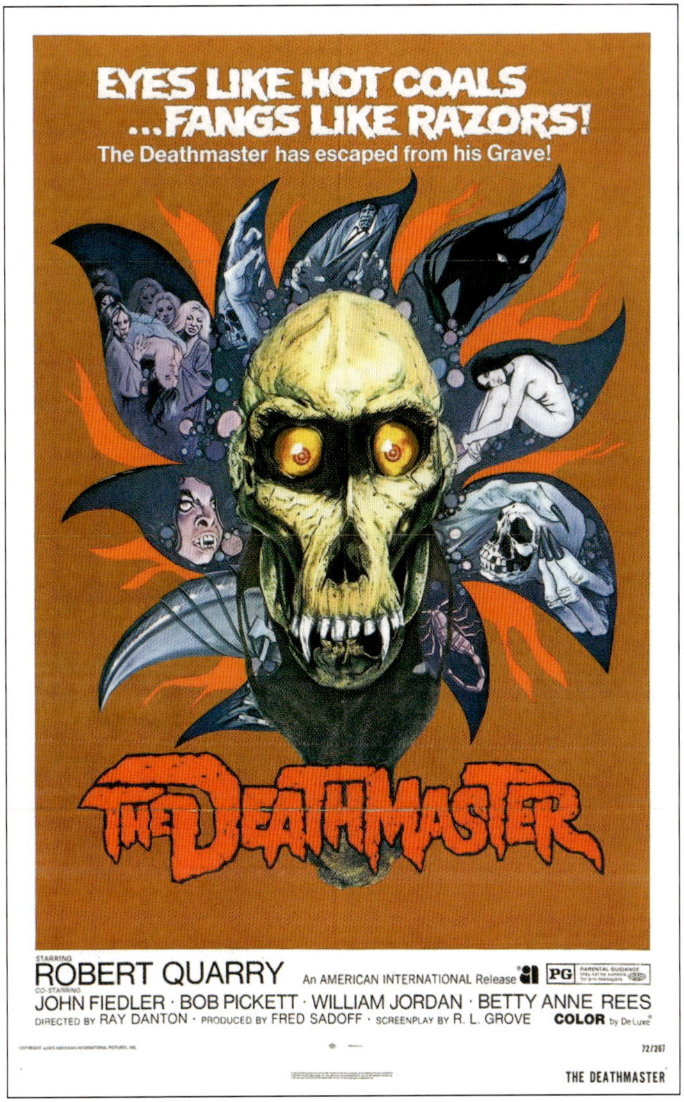

The Deathmaster was released in 1972.

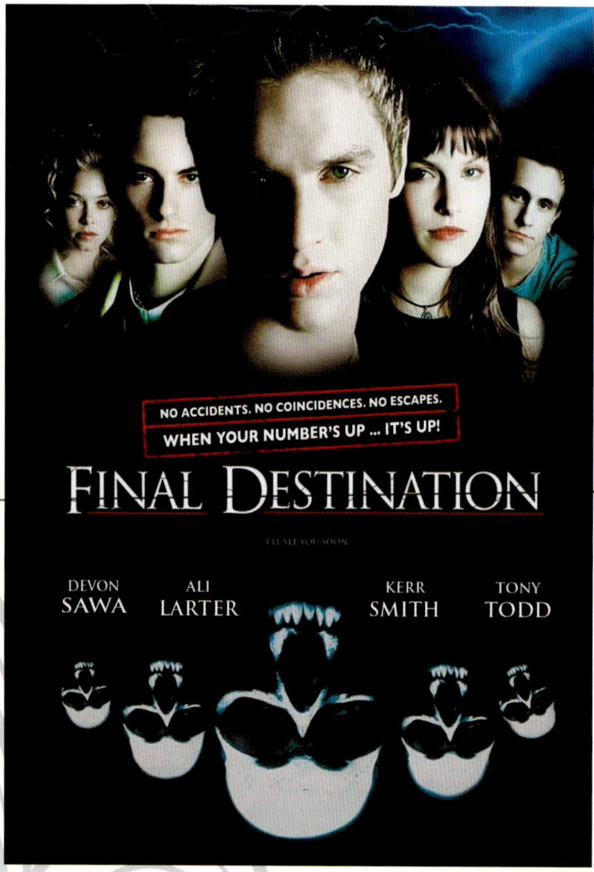

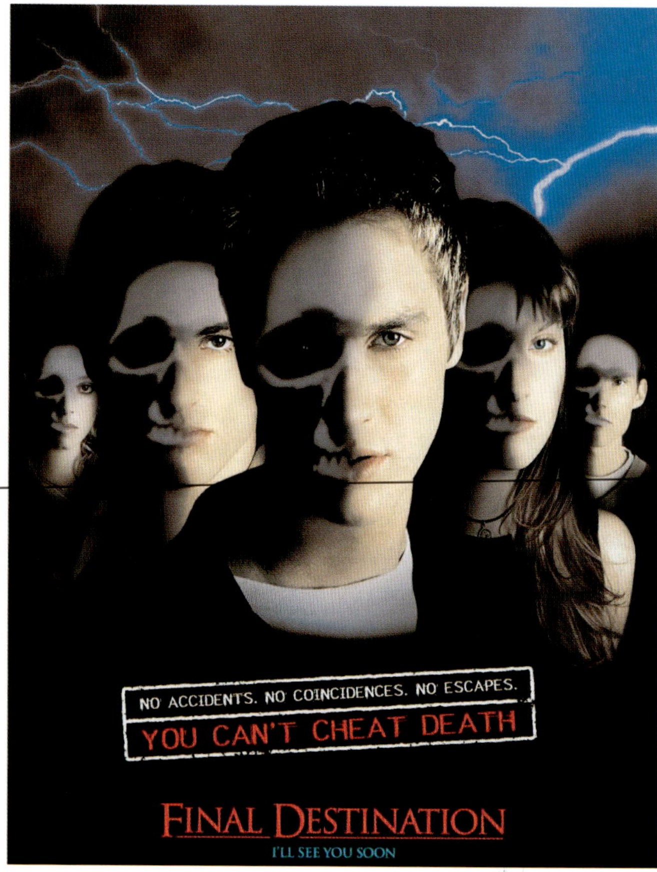

Movie posters advertising 2000's *Final Destination*.

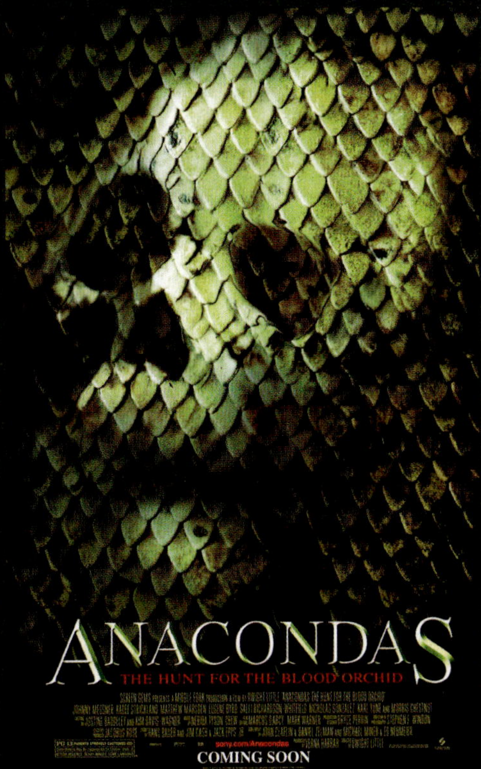

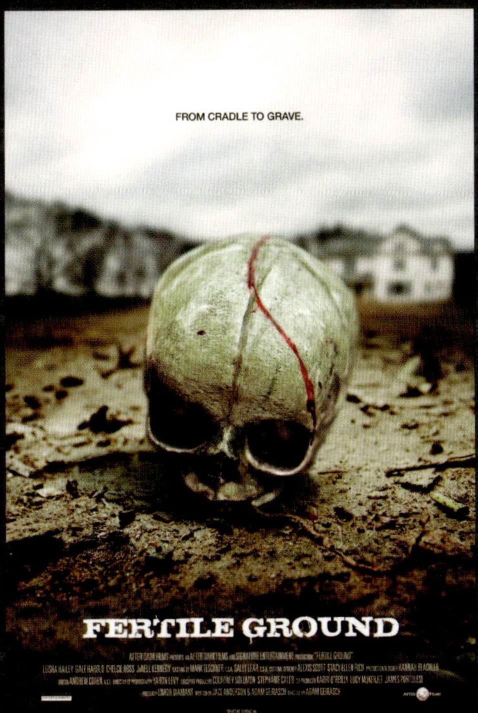

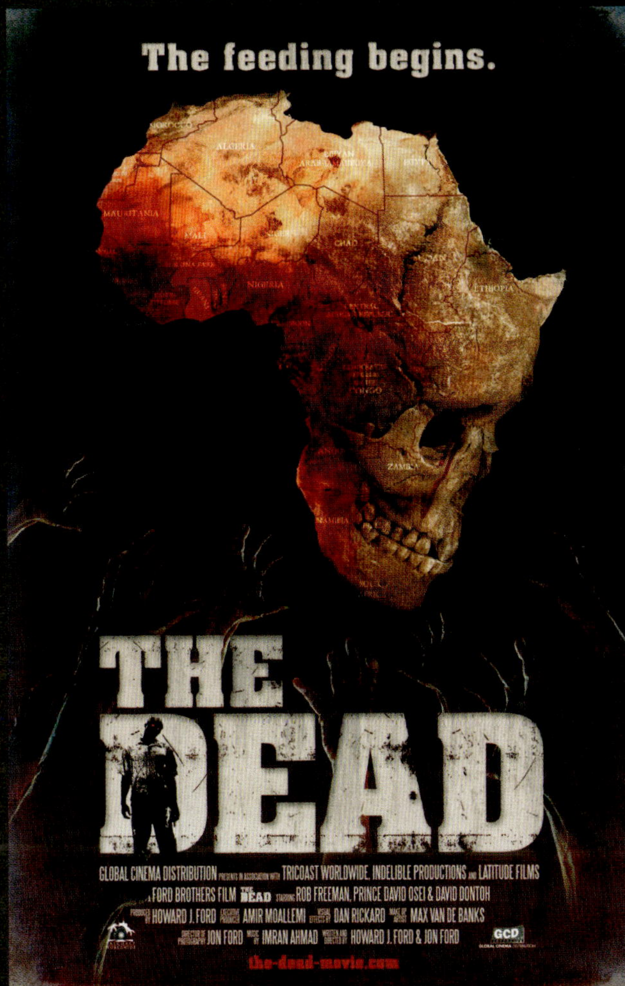

Above: This movie poster advertised 2010's *The Dead*.

Top left: Movie poster for 2004's *Anacondas.*

Bottom left: This movie poster advertised 2011's *Fertile Ground*.

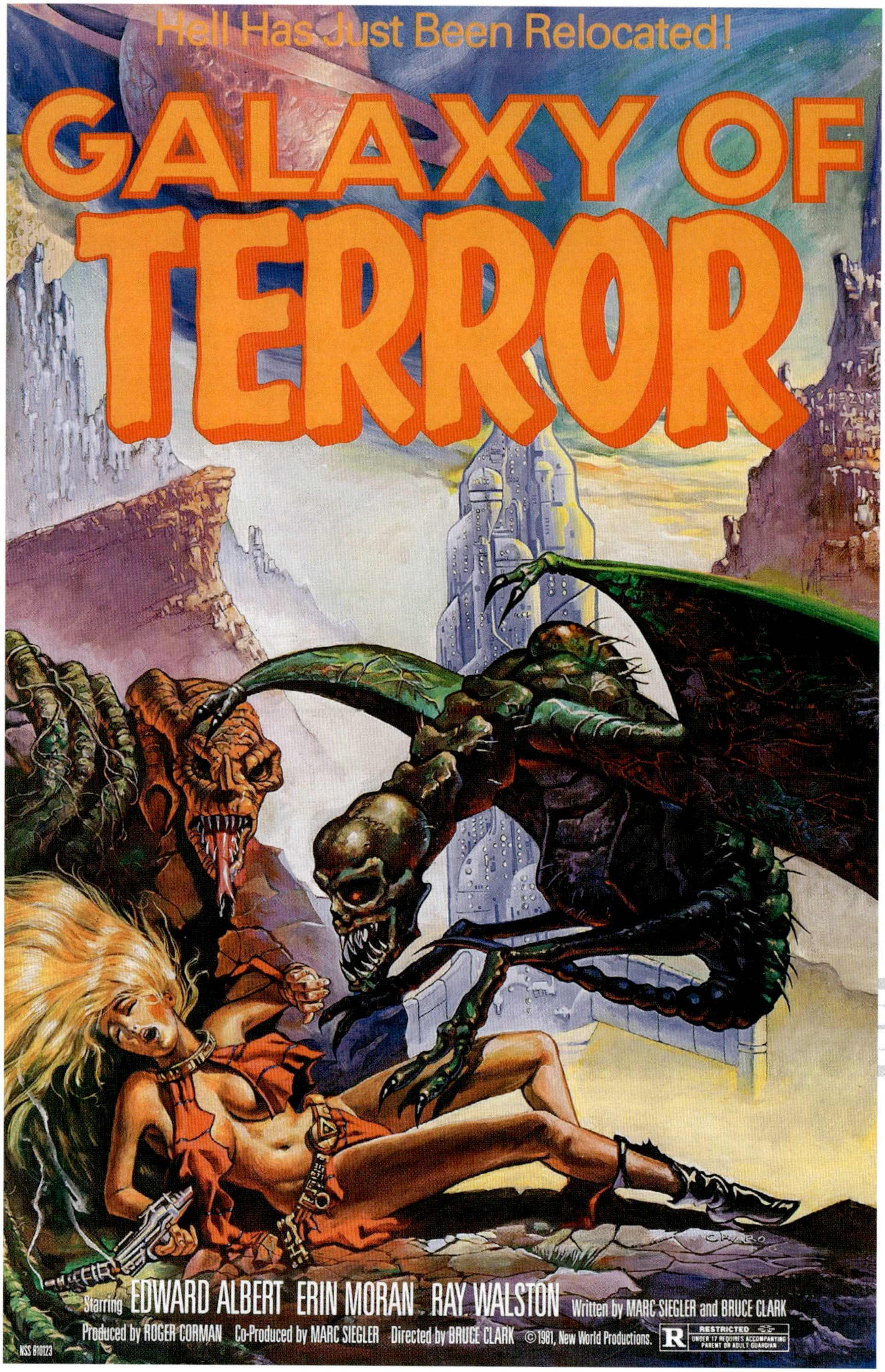

Movie poster advertising 1981's *Galaxy of Terror.*

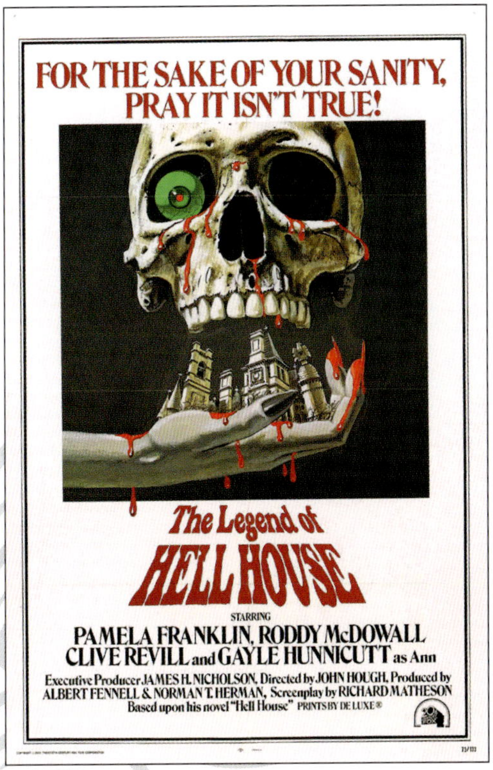

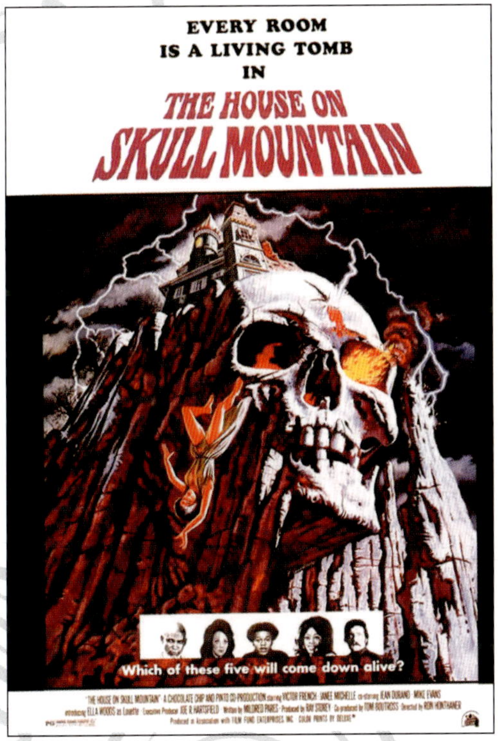

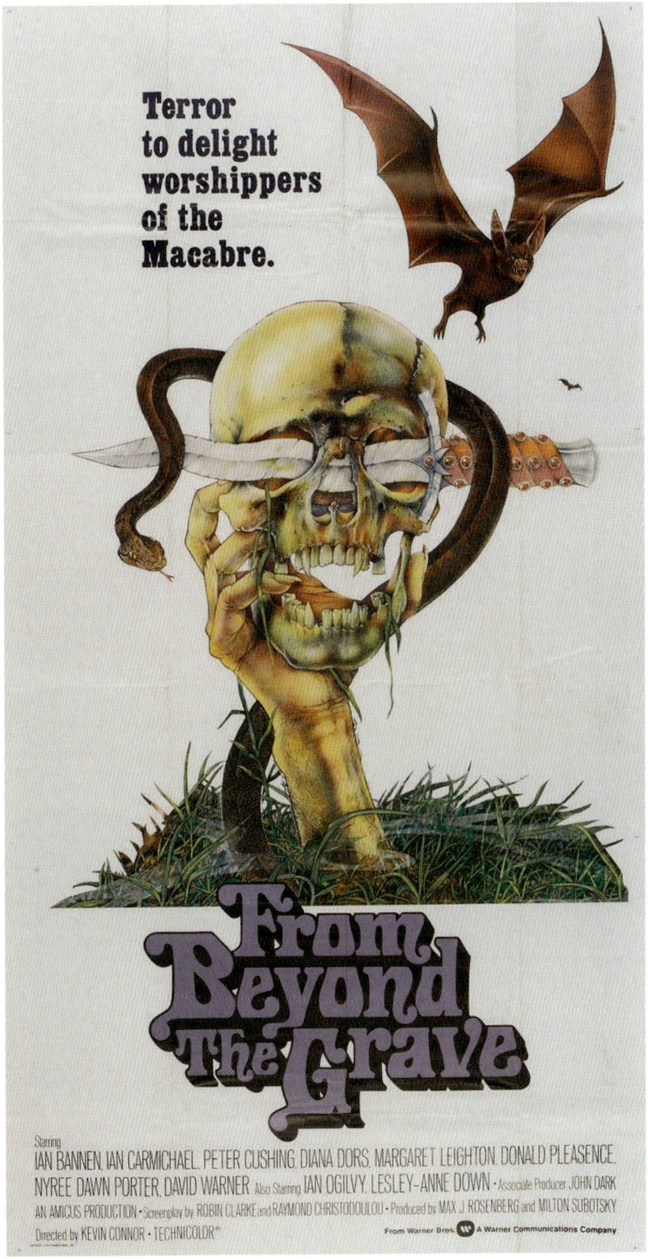

Above: A poster advertising the 1973 movie *From Beyond the Grave*.

Top left: A poster advertising 1973's *The Legend of Hell House.*

Bottom left: This poster advertised 1974's *The House on Skull Mountain*.

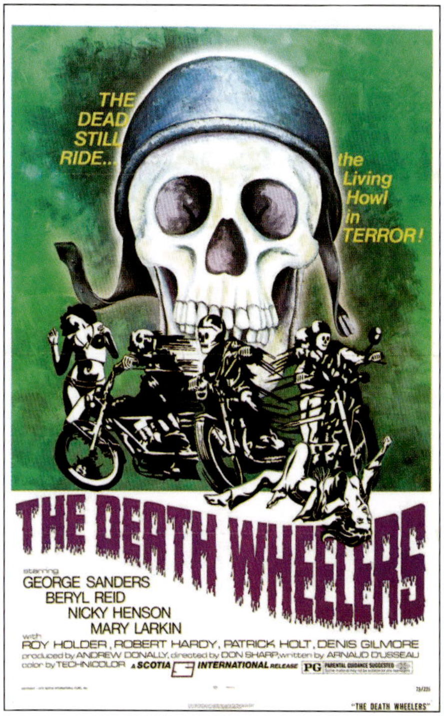

A movie poster from 1973's *The Death Wheelers*.

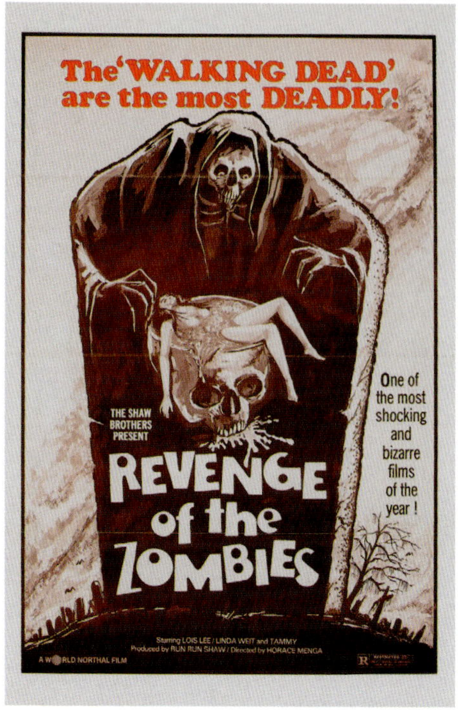

This poster advertised the 1981 movie *Revenge of the Zombies*.

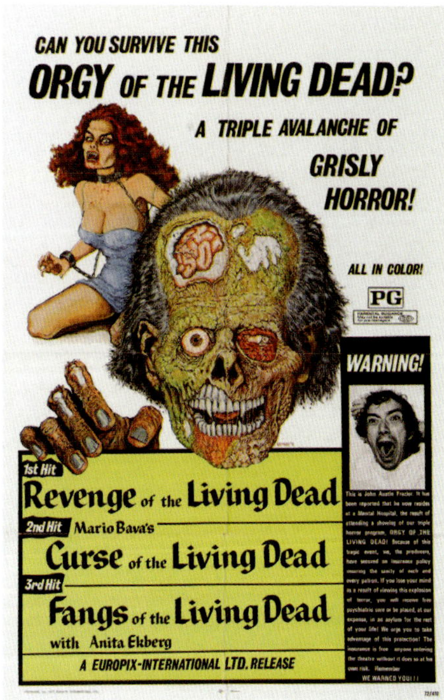

A poster advertising three hit zombie movies.

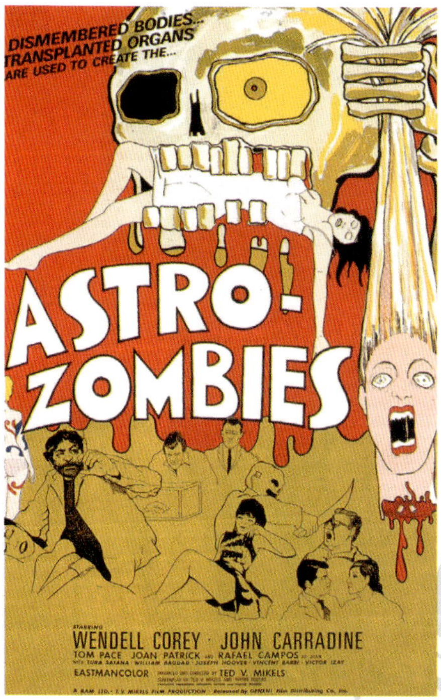

A movie poster for 1968's *Astro-Zombies*.

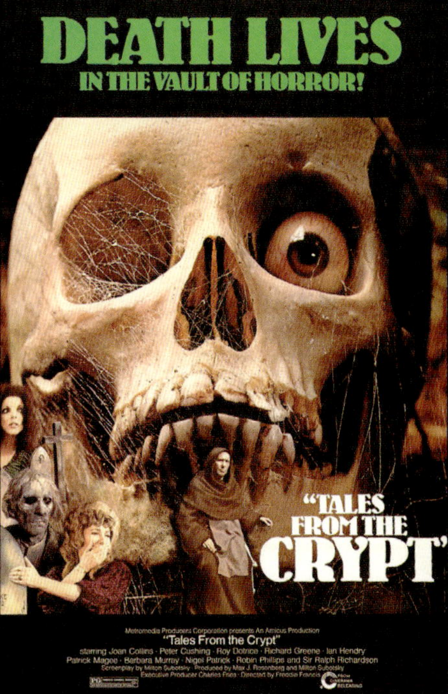

Movie poster advertising the 1972 movie *Tales from the Crypt*.

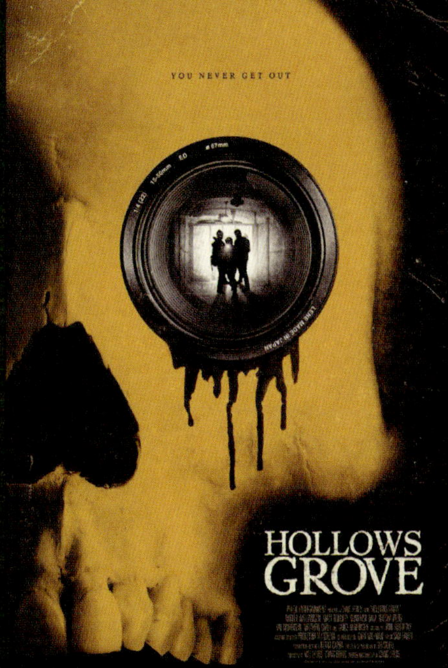

This movie poster advertised 2014's *Hollow Grove.*

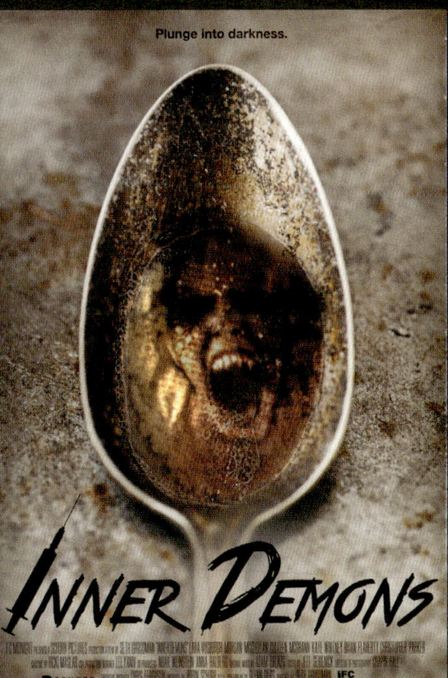

This movie poster advertised 2014's *Inner Demons.*

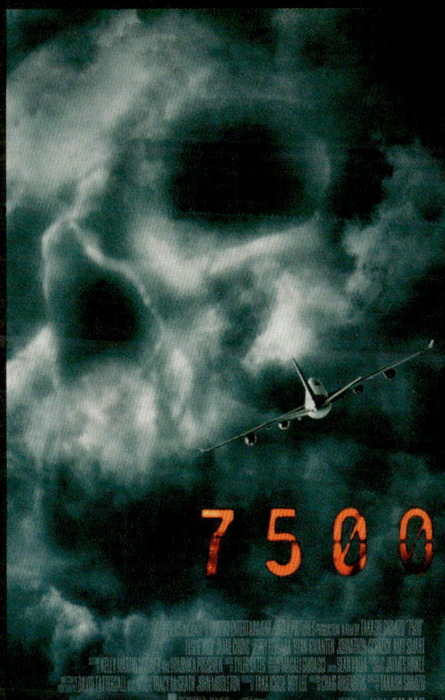

A poster for the 2014 movie *7500*.

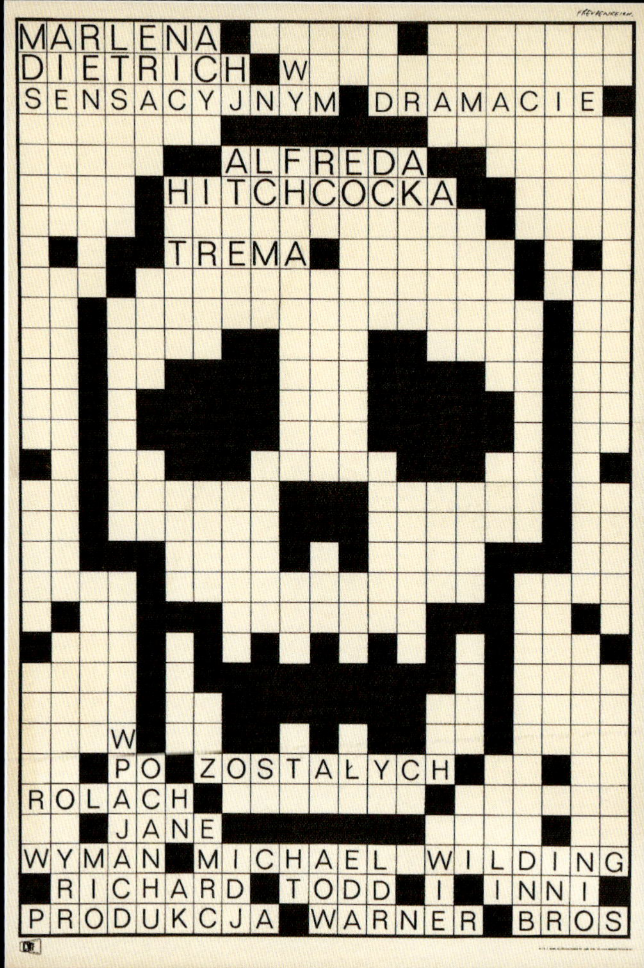

This Polish poster advertised Alfred Hitchcock's 1950 film *Stage Fright*.

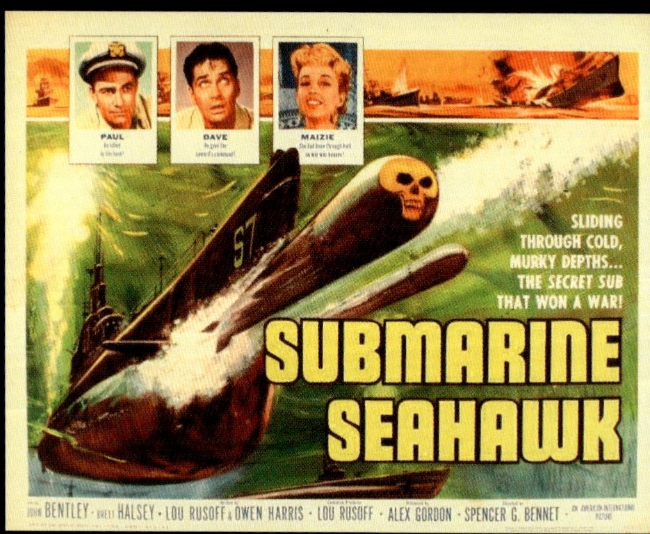

This poster advertised the 1958 movie *Submarine Seahawk*.

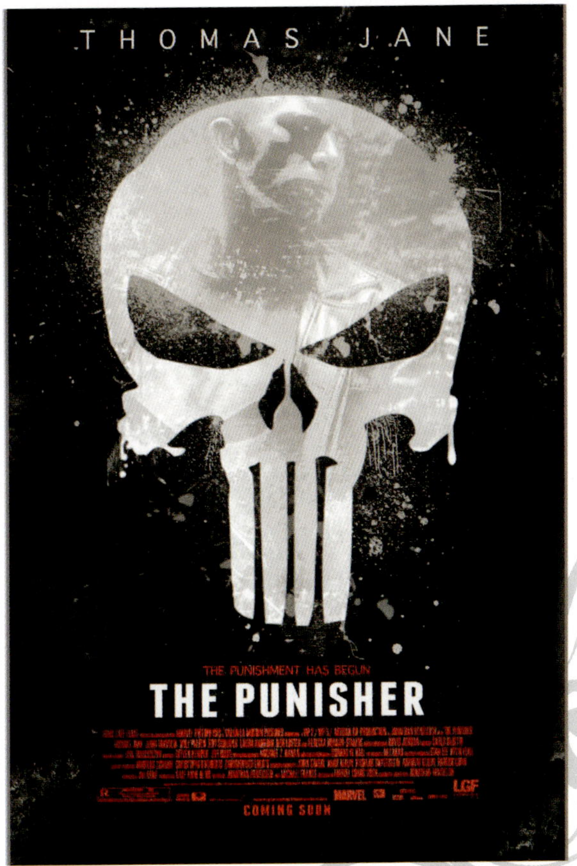

A movie poster advertising 2004's *The Punisher.*

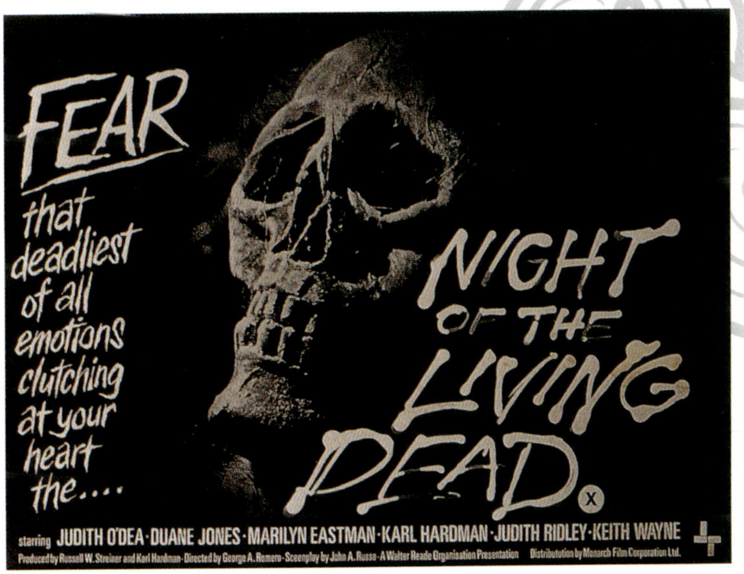

Movie posters from 1968's *Night of the Living Dead.*

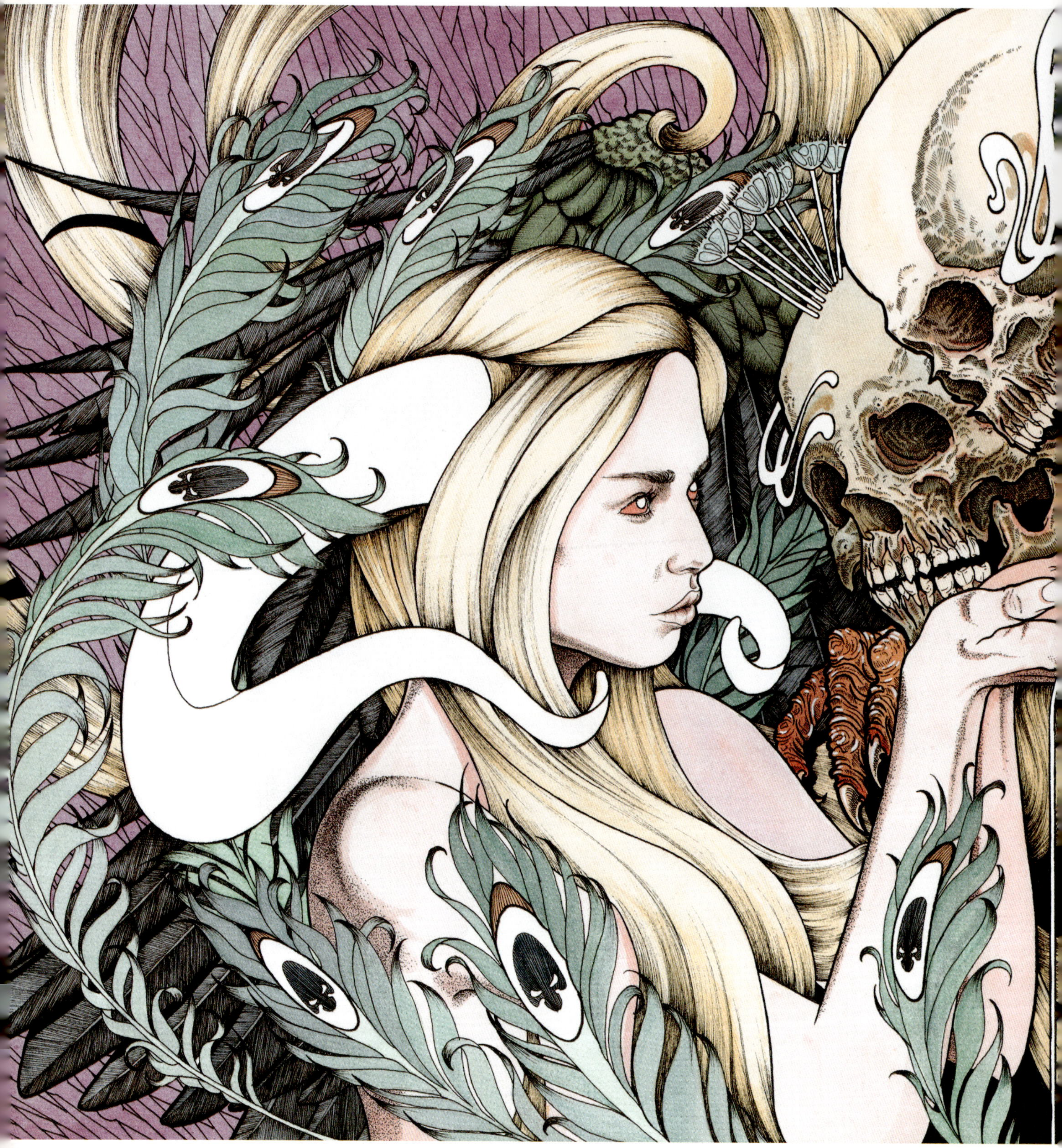

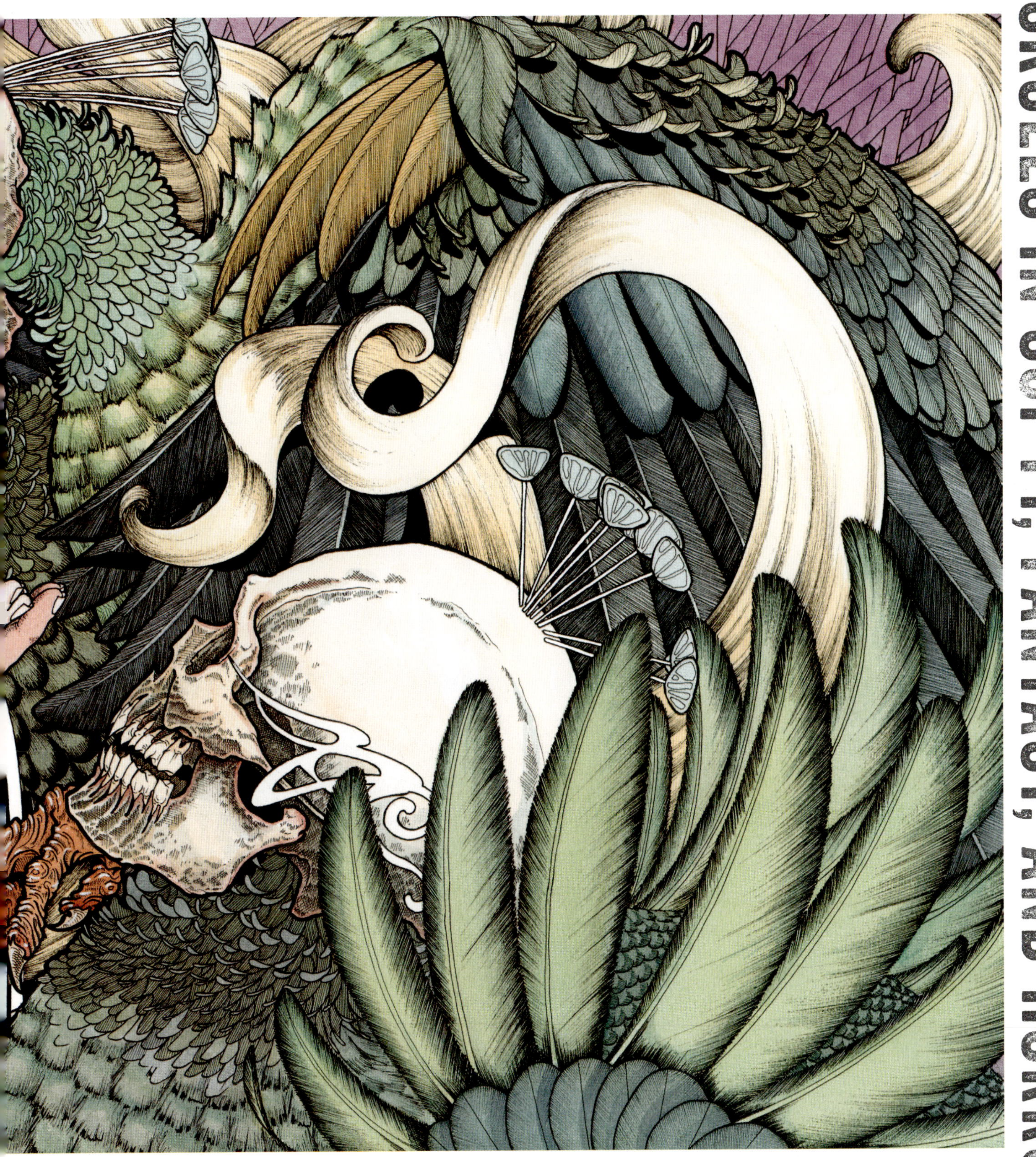

Communion, by artist Chris Panatier

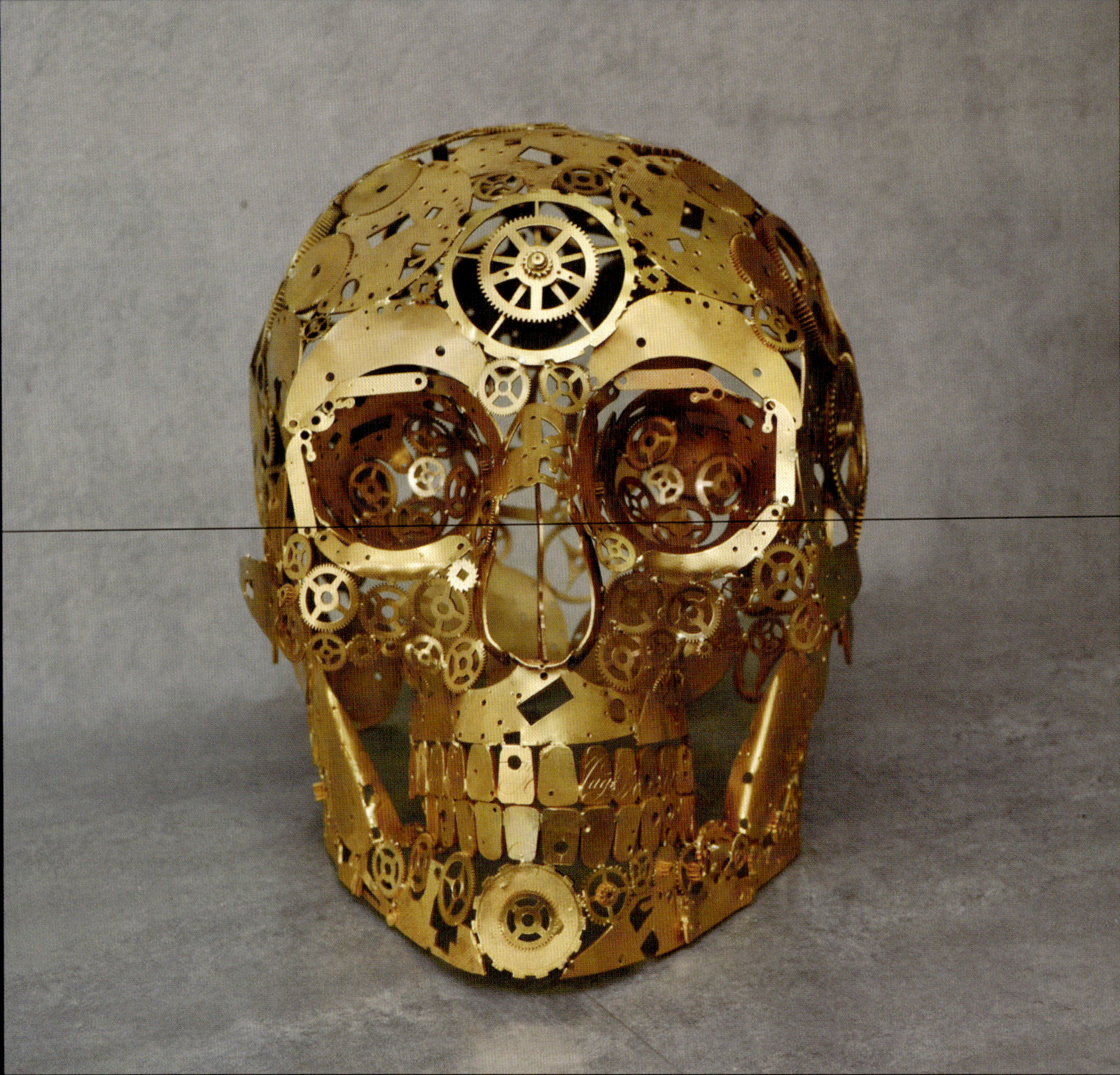

This spread: Steampunk skulls made from antique bronze ornaments by artist Alain Bellino.

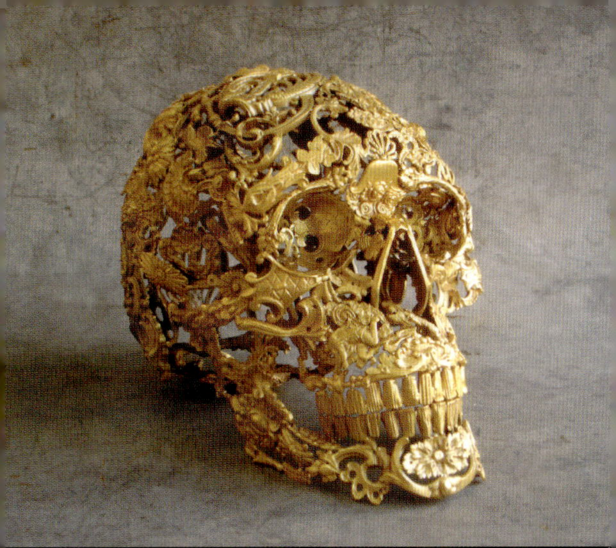

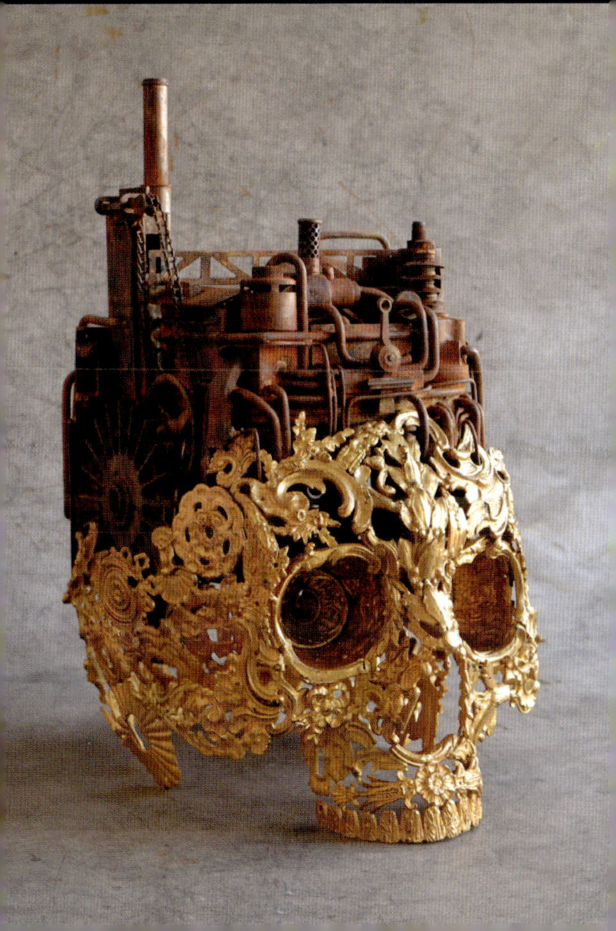

SKULLS IN RELIGION

ALTHOUGH ONE PERSON'S OCCULT PRACTICE MAY BE ANOTHER PERSON'S RELIGION, HERE WE USE "RELIGION" TO REFER TO CONVENTIONAL ORGANIZED PRACTICES.

You might think that the use of prayer beads, malas, or rosaries is common only to Catholicism but this is not the case. These are featured in many religions and spiritual practices; even those who aspire to no religious belief at all will use "worry beads" as a form of meditation or reflection as well as being something to occupy one's hands during idle moments. In some cases, each individual bead can be interpreted as representing a tiny human skull. For Catholics, these skulls act as "memento mori," a Latin phrase reminding one of mortality and impending death. Conversely, if you see a crucifix—a statue of Christ nailed to the cross—with a skull at the foot of the cross, this is a symbol of Christ's resurrection, and his victory over death.

As a term, "memento mori," also applies to sculptures, paintings, and other works of art, whether overtly religious or not that use the skull as a reminder of the inevitability of death. Similarly tombs and gravestones may also feature the skull as a reminder of how all life ends in death.

An ossuary is a container—either as small as a jewelry box or as large as a building—which is used to provide a final resting place for human skeletal remains. After being buried for some time, the skeleton is unearthed and removed to the ossuary. Skulls, as the most recognizable part of the human skeleton, feature prominently. Some ossuaries, such as the Sedlec Ossuary in the Czech Republic, use the bones to decorative effect. There, a chandelier is artfully constructed from bones and skulls. Ossuaries are common in both the Christian and Jewish traditions.

Some Catholic organizations, such as churches, convents, and monasteries, may boast about their possession of the relics of a saint which is associated with them. Such religious souvenirs generally take the form of the skull—or even the entire skeleton of the saint. Specialists may take years to elaborately cloth the skeletons in precious jewels and clothing. Some of these relics are on show for all to see; some are kept in private vaults. In art or in Scripture, some Catholic saints are depicted with skulls, too. These include St. Francis of Assisi, rarely seen without a skull in his hands. Reputedly, he sometimes put the skull on his breakfast table, keen to remind himself of the temporary nature of life on earth. Other saints kept skulls nearby for the same reason, including St. Gerard.

On other continents and in other places, the skull image also has a prominent part to play. The Hindu deities, Shiva and Kali, display their immortal powers by means of skull necklaces. Further, each of the human skulls around Kali's neck is said, by some, to represent a letter of the alphabet. Deities in Nepal and Tibet follow similar fashion precedents in dressing themselves in skulls. One of the more ferocious of these portraits belongs to the Tibetan god of death, Yama, who wears a crown of human skulls, symbolizing his control of the realm of the afterlife.

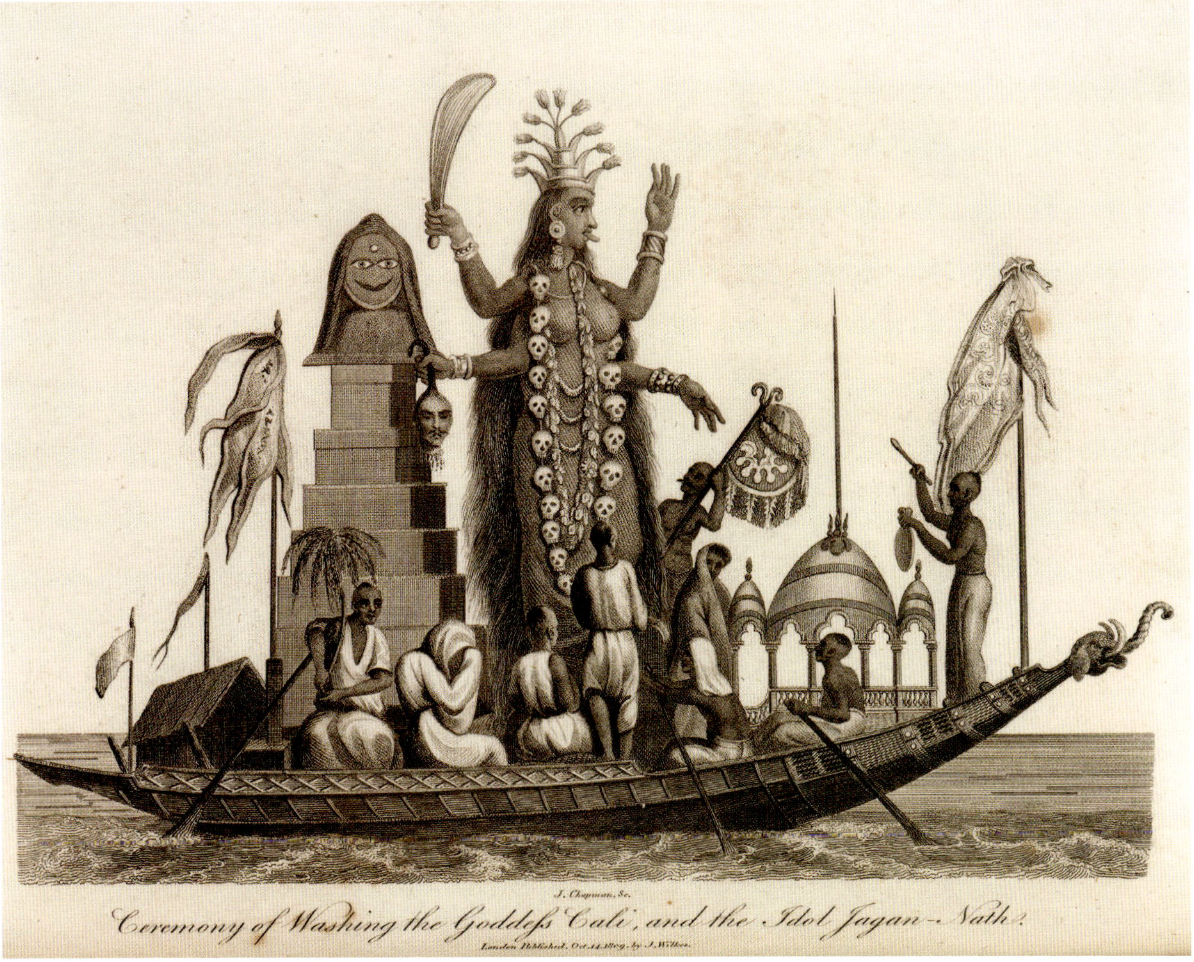

An engraving by John Chapman of the Hindu goddess Kali, wearing a necklace of skulls.

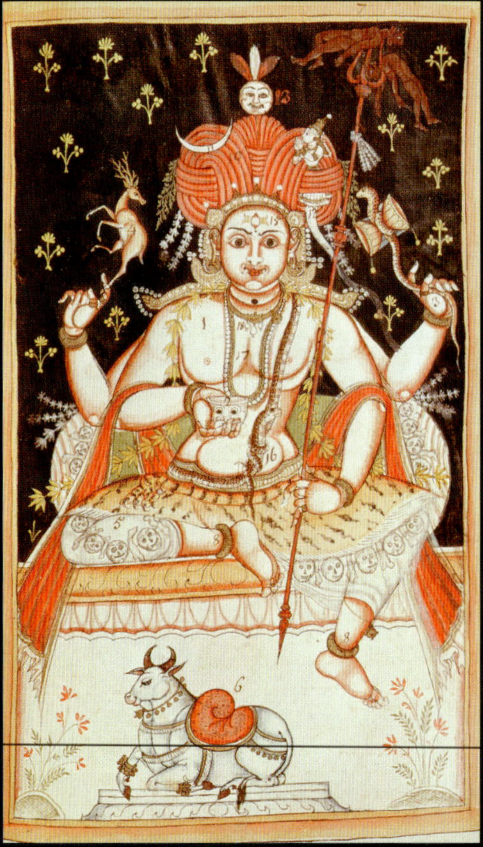

Artwork depicting Rutrim, the third of Hinduism's false trinity in the guise of Shiva.

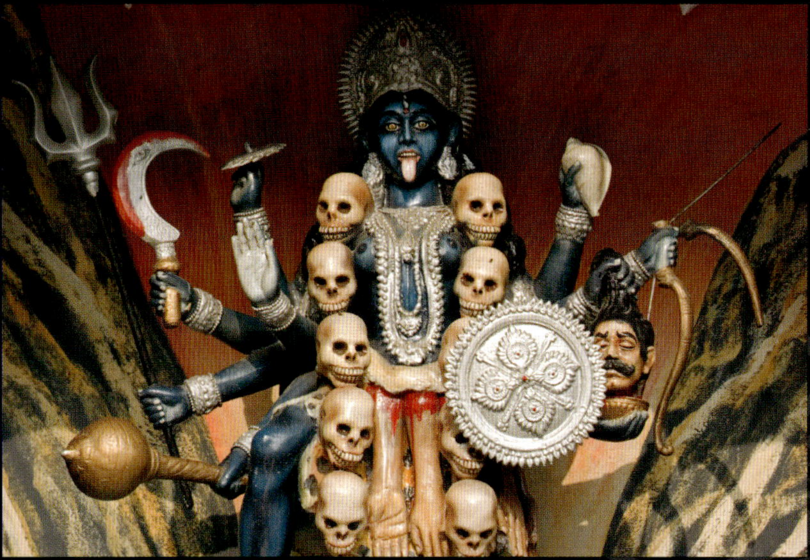

A statue of the Hindu goddess Kali, wearing a necklace of skulls.

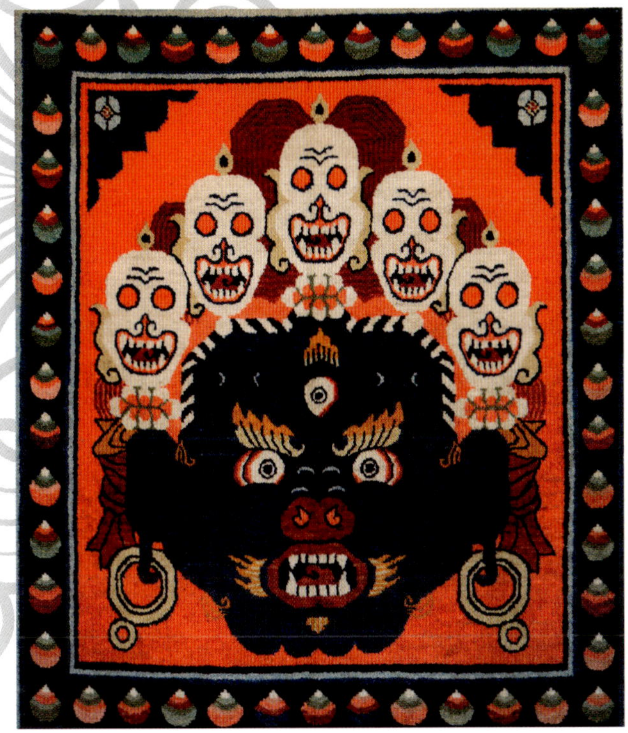

A depiction of Yama, Hindu god of death.

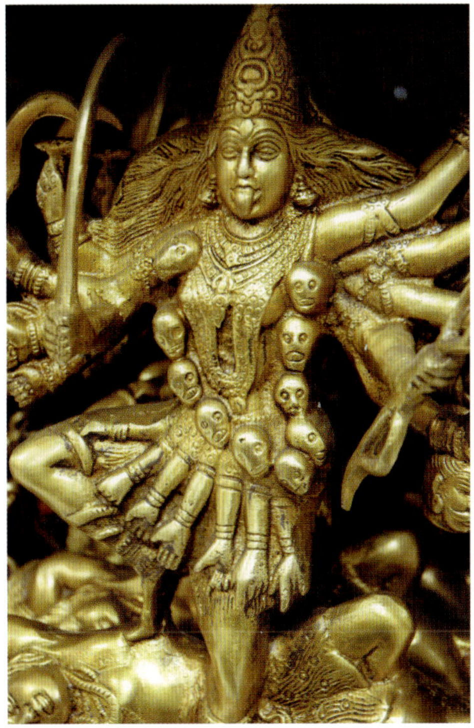

A golden statue of the Hindu goddess Kali, wearing a necklace of skulls.

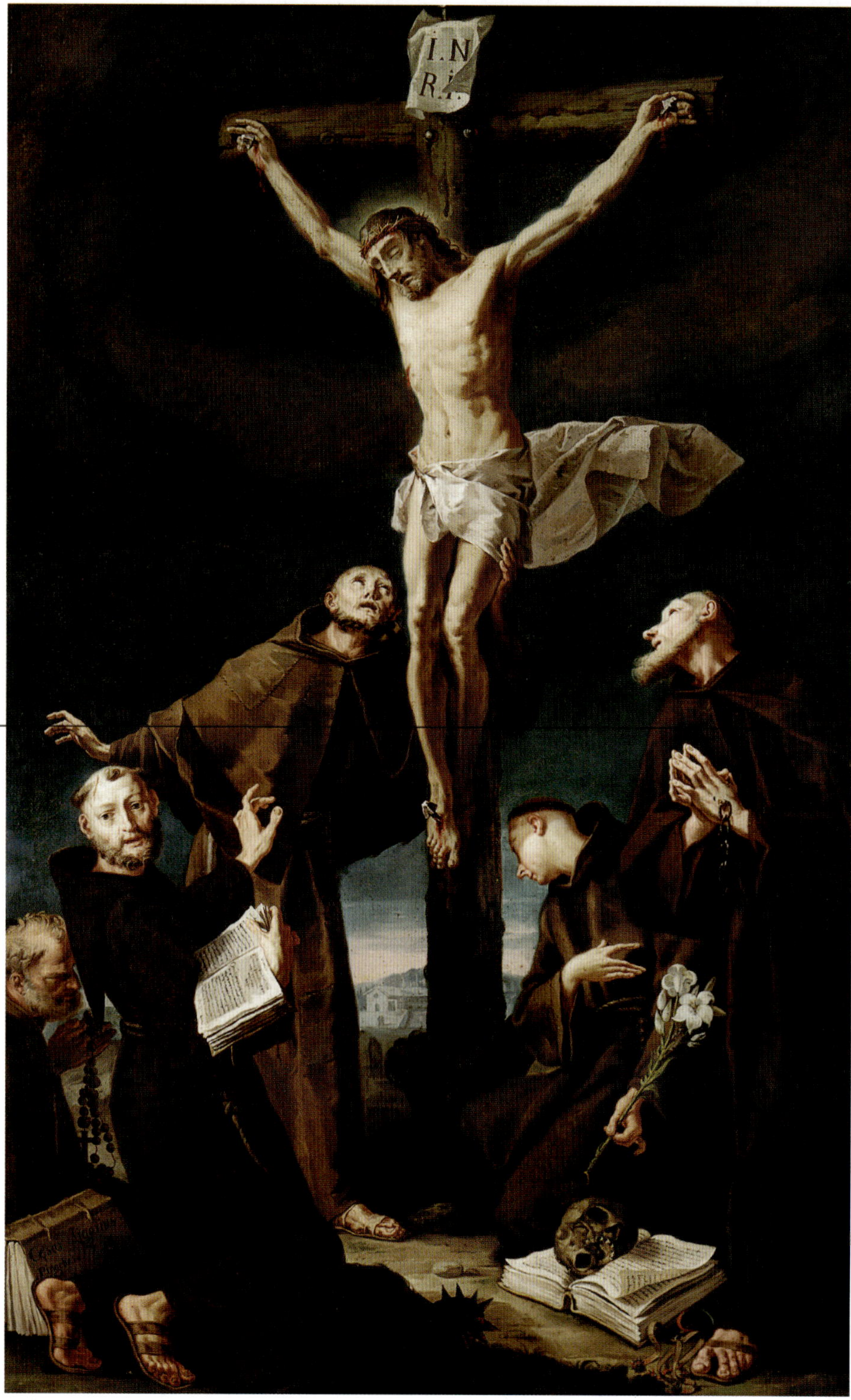

This unnamed artwork by an unknown artist depicts five saints around the cross; a skull lies near the foot of the cross.

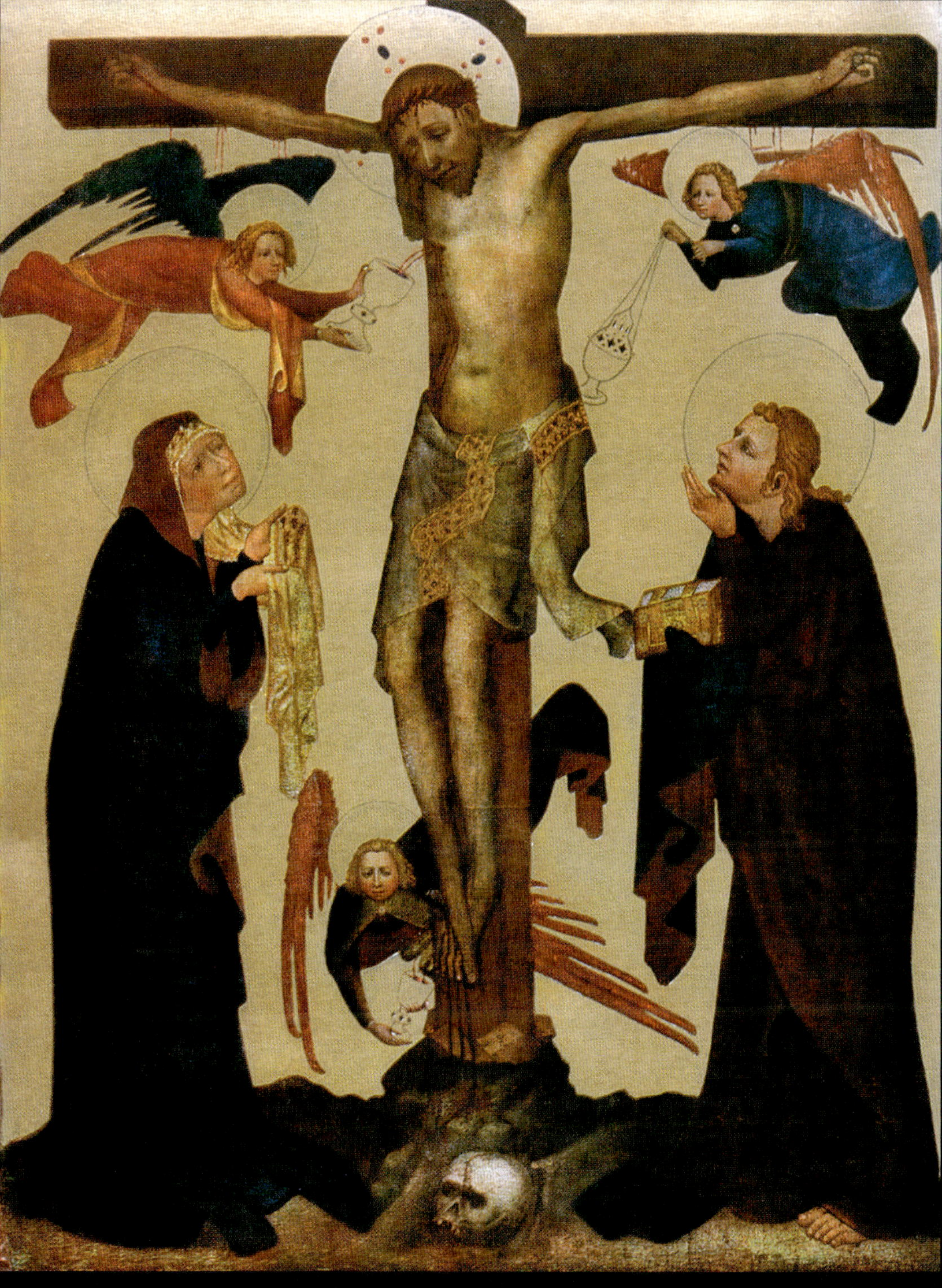

The Vyšší Brod *Crucifixion*, painted before 1400 by the Master of the Vyšší Brod Altar, shows a skull near the foot of the cross.

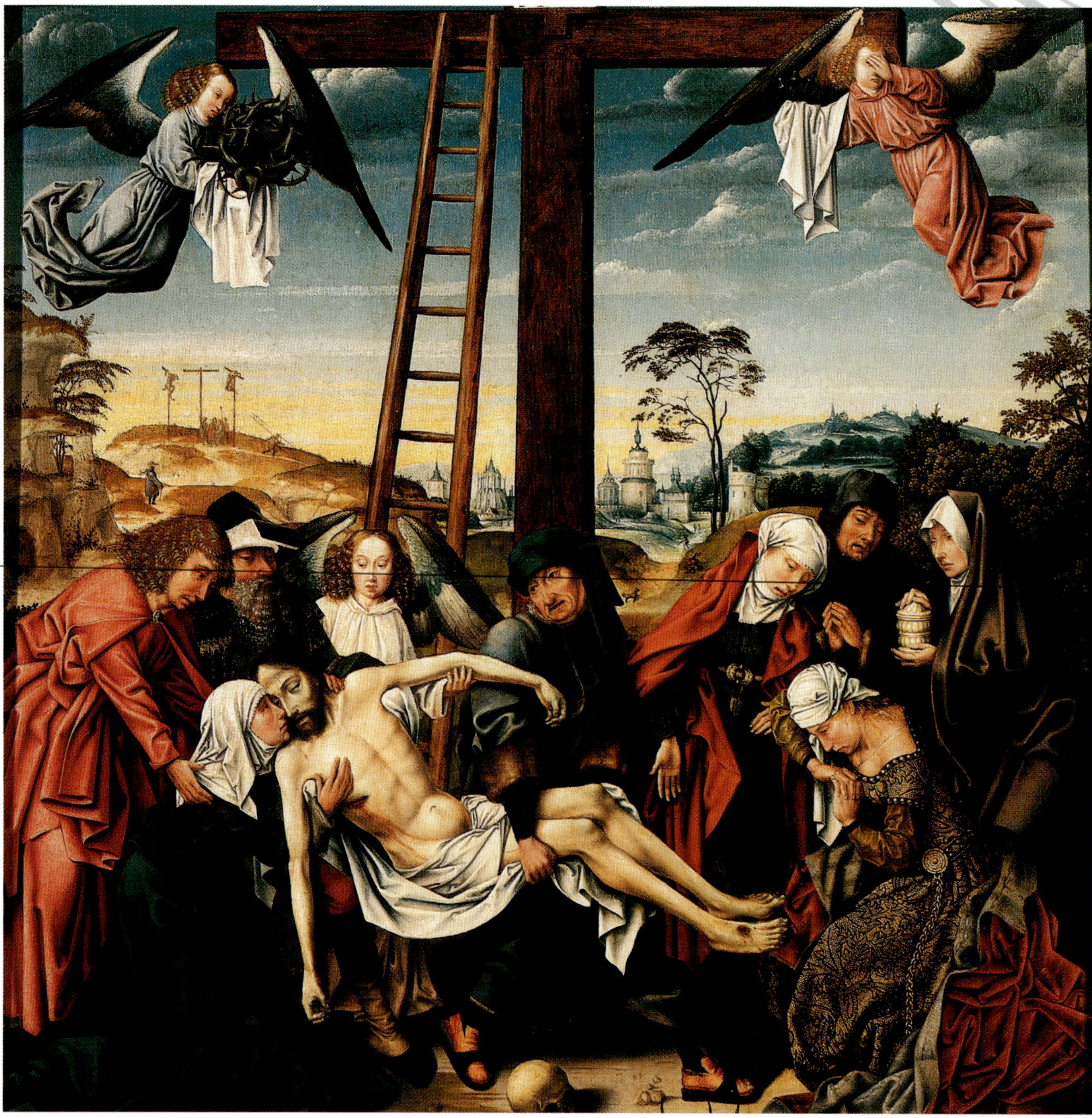

Pietà, by Rogier van der Weyden, was painted in approximately 1441.

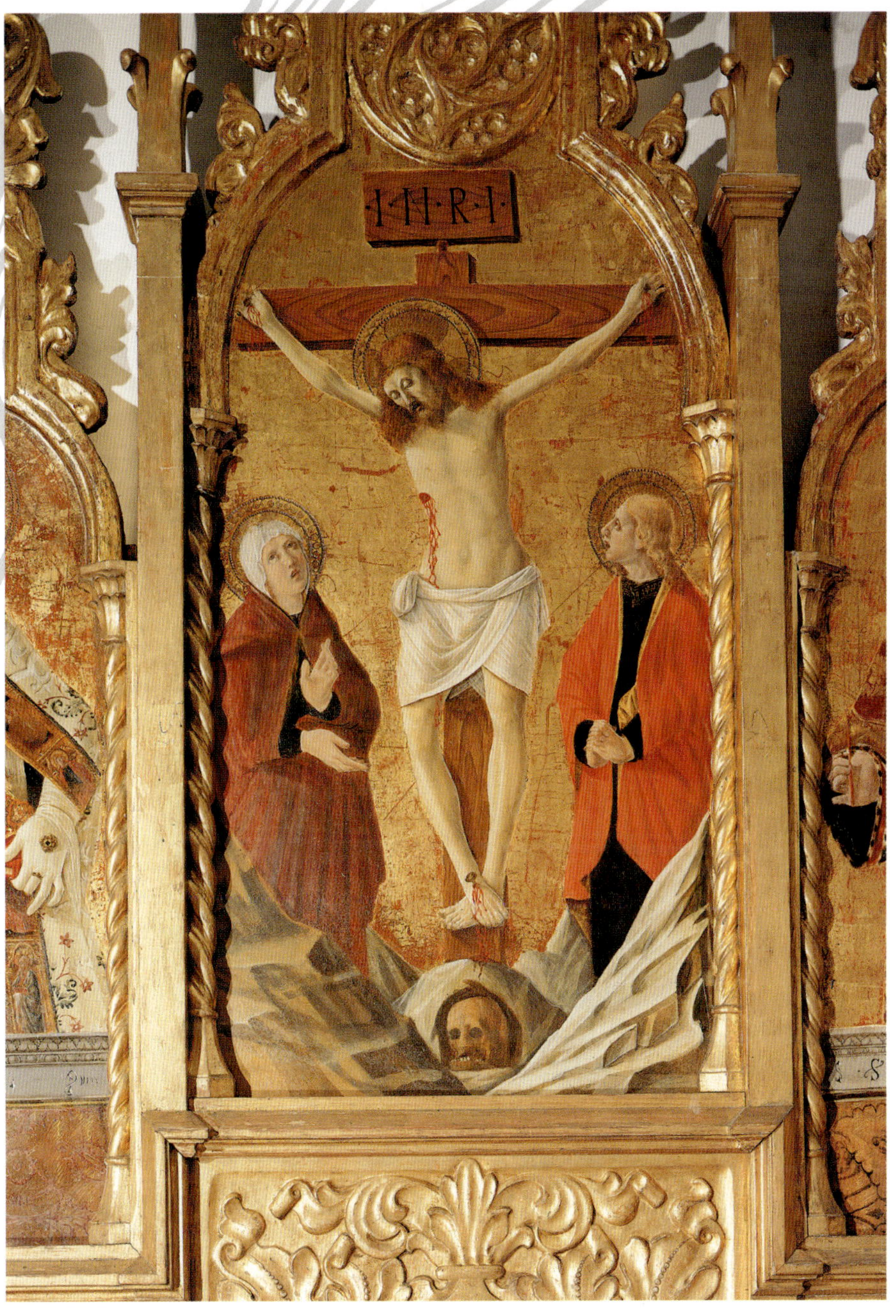

Polyptych of Gualdo Tadino, a painting by Girolamo di Giovanni, was completed in approximately 1462.

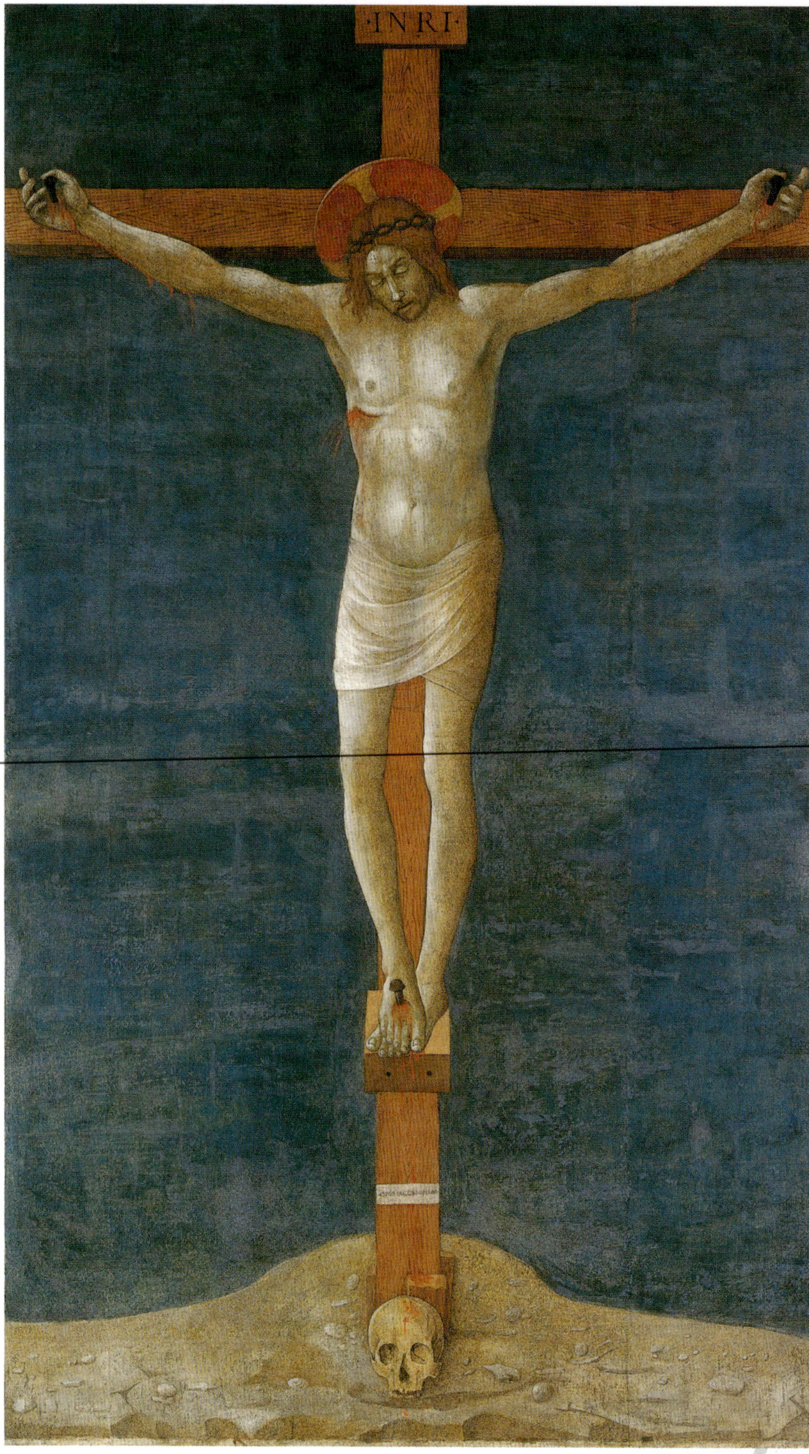

An unnamed painting showing the crucifixion by Jacopo Bellini, completed around 1436.

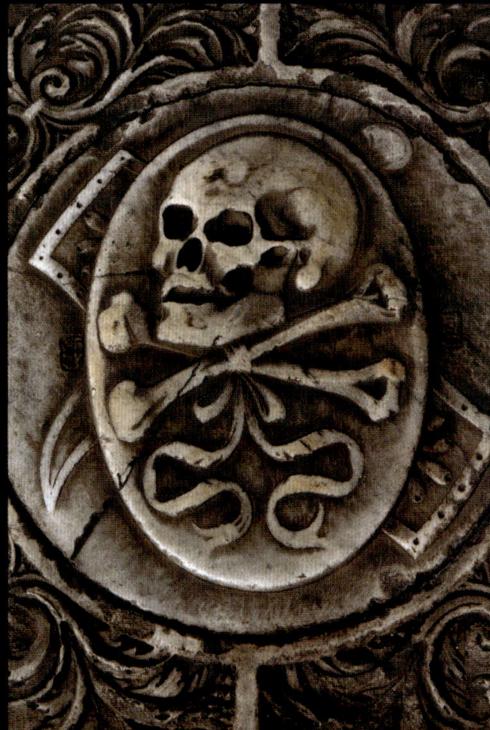

Skull and crossed bones motif on a tombstone inside the Camposanto Cemetery in Pisa, Italy.

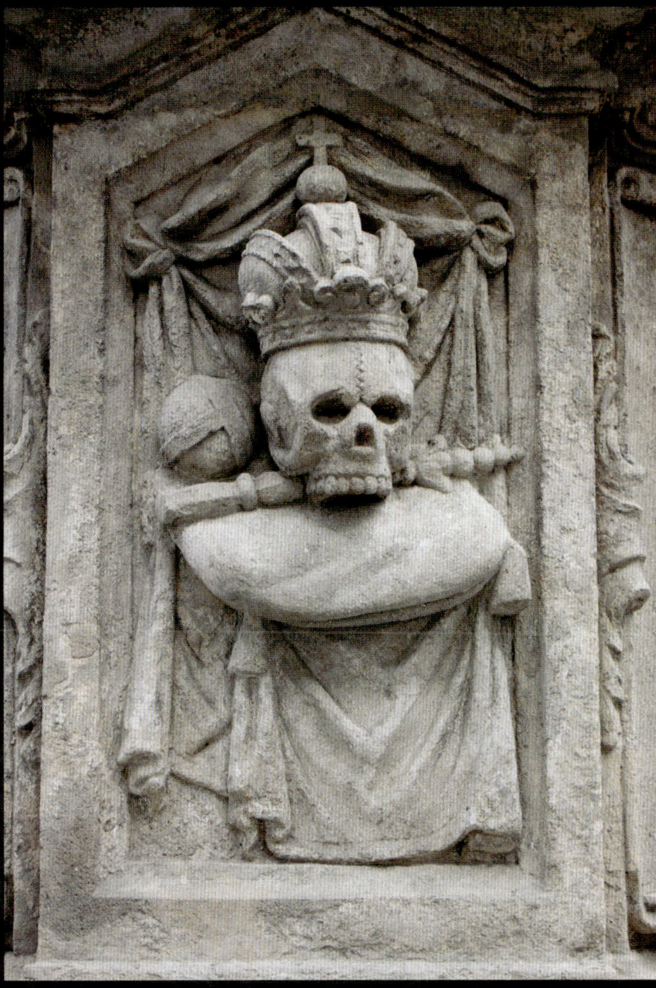

Skull sculpture in Kutná Hora, in the Czech Republic.

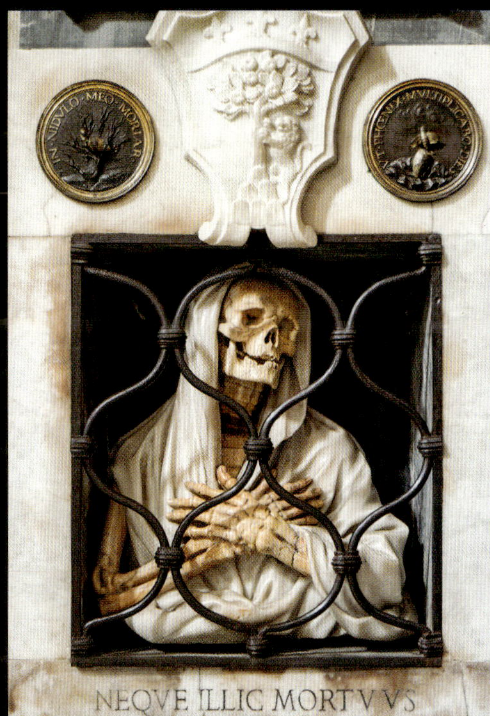

Funerary monument to Giovanni Battista Gisleni at the Basilica of Santa Maria del Popolo in Rome.

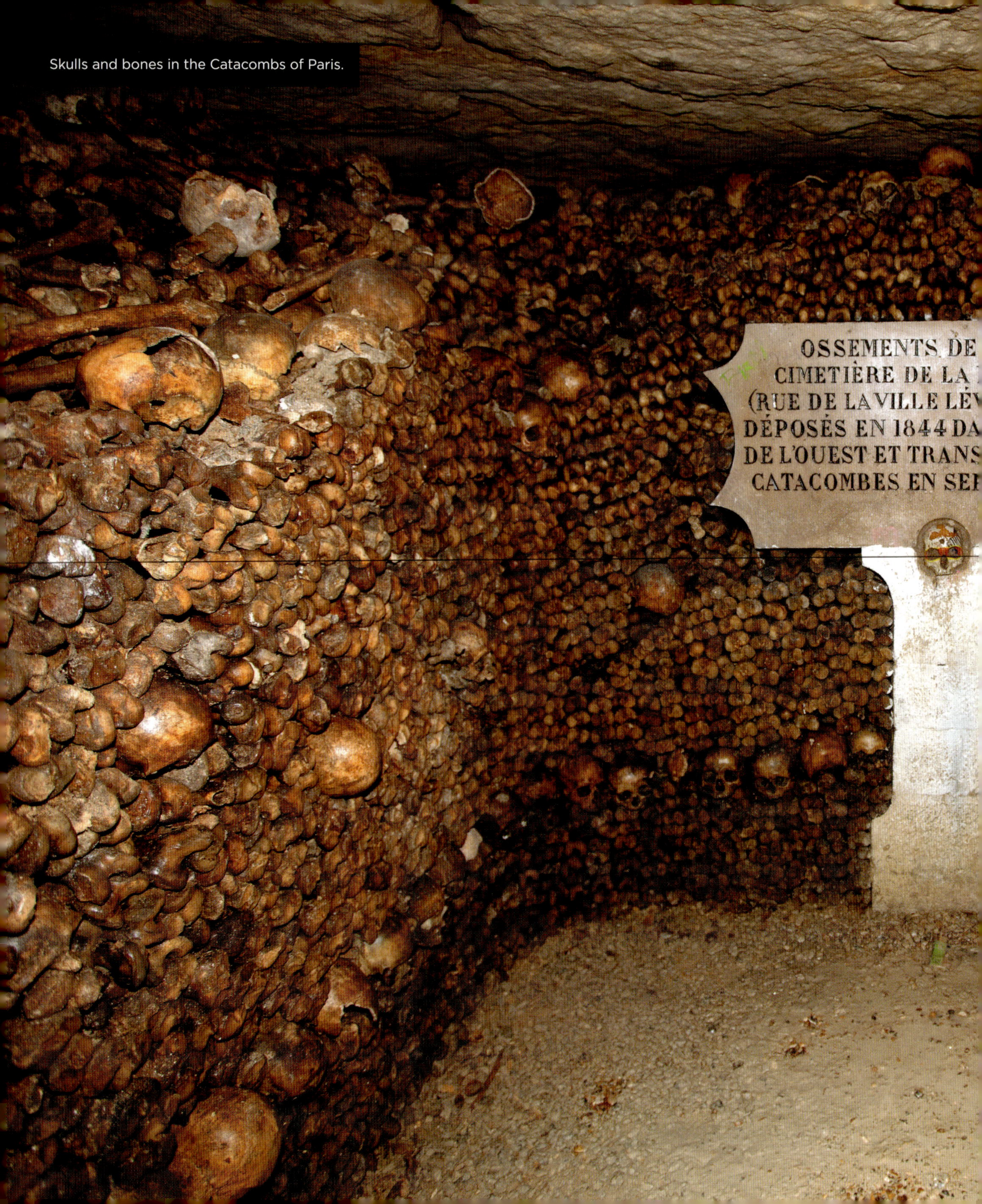

Skulls and bones in the Catacombs of Paris.

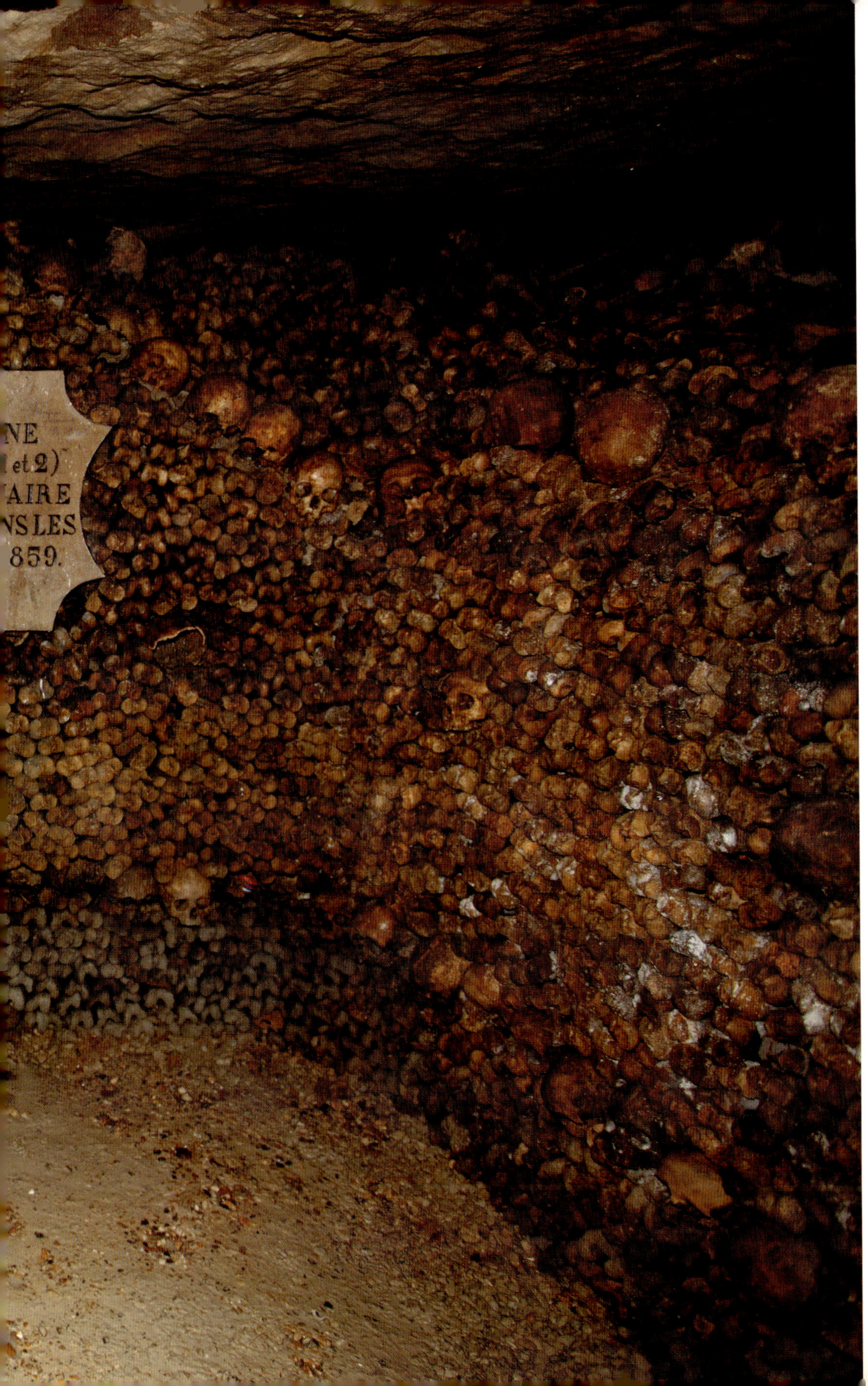

SKULLS IN RELIGION

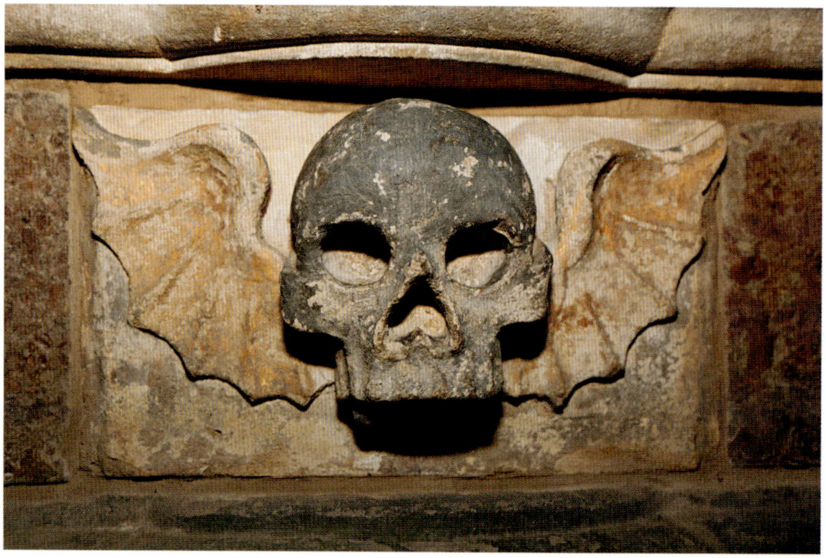

A carved skull with wings on a gravestone in Abergavenny, Wales.

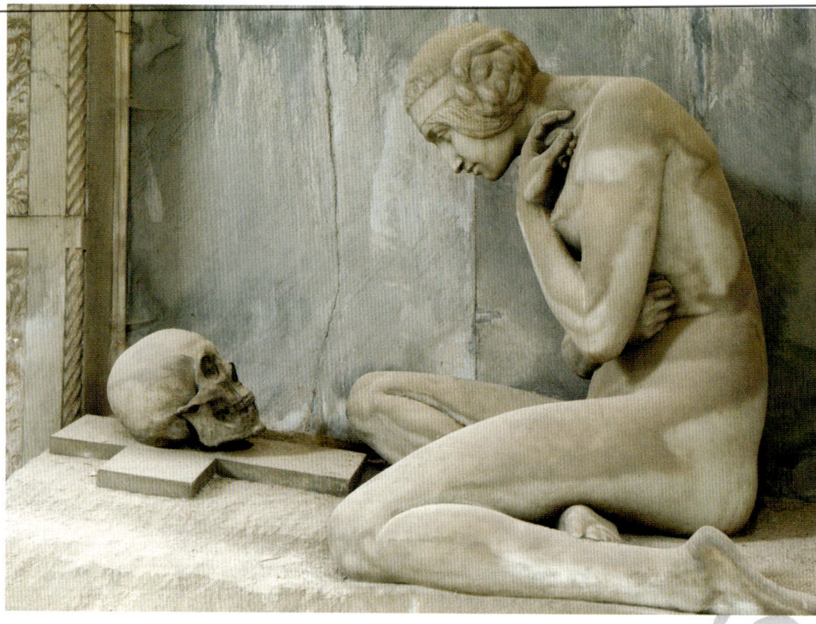

A grave sculpture at the Monumental Cemetery of Staglieno in Genoa, Italy.

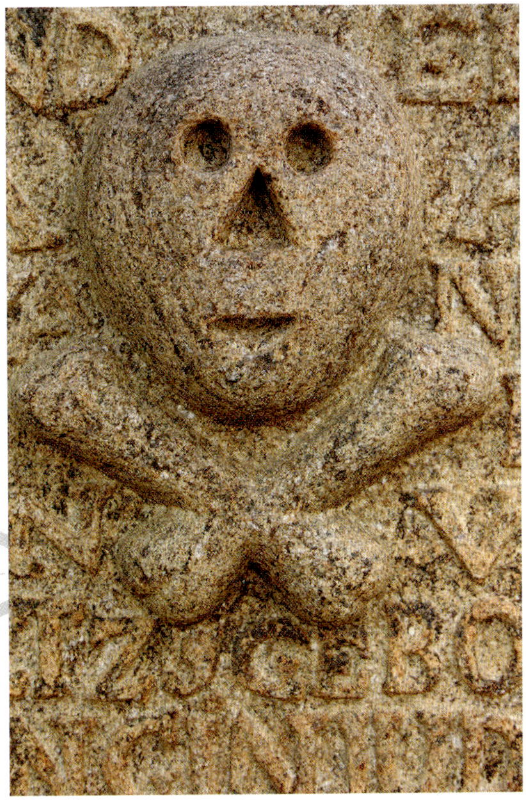

Skull and crossbones on a grave at the Dutch Reformed Church in Galle, Sri Lanka.

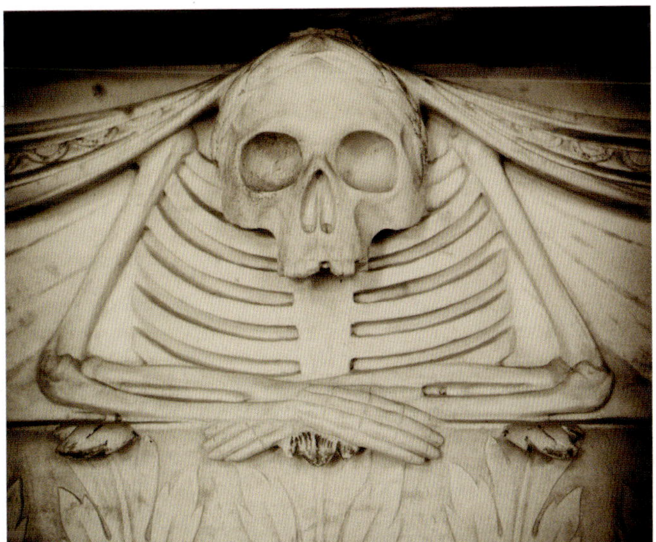

A skeletal carving, complete with skull, in a church relief.

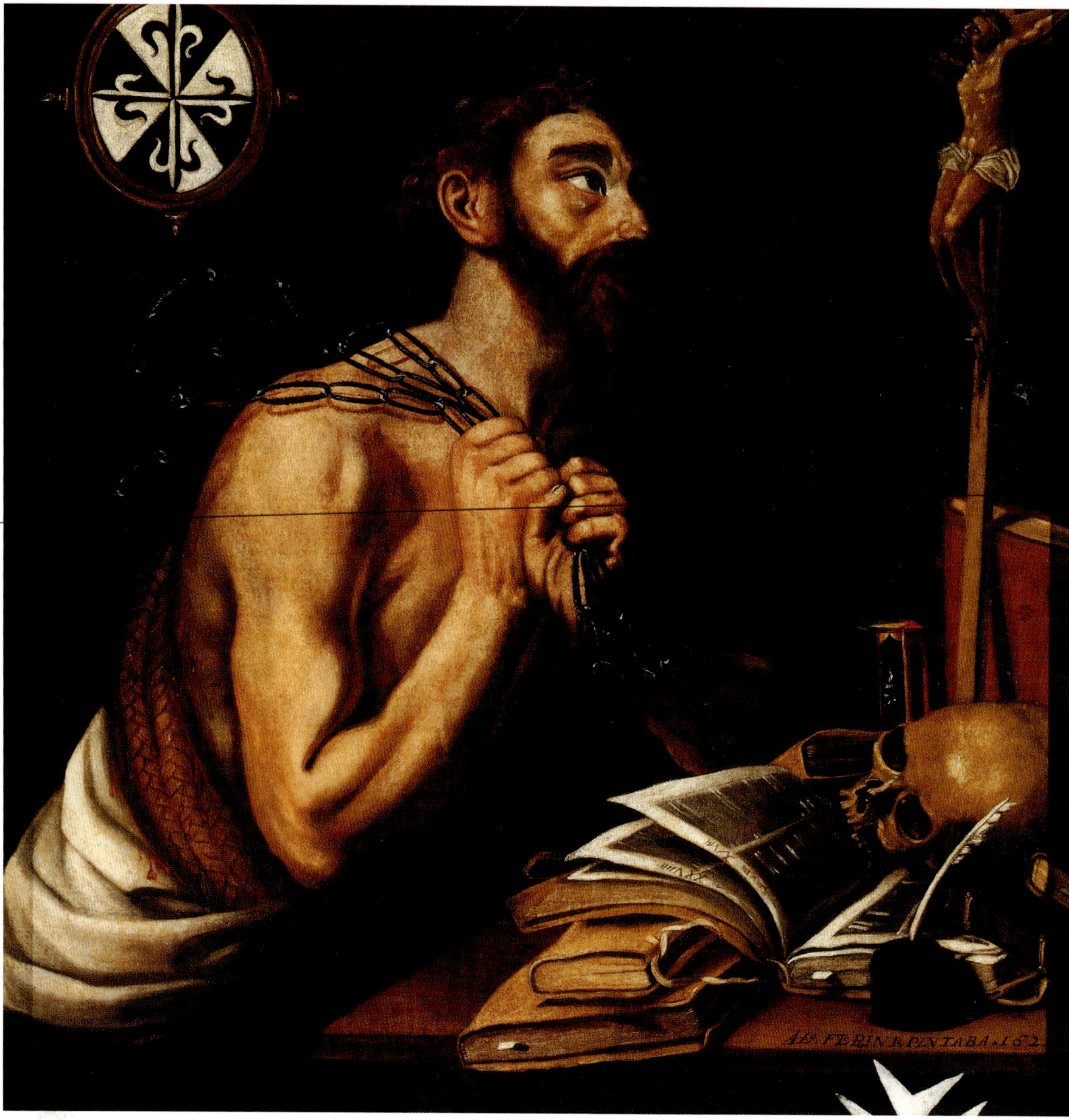

Penitent St. Dominic was painted by Alonso Florin in 1621.

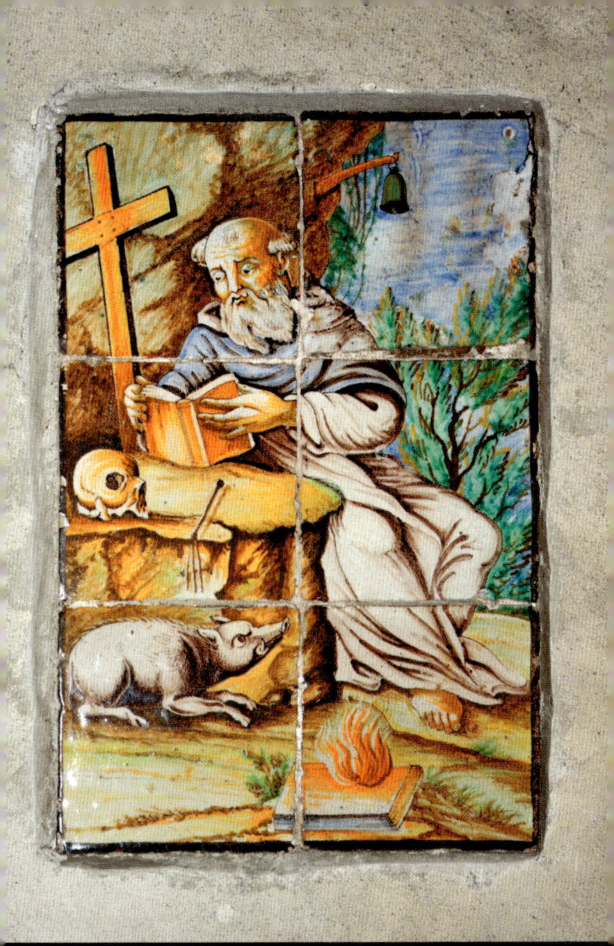

A tile portrait of Saint Anthony from the Church of Saint Anthony located in England's Lake District

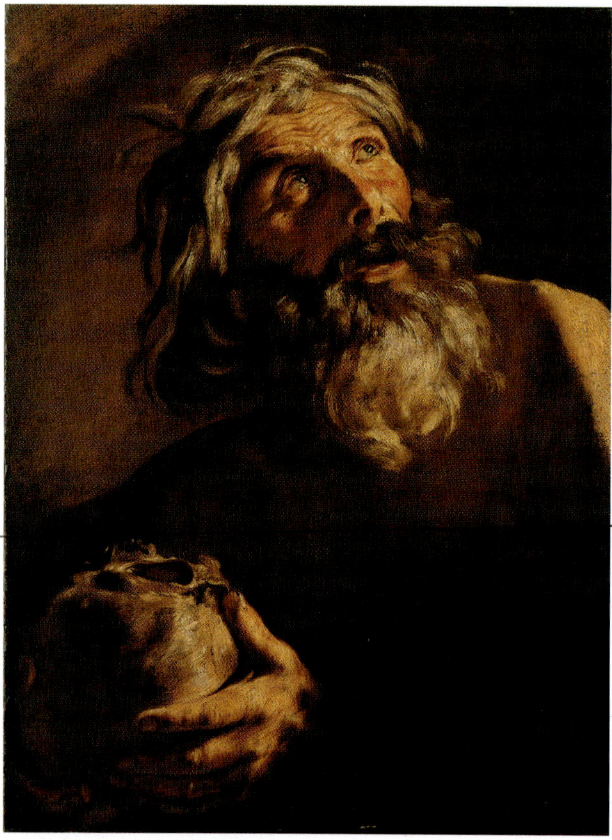

Saint Jerome was painted by Giovanni Serodine in the seventeenth century.

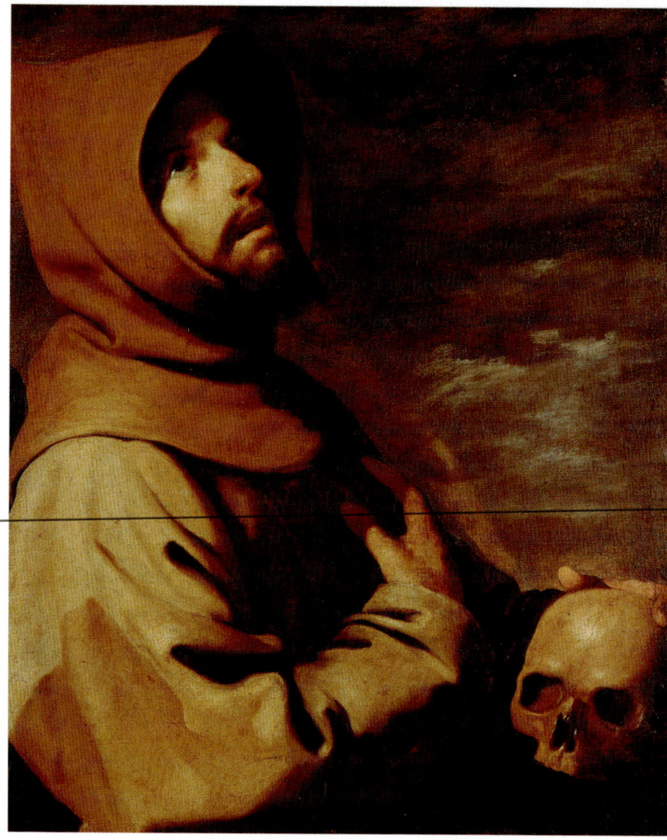

A painting of Saint Francis of Assisi.

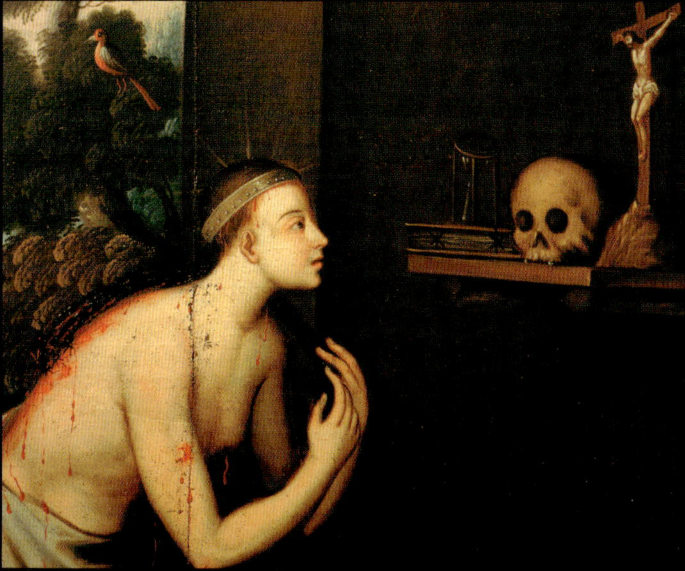

Saint Rose of Lima is depicted her committing self-flagellation as penance.

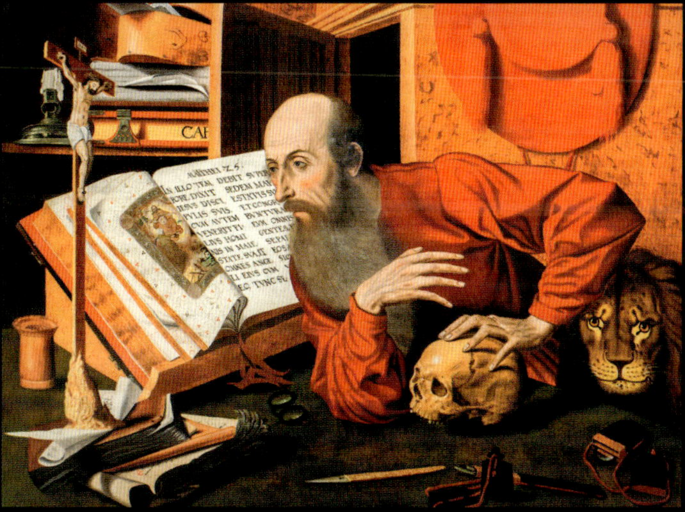

This image of Saint Jerome was painted in the late sixteenth century.

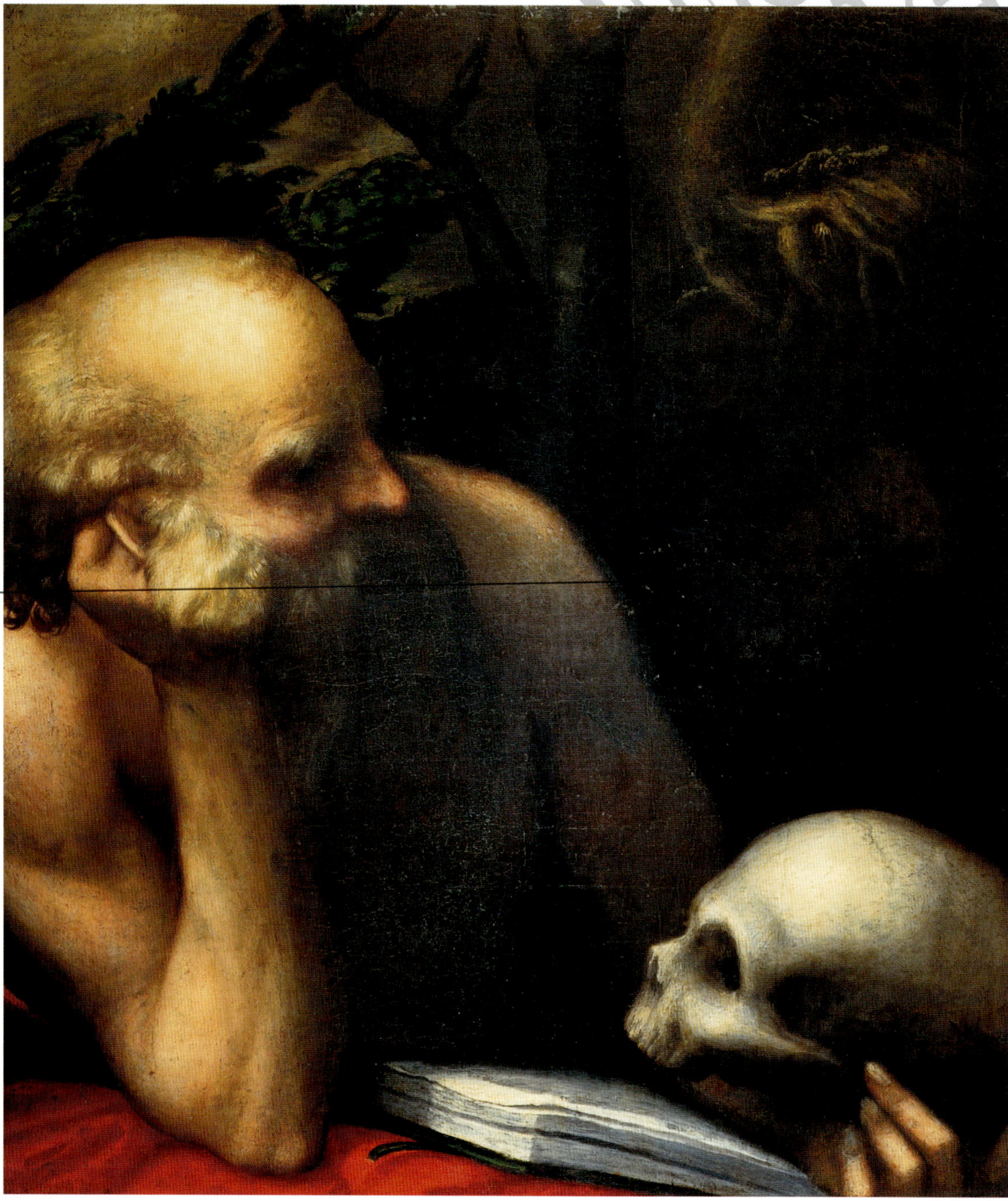

This image depicts Saint Jerome holding a skull.

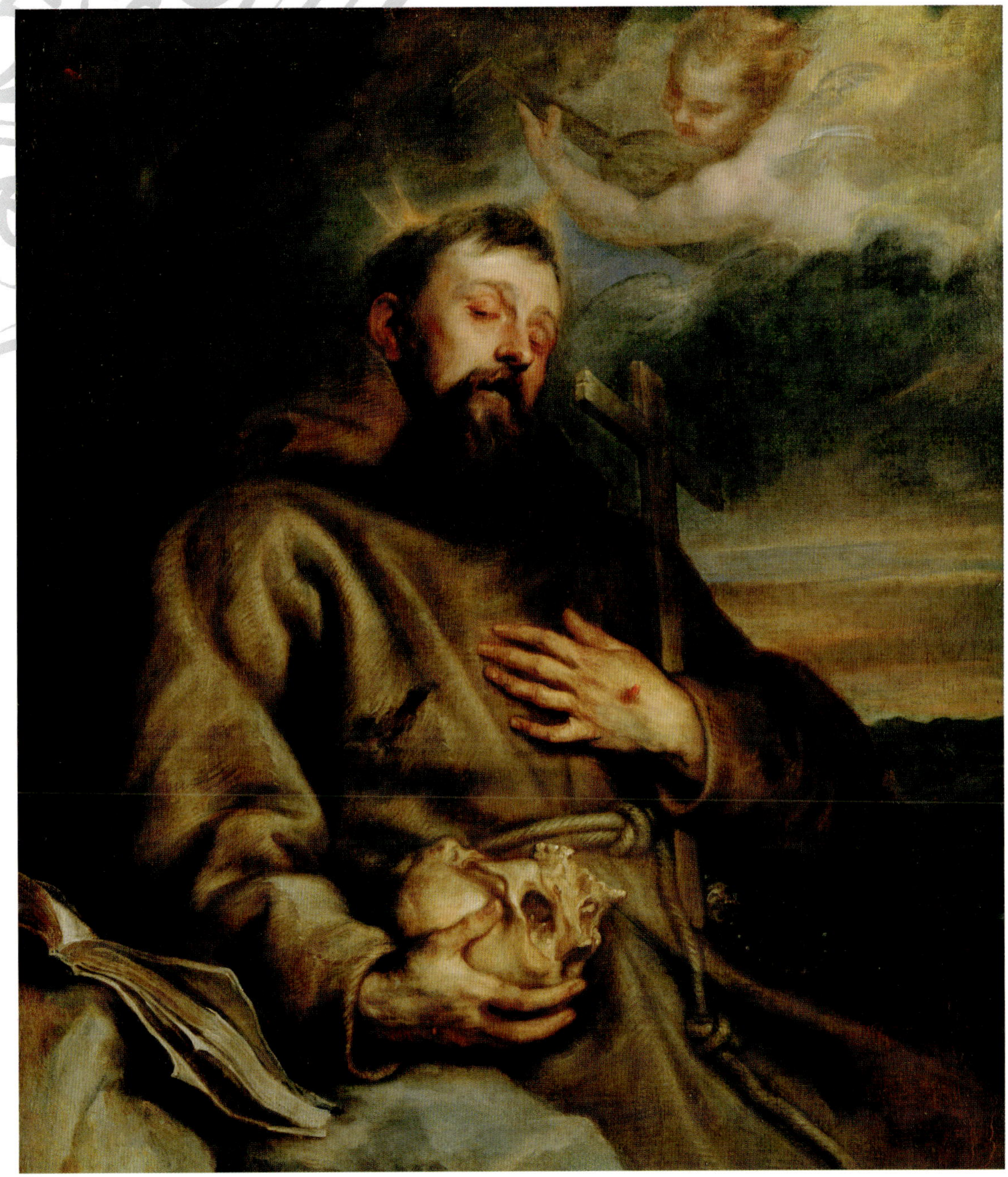

This image of Saint Francis of Assisi was painted in approximately 1627–1632.

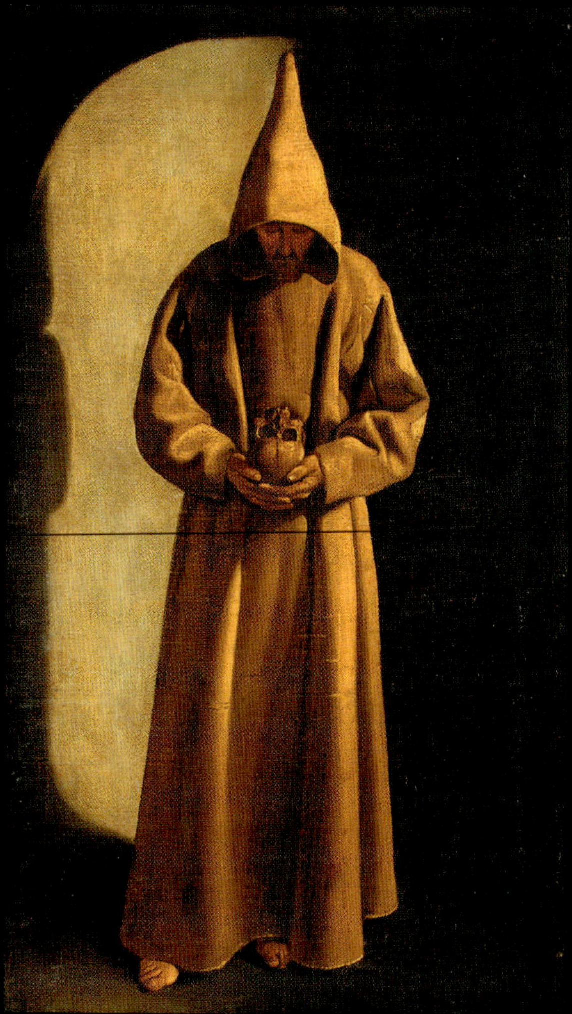

Francisco de Zurbarán painted *Saint Francis with a Skull in His Hands* in approximately 1630.

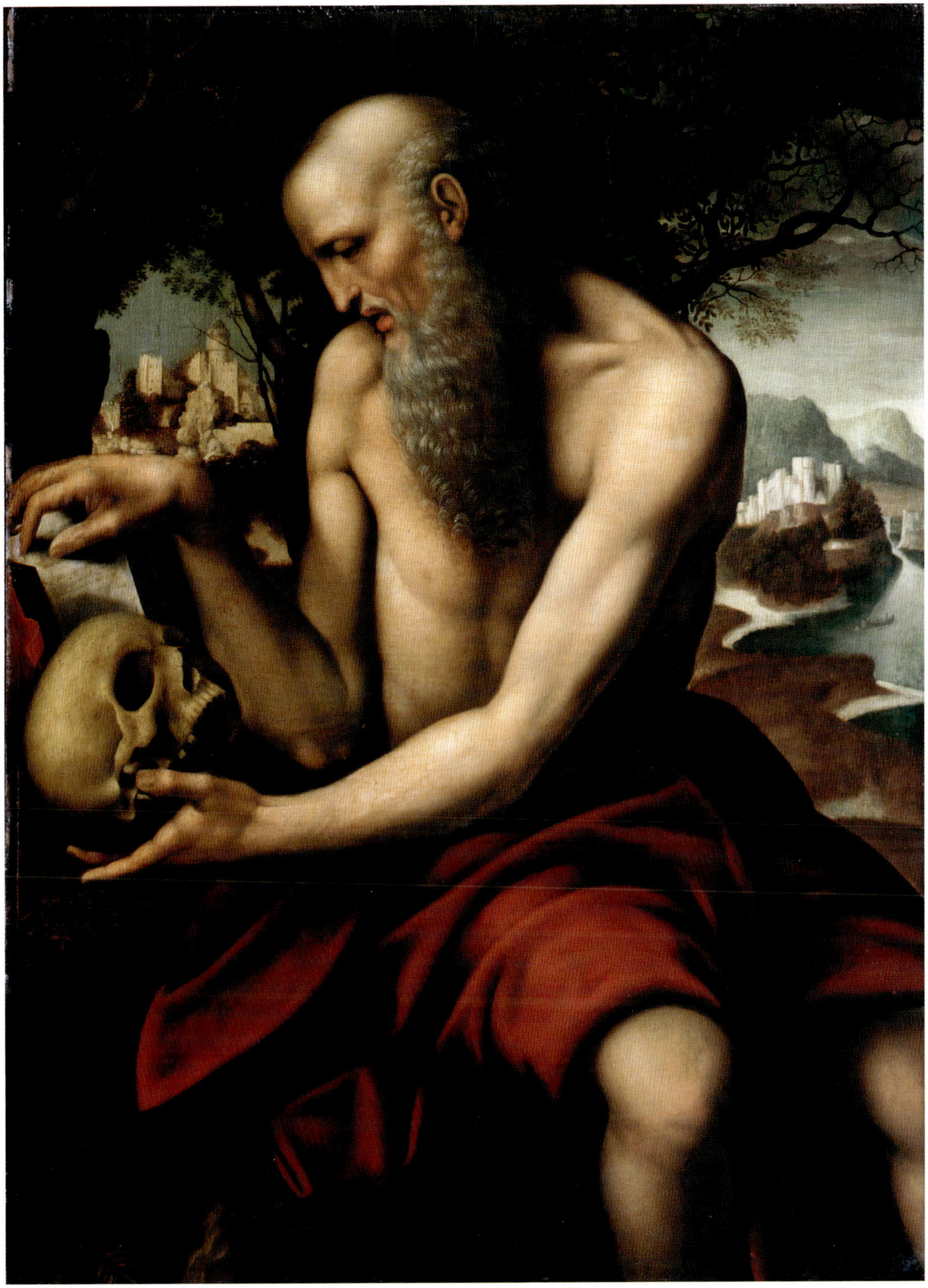

St. Jerome was painted by Cesare da Sesto around the turn of the sixteenth century.

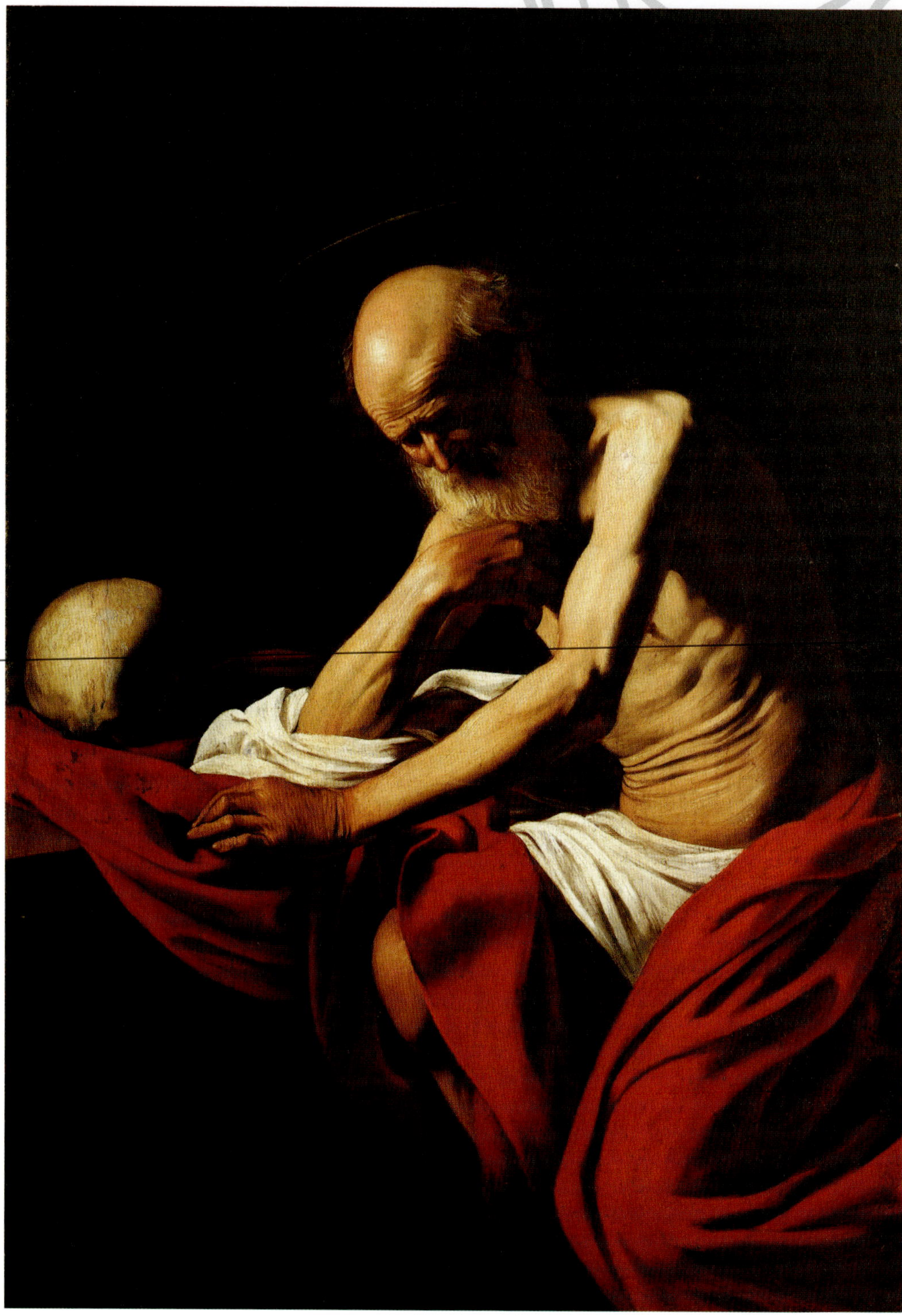

St. Jerome the Penitent was painted by Michelangelo Merisi da Caravaggio in 1606.

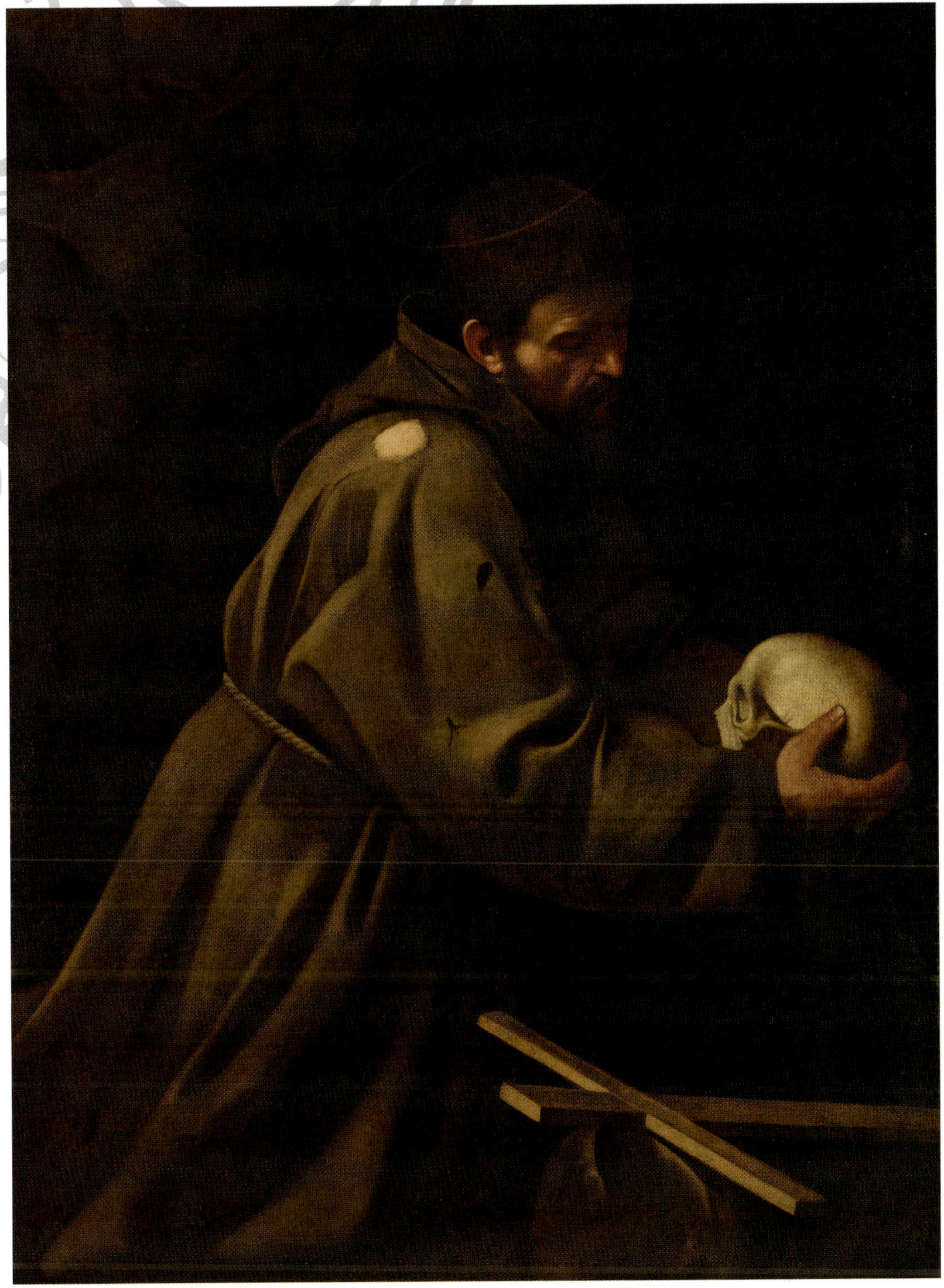

St. Francis in Prayer was painted by Michelangelo Merisi da Caravaggio in the early seventeenth century.

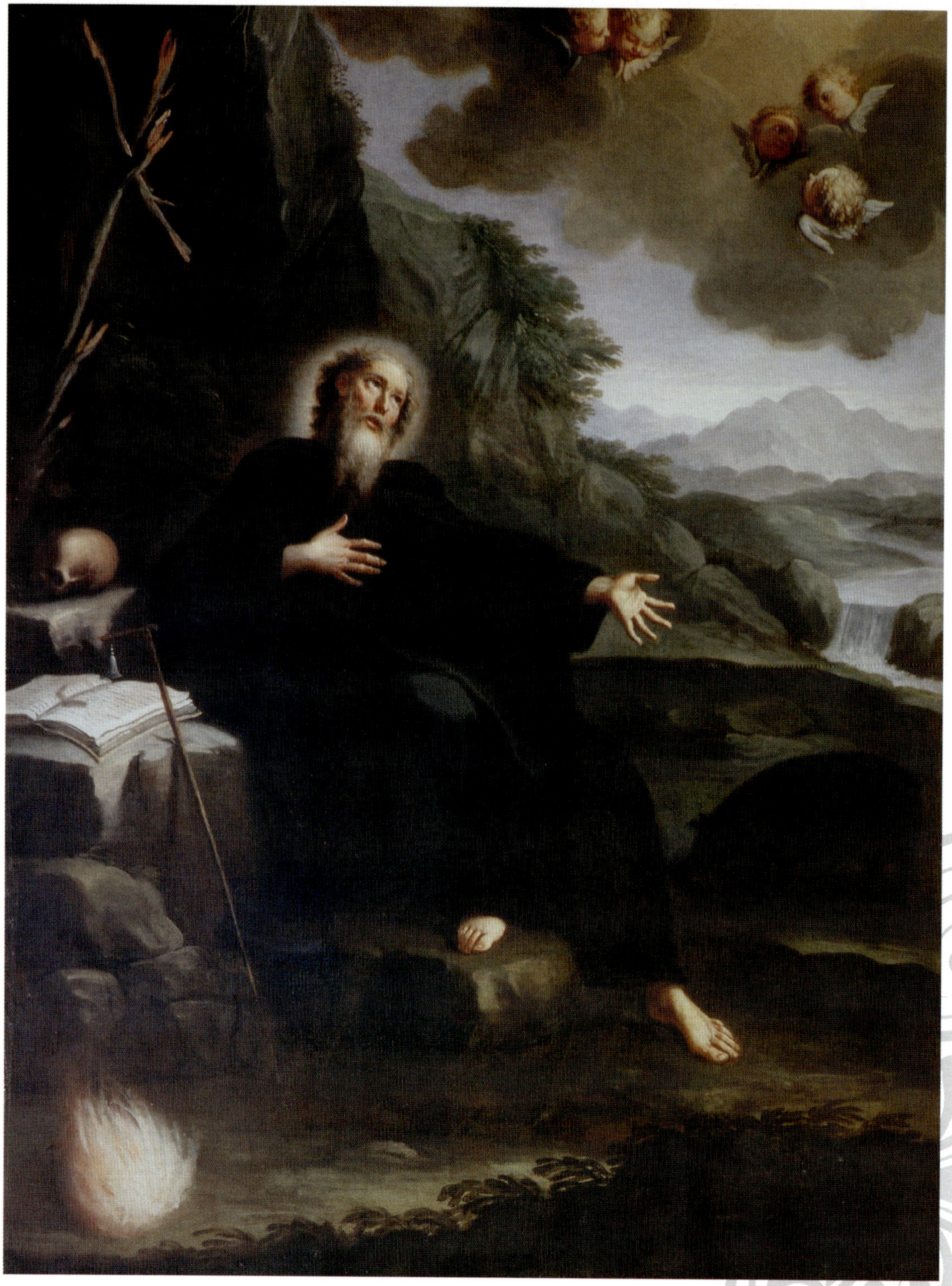

This seventeenth or eighteenth century painting, by an unknown artist, is *Saint Anthony the Abbot*.

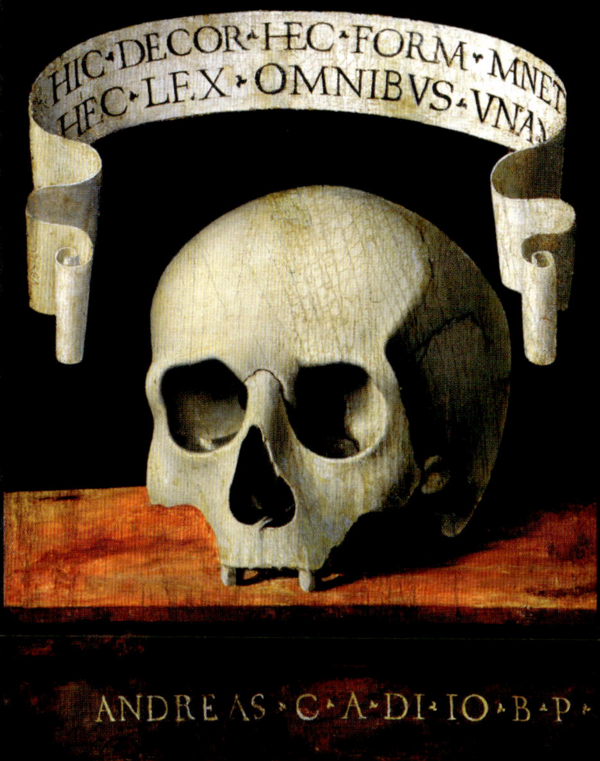

Skull of a Man, a painting by Andrea Previtali from approximately 1500, depicts a memento mori.

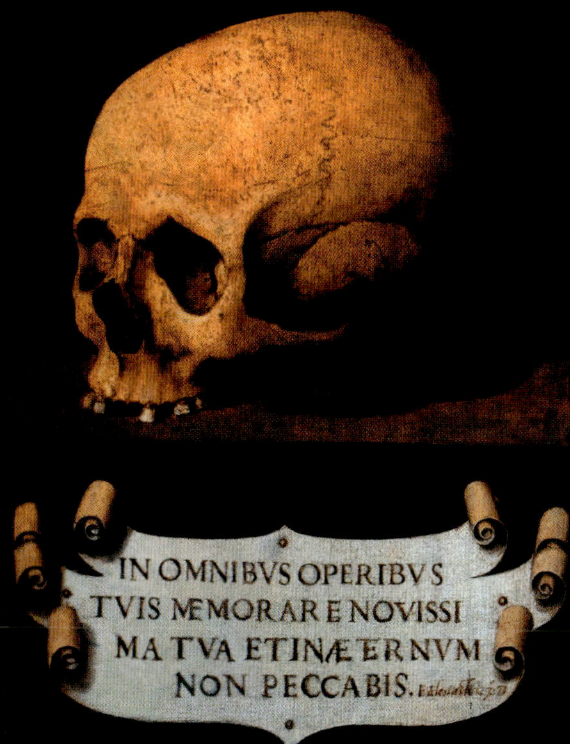

This skull memento mori is from Valencia and was painted around 1505.

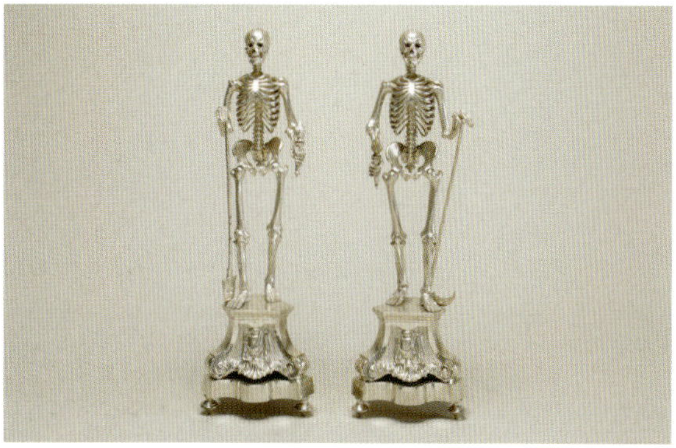

Two statues of *memento mori* skeletons.

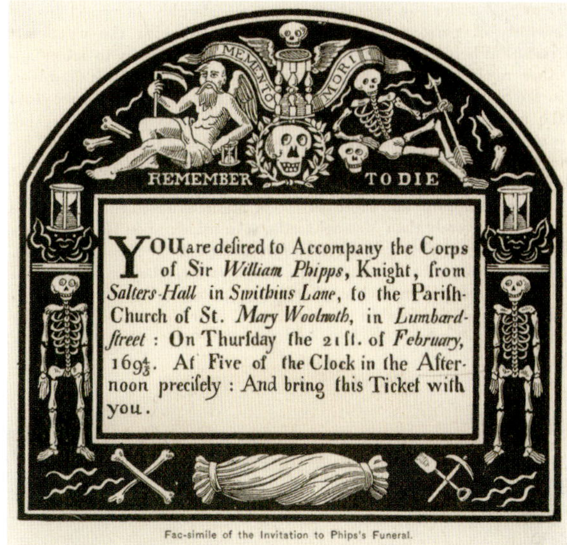

A 1694 funeral invitation with a *memento mori* message.

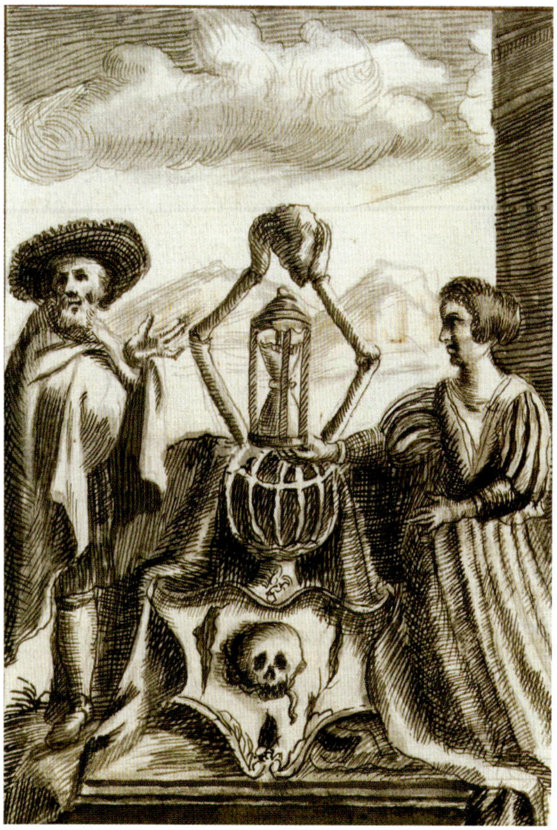

A *memento mori* drawing from a late-sixteenth century manuscript.

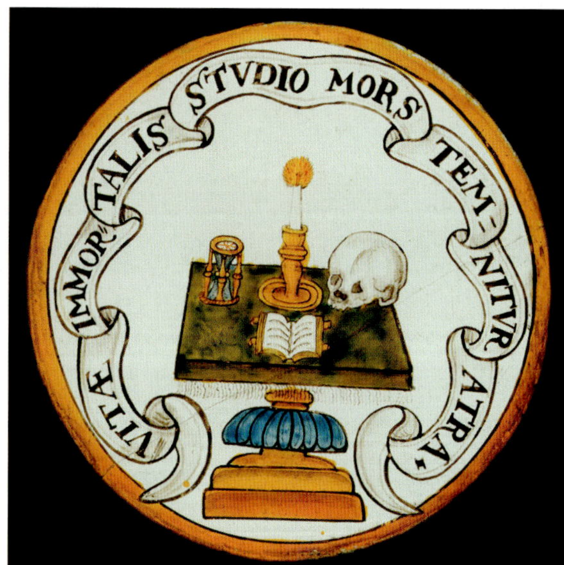

A stained glass roundel with a *memento mori* message.

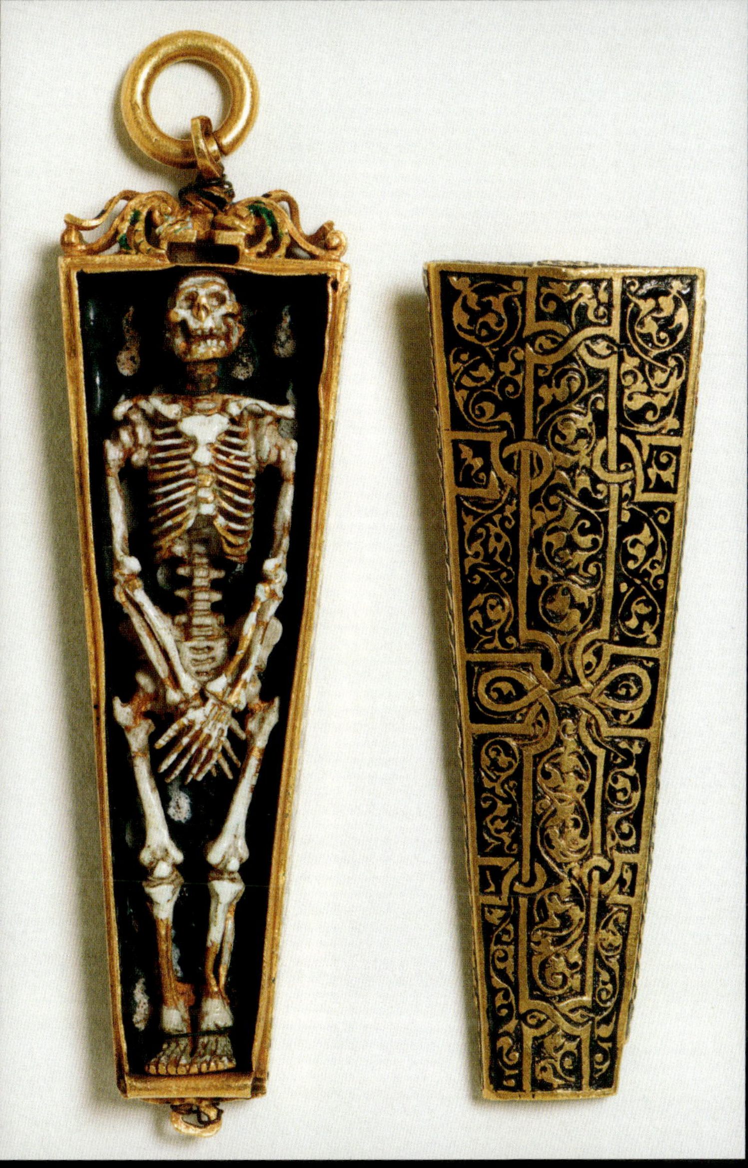

A *memento mori* pendant from mid-sixteenth century England.

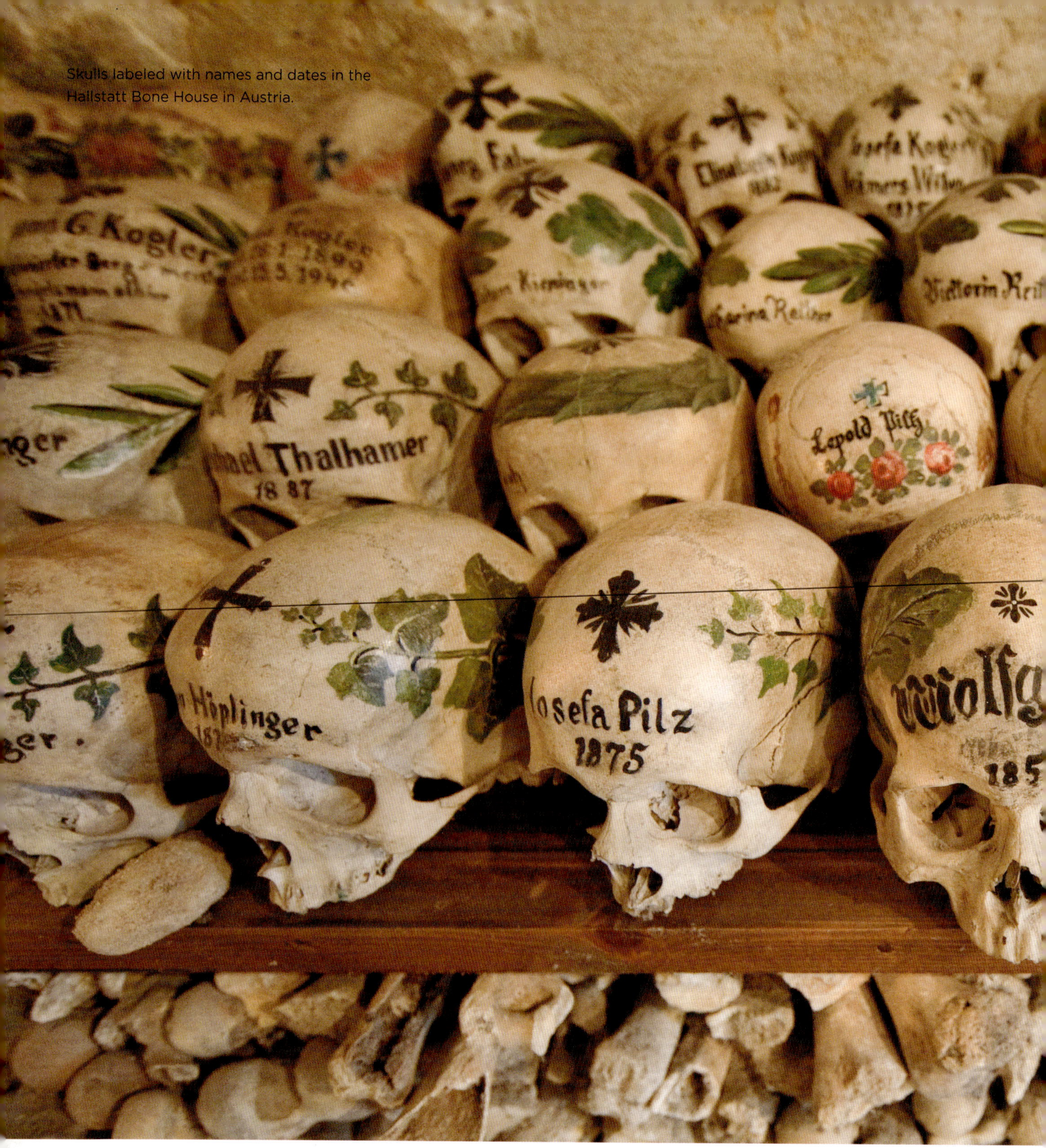

Skulls labeled with names and dates in the Hallstatt Bone House in Austria.

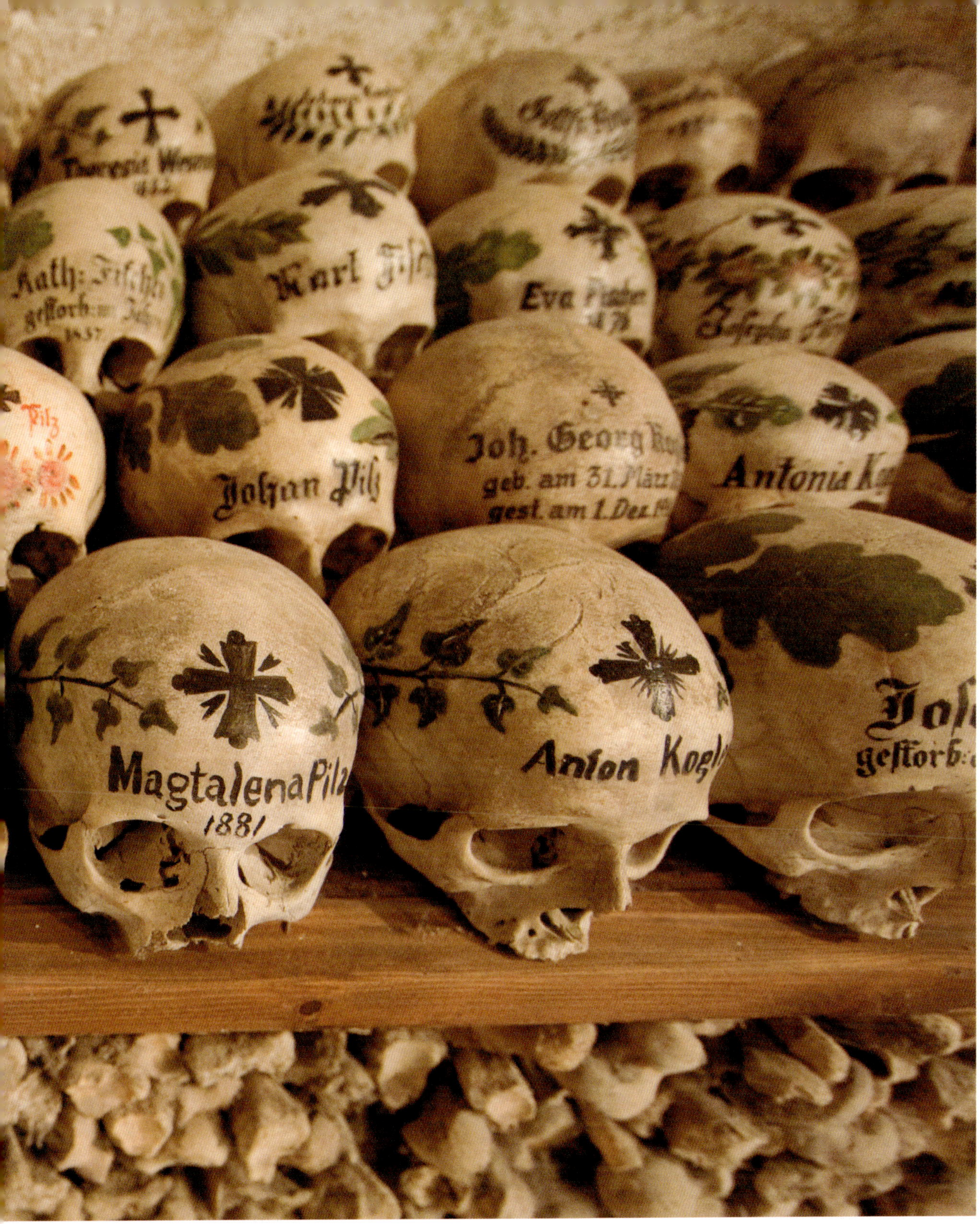

SKULLS IN RELIGION

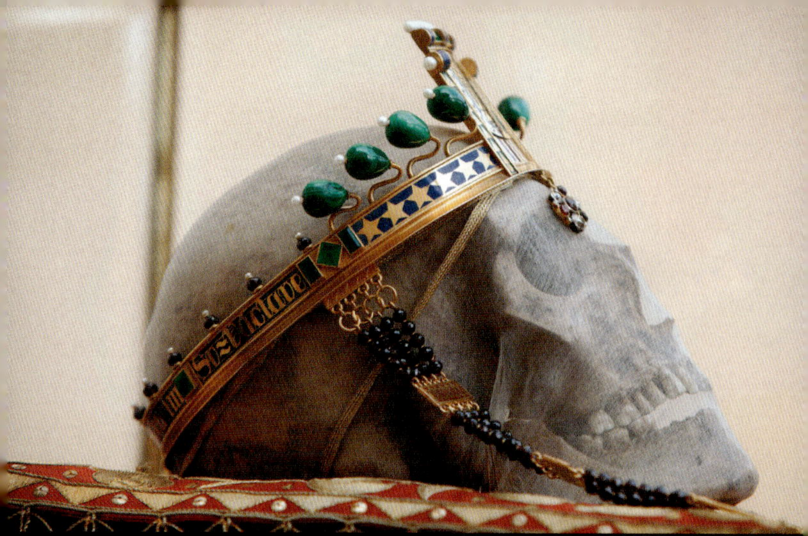

The Skull of St. Wenceslas is kept as a relic in Stara Bolelsav in the Czech Republic.

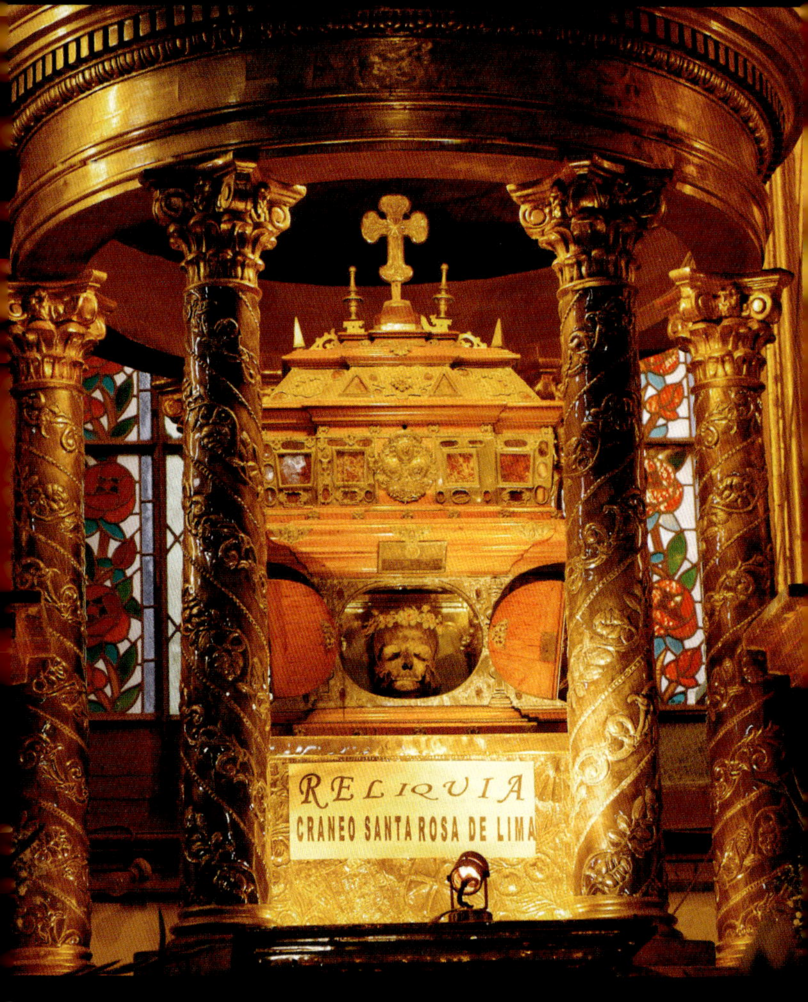

The skull of Saint Rose of Lima is held as a relic at a church in Lima, Peru.

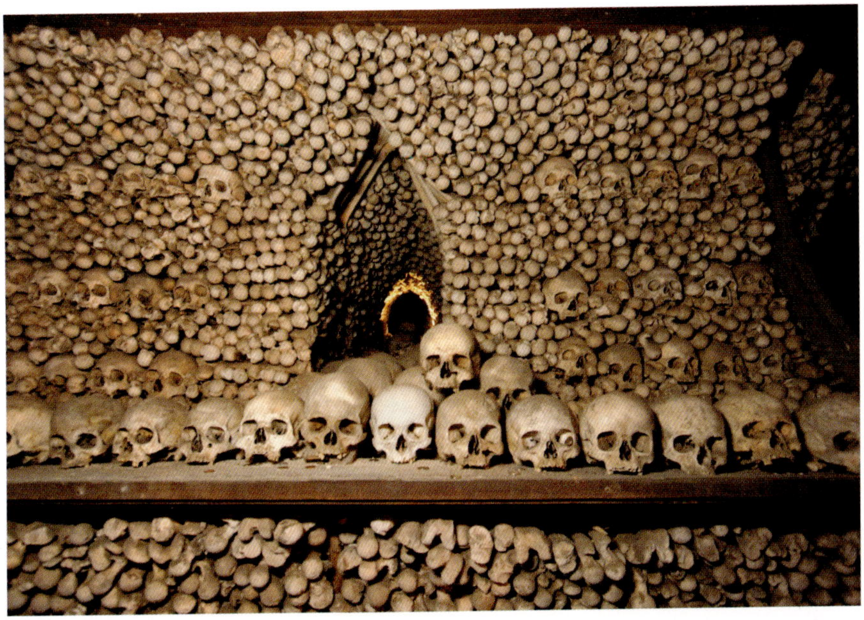

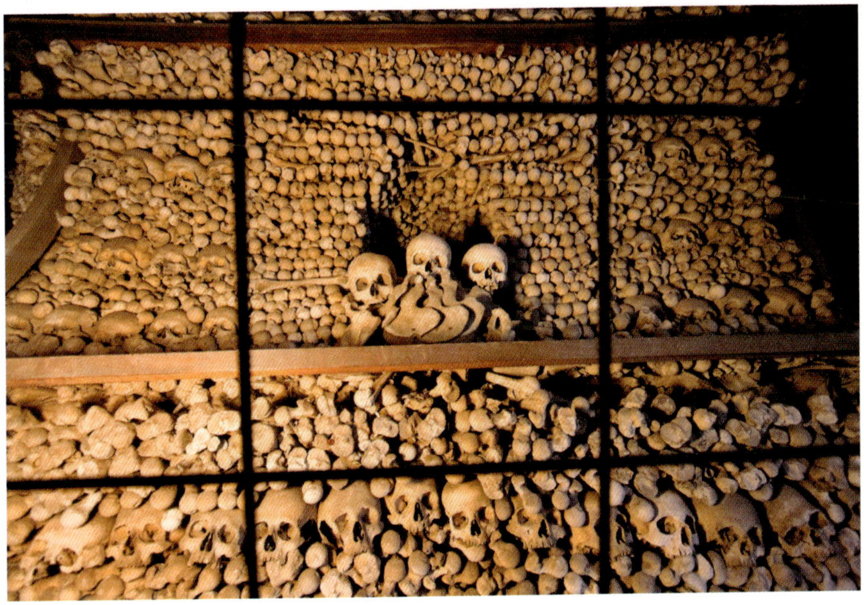

This page and through page 273: Photographs from within the Sedlec Ossuary in the Czech Republic.

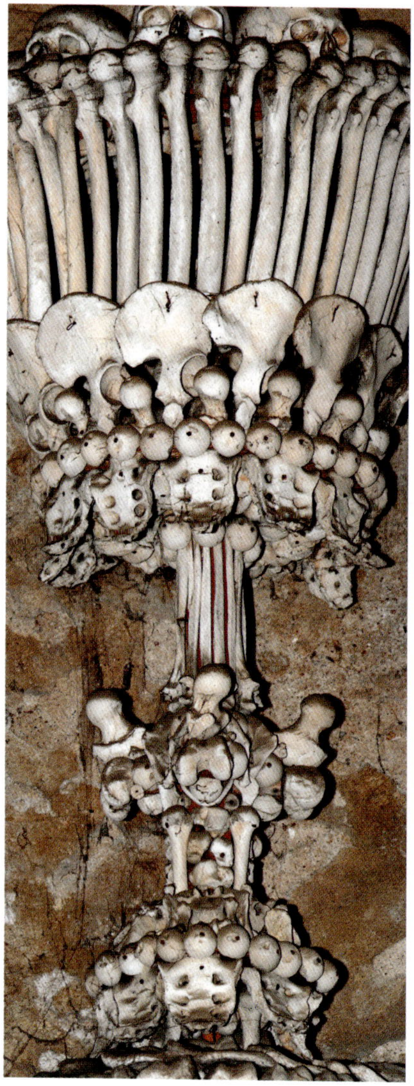

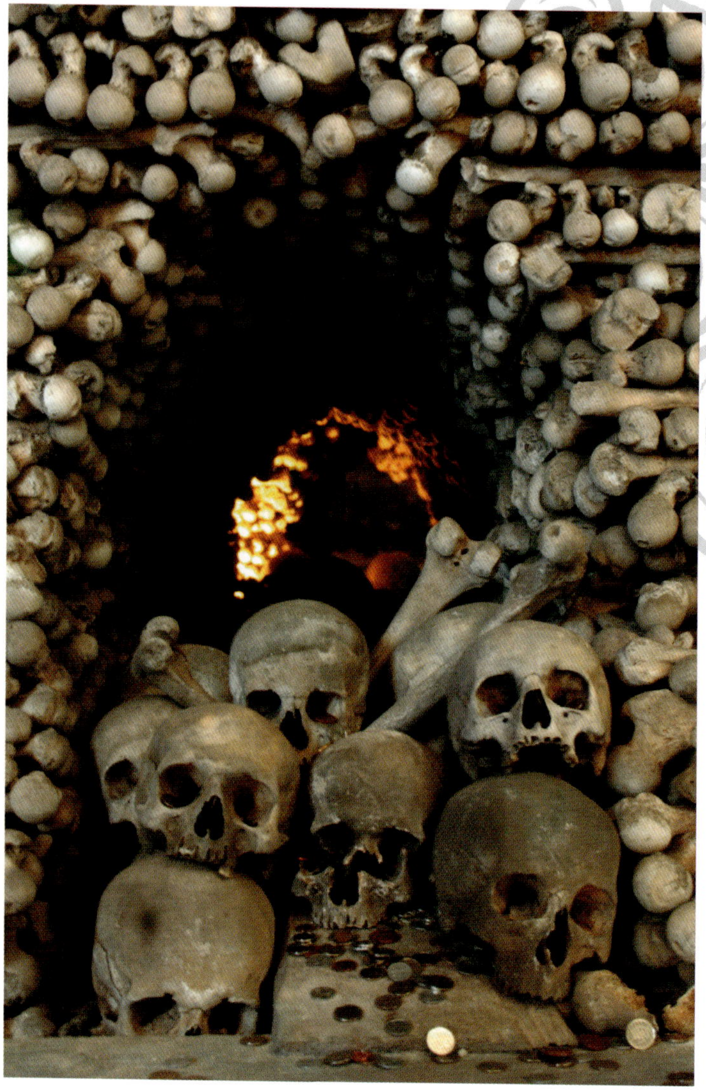

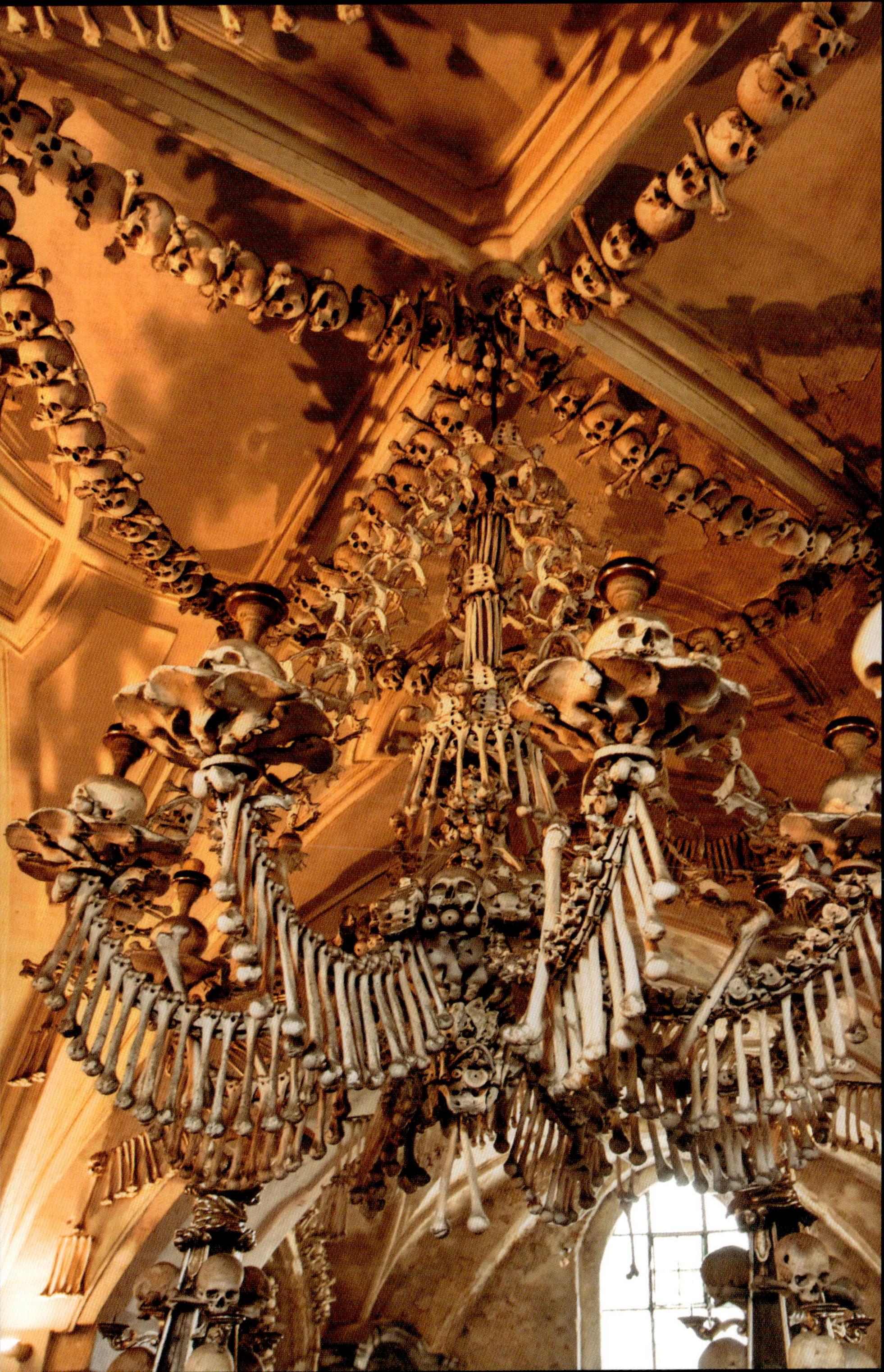

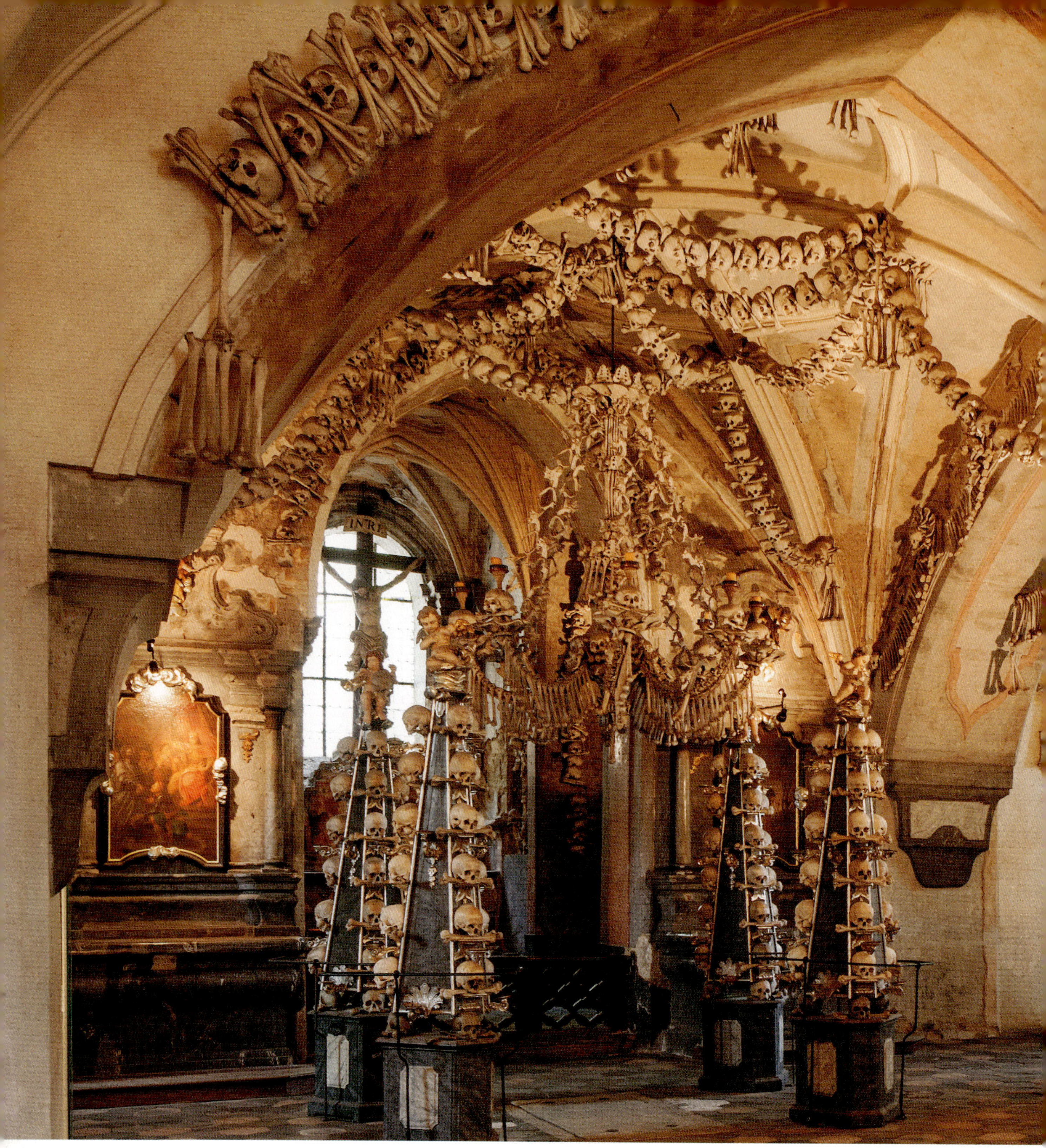
INRI

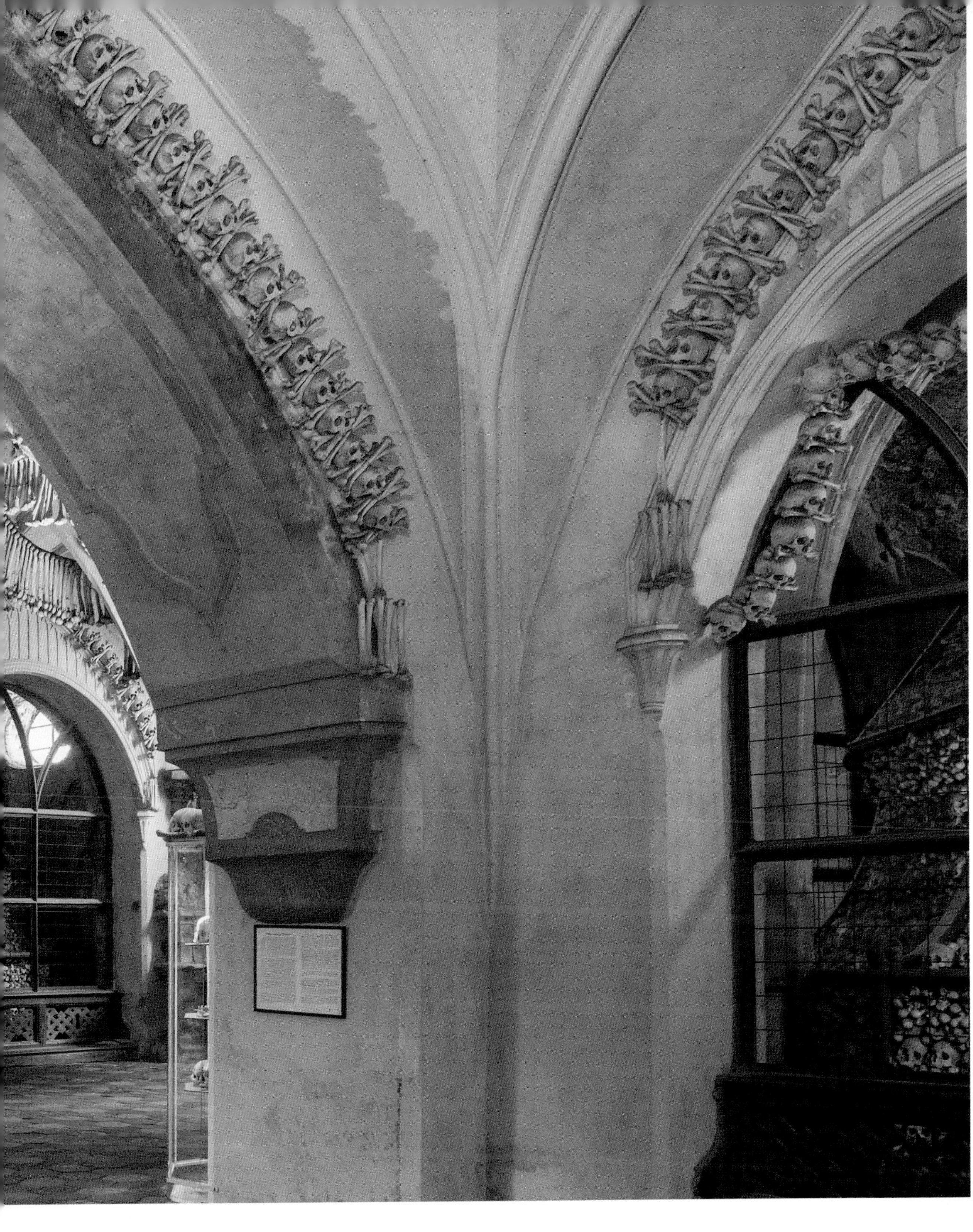

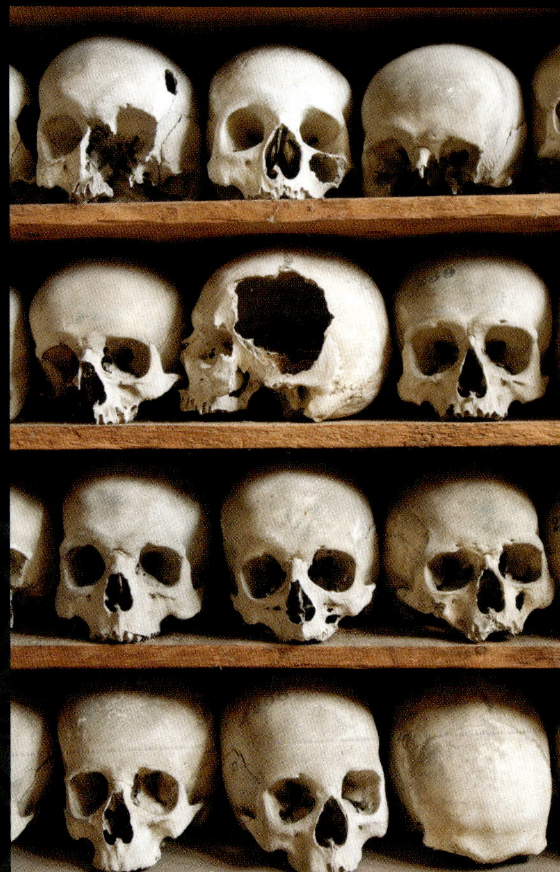

Medieval human skulls in the ossuary at St. Leonards Church in England.

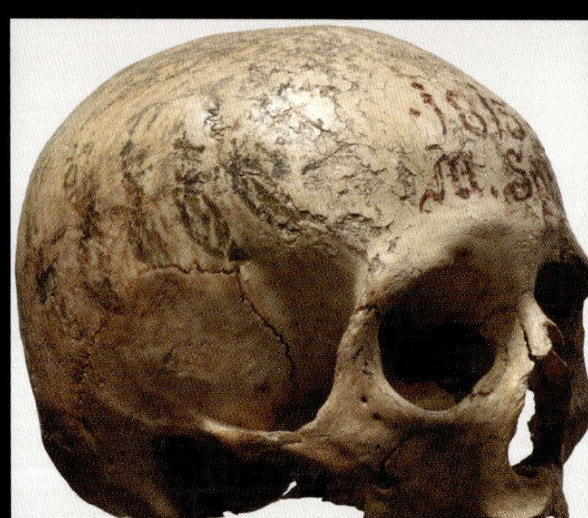

A skull stored in a South German ossuary missing the lower jaw; a name and year are painted on it.

SKULLS IN RELIGION

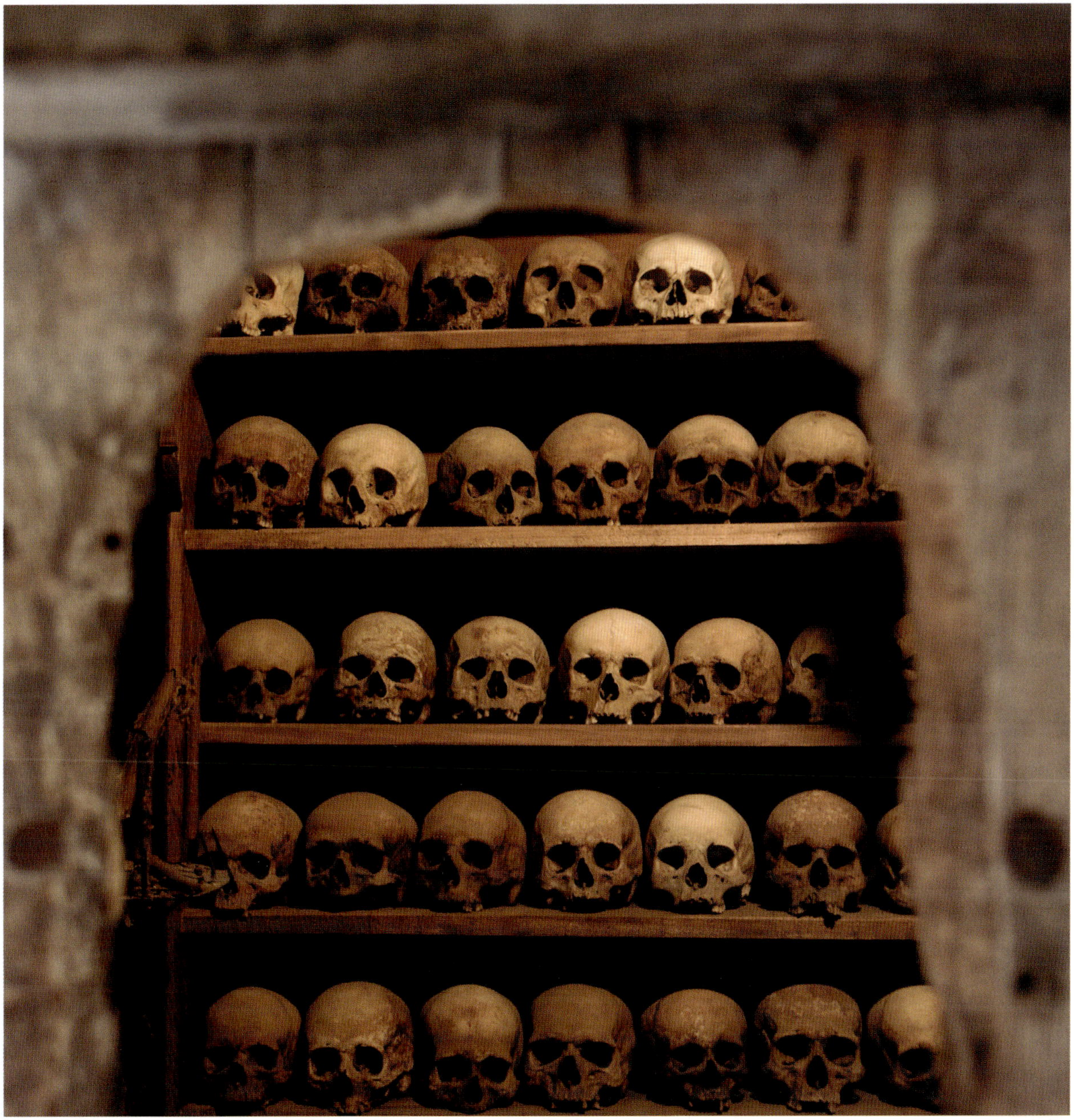

Skulls of deceased monks in an ossuary at the Megalo Meteoro, a monastery in Athens, Greece.

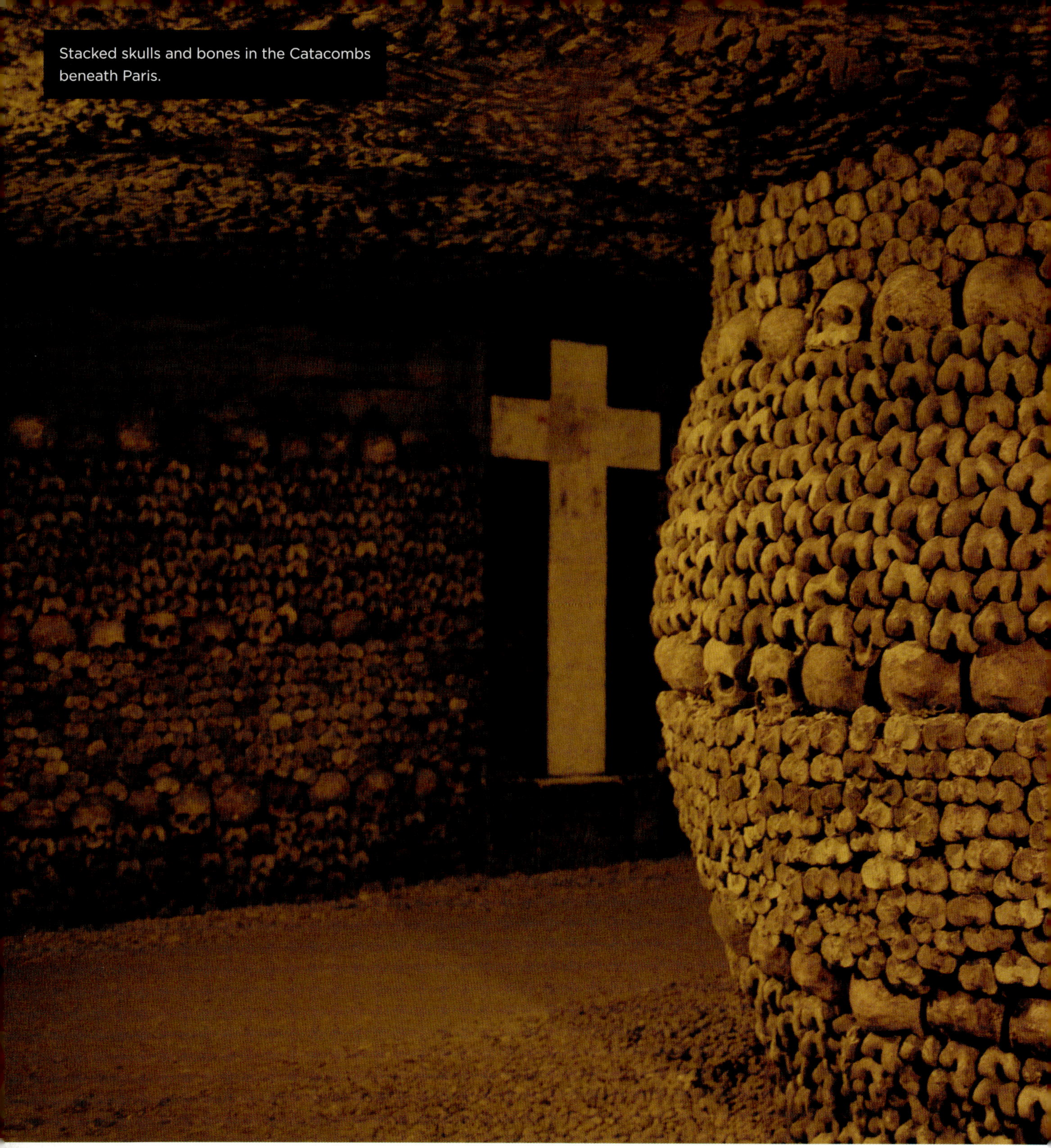

Stacked skulls and bones in the Catacombs beneath Paris.

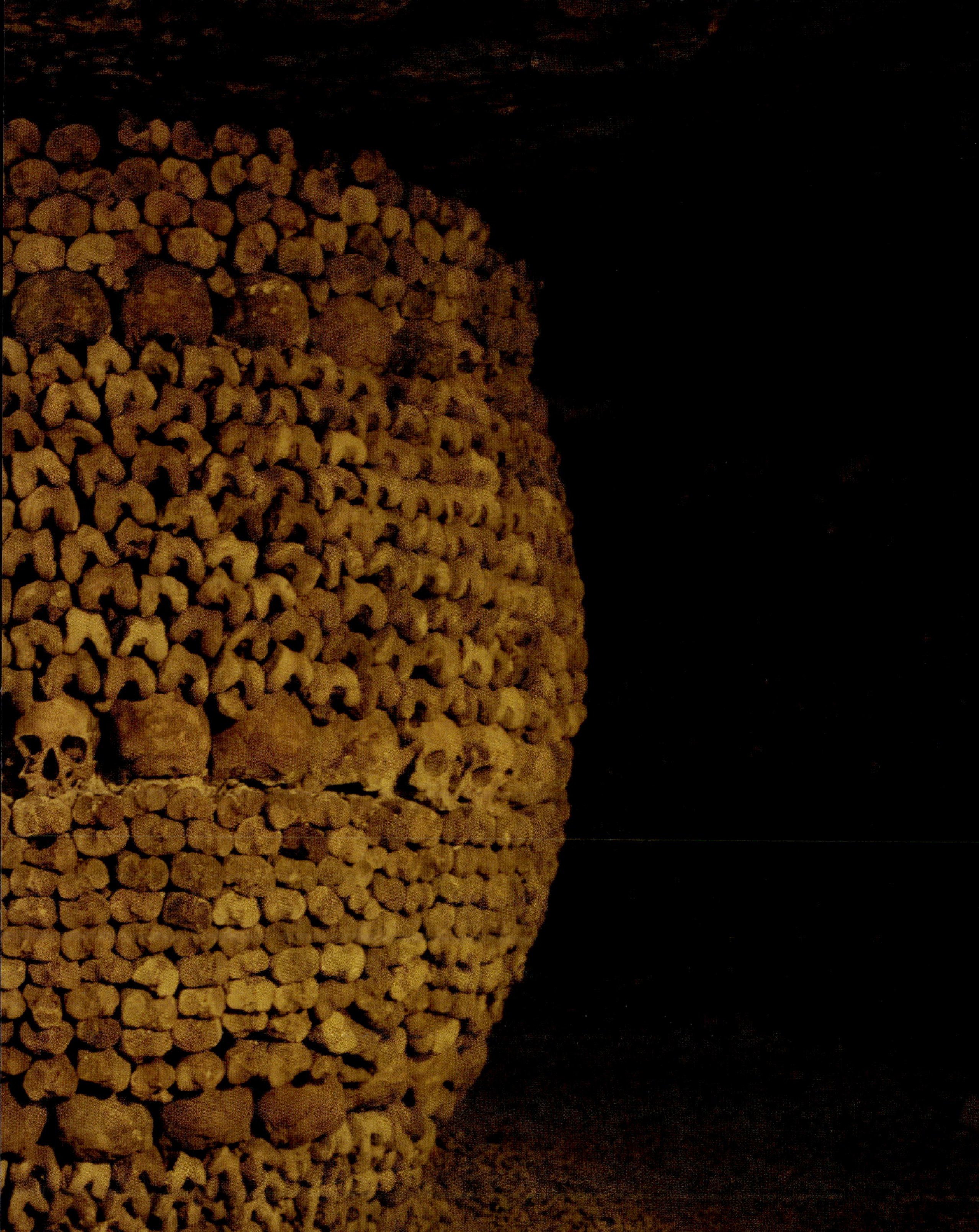

SKULLS IN RELIGION

IMAGE CREDITS

All reasonable efforts have been made to contact the copyright holders of all images.

HIDDEN SKULLS

p.11 (top): *Bridge to Eternity*, © G. Sarcone (www.giannisarcone.com). All rights reserved.

p.11 (bottom): *Deadly Dazzles*, © G. Sarcone (www.giannisarcone.com). All rights reserved.

p.12: © Brittany Riemer, *Hidden Between the Lines*, 2013.

p.13: © Scottish National Portrait Gallery, Edinburgh, Scotland/Bridgeman Images

p.14–16: © Museo Ilusionario

p.17: © Kerby Rosanes

p.18–22: © Christina Burch

p.23–25: © Kristy Patterson-D'Anna

p.26–28: © Istvan Orosz

p.29: © Vicky Sharma

p.30: © Joan Alturo/Bec Negre (www.joanalturo.com)

p.31: Courtesy of Mothers Against Drunk Driving, used with permission.

p.32: © The Print Collector/Getty Images

p.33: © The National Gallery, London/Google Art Project

SKULLS AND MUSIC

p.39: © Mark Noll

p.63 (top left): © Sam Dunn

p.63 (bottom left): © Sam Dunn

p.63 (bottom right): © Sam Dunn

SKULL TATTOOS

p.68 (top, bottom): © aquili-filius/CreateMyTattoo

p.68 (middle): © ladytat2/CreateMyTattoo

p.69 (top left): © lild_customs/CreateMyTattoo

p.69 (top right): © Justinwinter/CreateMyTattoo

p.69 (bottom): © Mythology/CreateMyTattoo

p.70 (top left): © PecuniamArt/CreateMyTattoo

p.70 (top right): © shady/CreateMyTattoo

p.70 (bottom left): © kasini/CreateMyTattoo

p.70 (bottom right): © chicktattoo/CreateMyTattoo

p.71 (top): © ShaunaHope/CreateMyTattoo

p.71 (bottom): © bmills/CreateMyTattoo

p.72 (top left): © Inksage/CreateMyTattoo

p.72 (top right): © KCool/CreateMyTattoo

p.72 (bottom): © kaneykreative/CreateMyTattoo

p.73 (top): © Dim-Dim/CreateMyTattoo

p.73 (bottom): © mfrank/CreateMyTattoo

p.74–75: © Barb Allen-Bolger

p.76 (top, bottom): © MissRita/CreateMyTattoo

p.76 (middle): © MacsTats/CreateMyTattoo

p.77 (top left): © Toddric76/CreateMyTattoo

p.77 (top right): © tat2Carson/CreateMyTattoo

p.77 (bottom): © chicktattoo/CreateMyTattoo

p.78 (top left): © Ghost13/CreateMyTattoo

p.78 (top right): © Gianluca/CreateMyTattoo

p.78 (bottom left): © rport35/CreateMyTattoo

p.78 (bottom right): © bcaso/CreateMyTattoo

p.79 (top left): © Justinwinter/CreateMyTattoo

p.79 (top right): © inkaholick/CreateMyTattoo

p.79 (bottom left): © achemical/CreateMyTattoo

p.79 (bottom right): © chicktattoo/CreateMyTattoo

p.80: © Gustavo Castro

p.81: © Ben T. Fiedler, Royal Tiger Tattoo

p.82 (top left): © Patricia/CreateMyTattoo

p.82 (top right): © wildheart/CreateMyTattoo

p.82 (center): © Dwiwilson/CreateMyTattoo

p.82 (bottom left): © ShyPersephone/CreateMyTattoo

p.82 (bottom right): © OMB/CreateMyTattoo

p.83 (top left, right): © MANCAVE/CreateMyTattoo

p.83 (bottom): © aquili-filius/CreateMyTattoo

p.84 (top): © hannonartink/CreateMyTattoo

p.84 (bottom): © achemical/CreateMyTattoo

p.85 (top left): © Porcelaine/CreateMyTattoo

p.85 (top right): © Mckayamber/CreateMyTattoo

p.85 (bottom): © devonsw/CreateMyTattoo

p.86–87: © Dino Tomic

p.88 (top): © vmslv/CreateMyTattoo

p.88 (bottom left): © gisty73/CreateMyTattoo

p.88 (bottom right): © Porcelaine/CreateMyTattoo

p.89 (top): © SeanG/CreateMyTattoo

p.89 (bottom): © GrechkaArt/CreateMyTattoo

p.90 (top): © Tattoodesigne/CreateMyTattoo

p.90 (bottom): © PatrickSellers/CreateMyTattoo

p.91 (top left): © carotdraws/CreateMyTattoo

p.91 (top right): © aquili-filius/CreateMyTattoo

p.91 (bottom): © mfrank/CreateMyTattoo

p.92–93: © Henning Jorgensen (www.royaltattoo.com); Instagram: @royaltattoodk

p.94 (top left): © achemical/CreateMyTattoo

p.94 (top right): © CHeRFe/CreateMyTattoo

p.94 (bottom): © L4_MU3RT3/CreateMyTattoo

p.95 (top): © bmills/CreateMyTattoo

p.95 (bottom): © AirbrushMike/CreateMyTattoo

p.96–97: © James Robinson

p.98: © Julien Lemoine

p.99: © Doug Hand; Instagram: @thughand

p.100–101: © Mario Delgado (www.mariodelgado.com)

p.102: © Salvatore Ffiti, Northern Liberty Tattoo

p.103 (top): © ribano2/CreateMyTattoo

p.103 (bottom): © Picasso2b/CreateMyTattoo

p.104–105: © Alasdair Roy (www.darkdesigngraphics.co.uk)

p.106–109: © Remi Larroque

p.110: © munkywrench/CreateMyTattoo

p.111 (top left): © mrsinister/CreateMyTattoo

p.111 (top right): © WhiteWhale/CreateMyTattoo

p.111 (bottom left): © inkaholick/CreateMyTattoo

p.111 (bottom right): © eraek/CreateMyTattoo

SKULLS IN ART

p.116–120: © Julien Lemoine

p.121: © Damien Hirst and Science Ltd. All rights reserved / DACS, London / ARS, NY 2015, Photo: Prudence Cuming Associates Ltd

p.122: © 2016 Georgia O'Keefe Museum / Artists Rights Society (ARS), New York

p.123: © Artists Rights Society (ARS), New York / VG Bild-Kunst, Bonn

p.124: © George Ioannou

p.125 (top): © Christie's Images/Corbis

p.125 (middle): © Detroit Institute of Arts, USA/ Bridgeman Images

p.125 (bottom): © Grand Palais, Paris, France/ Bridgeman Images

p.126–127: © Nancy Glenn-Nieto

p.128–129: © Noah Scalin

p.130–131: © Sasha Vinogradova

p.132 (top left): © age fotostock/Alamy

p.132 (bottom left): © INTERFOTO/Alamy

p.132 (right): © liszt collection/Alamy

p.133: © Salvador Dalí, Fundació Gala-Salvador Dalí, Artists Rights Society (ARS), New York 2016

p.134 (top): © Sebastien Micke/Paris Match/Getty Images

p.134 (bottom): © Nick Harvey/WireImage/Getty Images

p.135 (top): © Katina Houvouras/Corbis

p.135 (bottom): © Steve Russel/Toronto Star/ Getty Images

p.136–137: © Choccywoccydoodah, UK

p.139: © Fine Art Images/Heritage Images/Getty Images

p.140 (top): © DeAgostini/Getty Images

p.140 (bottom): © Leemage/UIG/Getty Images

p.141: © Fine Art Images/Heritage Images/Getty Images

p.142 (top): © Fine Art Images/Heritage Images/ Getty Images

p.142 (bottom): © Fine Art Images/Heritage Images/ Getty Images

p.143: © George Tatge for Alinari/Seat Archive/ Alinar Archives/Getty Images

p.144–146: © Isabelle Dalle

p.147: © Sam Dunn

p.148 (left): © Lidija Paradinovic Nagulov (www.celandinedesign.com) and Kimby Dawson (www.behance.net/heykimby)

p.148 (right): © Emily Marchesiello

p.149: © Alisa Grace

p.150: © Alice Dodard/Vesta Graphic Design/France

p.151: © Anneline Sophia (www.annelinesophiadesigns.com)

p.152: © 2014 Brian Allen (www.flylanddesigns.com)

p.153: © Lye Lawrence (www.inkloose.com)

p.154–155: © Kerby Rosanes

p.156: © Xrista Stavrou (www.xristastavrou.com)

p.157: © Jon Burgerman
p.158 (left): © Eric Pineda
p.158 (right): © Gabor Nagy
p.159: © Joseph Loughborough
p.160–161: © Jeremy Elder
p.162–163: © Francois Robert Photography (www.francoisrobertphotography.com)
p.164: © Michael Hacker (www.michaelhacker.at)
p.165: © Artwork by Philipp Rietz (www.badbugs.tumblr.com)
p.166–167: © Motohiro NEZU
p.168: © Rhys Owens/Hidden Moves
p.169: © Ryan Conners/KilkennycatArt
p.170–171: © Yuriy Skorohod
p.172–173: © Tommy Fiendish
p.174–175: © Vadim Zhulanov (www.behance.net/artbyvad)
p.176–177: © Wilhelm Steiner/Deaddreamer
p.178: © Craig Watkins
p.179: © Christopher Beikmann

SKULLS IN HISTORY

p.184 (top): © Brenda Kean
p.184 (bottom left): © David Gallimore/Alamy
p.184 (bottom right): © World History Archive/Alamy
p.185: © DeAgostini/Getty Images
p.186: © GraphicaArtis/Getty Images
p.187: © quisp65/Getty Images
p.188: © Peter Horree/Alamy
p.189 (top): © DeAgostini/Getty Images
p.189 (bottom): © University History Archive/UIG/Getty Images
p.190: © The Art Archive/Alamy
p.191: © Dea/G. Dagli Orti/Getty Images
p.192: © Bibliotheque de l'Assemblée Nationale, Paris, France/Bridgeman Images
p.193 (top): © General Photographic Agency/Getty Images
p.193 (bottom): © Evans/Three Lions/Getty Images
p.194 (top): © PHAS/UIG/Getty Images
p.194 (bottom left): © Sabena Jane Blackbird/Alamy
p.196 (top): © Judy Bellah/Alamy
p.196 (bottom left): © David McGlynn/Alamy
p.196 (bottom right): © RMX/Alamy
p.197: © Paige E.Sosebee
p.198: © National Geographic Image Collection/Alamy Stock Photo
p.199: © Eye Ubiquitous/Alamy
p.200 (top): © DeAgostini/Getty Images
p.200 (bottom): © Mary Evans Picture Library/Alamy
p. 201 (left): © Galerie Bilderwelt/Getty Images
p.201 (right): © Transcendental Graphics/Getty Images
p.202: © Leemage/Getty Images
p.203: © DeAgostini/Getty Images

SKULLS IN FANTASY, SCI-FI, AND HORROR

p.208: © SSPL/Getty Images
p.209: © Leemage/Getty Images
p.210: © Hulton Archive/Getty Images
p.211: © Fine Art Images/Getty Images
p.212: © Bertrand Demee/Getty Images
p.213: © amdandy/Getty Images
p.214 (left): © CSA Images/Printstock Collection/Getty Images
p.214 (right): © CSA Images/Getty Images

p.215–216: © CSA Images/Printstock Collection/ Getty Images

p.217: © Duncan Walker/Getty Images

p.218: © Man_Half-tube/Getty Images

p.219 (top): © Mart Klein/Getty Images

p.219 (bottom): © Igor Djurovic/Getty Images

p.220 (bottom left): © Pictorial Press Ltd/Alamy

p.232–233: © Chris Panatier

p.234–235: © Alain Bellino

SKULLS IN RELIGION

p.239: © SSPL/Getty Images

p.240 (top): © The Art Archive/Alamy

p.240 (bottom): © Martin Broeze/Alamy

p.241 (left): © Thomas Boehm/Alamy

p.241 (right): © age fotostock/Alamy

p.242: © Sergio Anelli/Mondadori Portfolio/ Getty Images

p.243: © The Print Collector/Print Collector/ Getty Images

p.244: © DeAgostini/Getty Images

p.245: © Sergio Anelli/Electa/Mondadori Portfolio/ Getty Images

p.246: © Paolo e Federico Manusardi/Electa/ Mondadori Portfolio/Getty Images

p.247 (left): © Alessandro0700/Alamy

p.247 (top right): © David Lyons/Alamy

p.247 (bottom right): © John Greim/Light Rocket/ Getty Images

p.248–249: © Eye Ubiquitous/UIG/Getty Images

p.250 (top): © Chris Howes/Wild Places Photography/ Alamy

p.250 (bottom): © Agencja Fotograficzna Caro/Alamy

p.251 (top): © Sabena Jane Blackbird/Alamy

p.251 (bottom): © Stuart Cox/Alamy

p.252: © DeAgostini/Getty Images

p.253: © Stan Pritchard/Alamy

p.254 (left): © Gonella/Electa/Mondadori Portfolio/ Getty Images

p.254 (right): © INTERFOTO/Alamy

p.255 (top): © The Art Archive/Alamy

p.255 (bottom): © Active Museum/Alamy

p.256: © DEA Picture Library/DeAgostini/Getty Images

p.257: © Imagno/Getty Images

p.258: © Heritage Image Partnership/Alamy

p.259: © Fine Art Images/Heritage Images/Getty Images

p.260: © Dea/G. Nimatallah/Getty Images

p.261: © Mondadori Portfolio/Getty Images

p.262: © DeAgostini/V. Pirozzi/Getty Images

p.263 (left): © Corbis

p.263 (right): © Peter Horree/Alamy

p.264 (top left): © Gallery Stock

p.264 (top right): © Mary Evans Picture Library/Alamy

p.264 (bottom left): © The Art Archive/Alamy

p.264 (bottom right): © V&A Images/Alamy Stock Photo

p.265: © V&A Images/Alamy Stock Photo

p.266–267: © Franz Pritz/Getty Images

p.268 (top): © Radim Beznoska/Alamy

p.268 (bottom): © age fotostock/Alamy

p.269: © Louise Heusinkveld/Alamy

p.270 (left): © Michal Boubin/Alamy

p.270 (right): © Michal Cizek/AFP/Getty Images

p.271: © Miroslav Beneda/Alamy

p.272–273: © Maremagnum/Getty Images

p.274 (top): © P. Tomlins/Alamy

p.274 (bottom): © INTERFOTO/Alamy

p.275: © Milos Bicanski/Getty Images

p. 276–277: © Panoramic Images/Getty Images

ABOUT THE ARTISTS

Brian Allen is a freelance artist and creative director who lives in the State College/Bellefonte area of Pennsylvania with his wife and two children. Explore more of his work at www.FlylandDesigns.com or contact him at brian@FlylandDesigns.com.

Barb Allen-Bolger is a tattoo artist, based in Brighton, United Kingdom. Explore more of her work on Facebook (Tattoos by Barb), Instagram (@Barboo72), Pinterest (@Barbinka80c), or at www.InkaTattoos.co.uk.

Joan Alturo, who goes by the artist name Bec Negre, is an artist based in Bournemouth, United Kingdom. He is originally from Barcelona, Spain. He is interested in illustrating many topics, but focuses mainly on dark subject matter, often using skeletons and skulls in his work. He finds them aesthetically interesting because of the spaces that the skeleton-shape creates, making them versatile object to draw. Explore more of his work at www.JoanAlturo.com.

Christopher Beikmann is a professional digital and mixed media artist from New Mexico. His artwork blends collage techniques with illustration. His subject matter fuses ancient cultural motifs with botanical elements and modern designs. Explore more of his work at www.FusionIdol.com.

Alain Bellino is a sculptor based in Nice, France. Explore more of his work at www.alain-bellino.net.

Vicky Bharadwaj is an artist who expresses herself through illusions. She can be contacted at vickybharadwaj007@gmail.com.

Christina Burch's work plumbs the imagination to penetrate life's deeper meanings and inherent poetry. A landscape of magical elements, a universe of pure mystery, divine energies at play: Burch's work is a lyrical exploration of the sublime dimension of figure in painting. Explore more of her work at www.ChristinaBurch.com.

Jon Burgerman is a British artist based in Brooklyn, New York. Explore more of his work on Facebook, Instagram, and Twitter (@jonburgerman) or at www.JonBurgerman.com.

Gustavo Castro is a Brazilian tattoo artist with experience in many countries around the world. He does illustrational tattoo art with solid colors or classic black and grey designs with a modern feel. Explore more of his work on Facebook (Gustavo Castro) or Instagram (@gcastrotattoo).

Ryan Conners is a Pennsylvania-based folk artist. Explore more of her work on Instagram (@KilkennycatArt) or at www.kilkennycat.wordpress.com.

CreateMyTattoo brings custom tattoo design ideas to life by connecting you to a community of over 20,000 custom tattoo designers from all over the world. The designers compete to create a custom tattoo design based on your specifications. With CreateMyTattoo, you're not limited to the imagination of one tattoo designer. Instead, you'll receive over ten variations (the average is twenty-two submissions) of your dream tattoo design from multiple tattoo designers. With CreateMyTattoo, you get: a 100 percent money back guarantee; an original high resolution design and stencil for your local tattoo shop; and pricing starting at just $20 with a $2.99 listing fee. Explore more tattoo designs or start a contest of your own at www.CreateMyTattoo.com.

Alice Dodard is a French graphic designer working under the name Vesta. She loves to find old newspapers and create hybrid figures with them. Explore more of her work at www.vesta-graphiste.com.

Sam Dunn is a freelance Illustrator based in East London. Her work is delicately crafted by hand with pen and ink and then digitally colored with many layers of found and forged textures. Explore more of her work at www.sam-dunn.com, or contact her at sam_dunn@live.co.uk.

Jeremy Elder is a Virginia-based artist. His work can be found gracing gallery walls, the bottoms of skateboards, or in dozens of publications. His style combines traditional and digital techniques, creating images that are unique and personal, yet clean and contemporary. Explore more of his work at www.elderhousearts.com.

Isabelle Dalle is a Paris-based digital artist who showcases an astonishing series of beautiful anatomical portraits. The imagery is inspired by archaic medical journals, and how they can be translated into pieces of art. Using areas of the body to infuse her digital creativity, she creates something that goes beyond surrealism. She uses a diverse range of colors, prints, and texts to redefine the human form. Explore more of her work at www.Isabelle DalleStore.com.

Kimby Dawson is an Australia-based graphic designer. Explore more of her work at www.behance.net/heykimby.

Mario Delgado is a San Francisco–based tattoo artist. Explore more of his work on Instagram (@mariod1979) or at www.MarioDelgado.com.

Salvatore Ffiti is a tattoo artist based in Philadelphia, Pennsylvania. He works at Northern Liberty Tattoo. Explore more of his work on Instagram (@ffititattoo) or contact him at salffiti@gmail.com.

Ben Fiedler is a tattoo artist, painter, 3D animator, and owner of Royal Tiger Tattoo in Austin Texas. Explore more of his work on Instagram (@benfiedler, @royaltigertattoo) or at www.RoyalTigerTattoo.com

Tommy Fiendish is a London-based artist who expresses himself through design. Explore more of his work at www.TommyFiendish.com.

Alisa Grace is an artist who can be contacted at AlisaGrace@mail.ru.

Michael Hacker is an illustrator based in Vienna, Austria. Other than illustrating and self-publishing comics, he designs gig posters for bands like Green Day, QOTSA, The Melvins, Sonic Youth, Mastodon, and Dinosaur Jr. Explore more of his work at www.MichaelHacker.at.

Doug Hand is a tattoo artist based in Philadelphia, Pennsylvania. Explore more of his work on Instagram (@thughand) or at www.NorthernLibertyTattoo.com.

George Ioannou is a self-taught artist based in the United Kingdom. He uses a variety of techniques on his artwork to achieve the desired look, including spray and acrylic paint, metal fabrication, and even a shotgun. Explore more of his work at www.GeorgeIoannou.com.

Henning Jørgensen is a tattoo artist based in Denmark. He is very happy that he was able to make his hobby into a career as a tattoo artist at Royal Tattoo. Explore more of his work on Instagram (@royaltattoodk) or at www.RoyalTattoo.com.

Remi Larroque is a Montreal-based tattoo artist and illustrator. Explore more of his work on Tumblr (RemiIsMeltingDots).

Lye Lawrence is an illustrator, cartoonist, and tattoo artist. Explore more work at www.inkloose.com.

Julien Lemoine is a Paris-based art director who loves rock and roll. Explore more of his work on Instagram (@Julien_Lemoine_Art) or at www.JulienLemoine.com.

Joseph Loughborough is a British artist. Explore more of his work at www.JosephLoughborough.co.uk.

Emily Marchesiello is a graduate of the Masters of Fine Arts program at Queens College CUNY. She is a visual artist from New York who specializes in fine point line drawing, oil painting, and watercolor painting. Emily derives her inspiration from the natural world and the fantasy realm. The process of intentional mark making is a meditation—each line, dot, and application of vibrant color is a physical manifestation of thought, while at the same time a product of mindless release. Explore more of her work at www.EmilyMarchesiello.com.

Gábor Nagy is an artist. Explore more work at www.ephynephryn.deviantart.com.

Motohiro Nezu is an illustrator and graphic designer based in Tokyo. Explore more of his work at www.Motohiro-Nezu.com.

Mark Noll is an avid carver and skull-lover. He is fascinated by skulls because they determine what you look like, how people see you, and how you are recognized. Skulls also represent the temporary nature of life. He is based in Saint Petersburg, Florida. Explore more of his work at www.MarkNollDesigns.com.

István Orosz experiments with the paradoxes of geometry and optics. His pseudonym Utisz (pronounced: *Outis*, which means "nobody"), was also Odysseus's pseudonym in the affair with the Cyclops that ended in the blinding of the monster. In accordance with the name, his art is a kind of attack on the eye. Explore more of his art at www.utisz.blogspot.hu.

Rhys Owens is a digital artist based in South Wales, United Kingdom. Explore more of his work at www.HiddenMoves.com.

Chris Panatier is an artist based in Dallas, Texas. He likes to listen to metal music and draw album covers. Explore his work on Facebook (ChrisPanatier), Tumblr (PrevailAndRide), or at www.panatier.com.

Lidija Paradinovic Nagulov is a Serbia-based surface and pattern designer. Explore more of her work at www.CelandineDesign.com or www.behance.net/celandine.

Kristy Patterson D'Anna is an illustrator based in Baltimore, Maryland. Explore more artwork at www.society6.com/kristypattersondesign.

Eric Pineda is a Los Angeles–based artist. Explore more of his work at www.EricPineda.com.

Brittany Riemer is a commission-based artist who is in the process of launching an online web comic. Her art can be followed online at www.BrittanyRiemer.com, www.gamesoffate.com, and www.skull-splinters.deviantart.com.

Philipp Rietz is a German-based digital artist working under the name Badbugs Art. Explore more of his work on Instagram (@badbugs_art) or www.Badbugs.de, or contact him at info@Badbugs.de.

Francois Robert is a photographer working in Chicago, Illinois, and Tucson, Arizona. Explore more of his work at www.FrancoisRobertPhotography.com.

James Robinson is a tattooist and the owner of the Gilded Cage Tattoo Studio in Brighton, United Kingdom. Explore more of his work at www.GildedCageTattooStudio.com.

Kerby Rosanes, a Philippines-based illustrator, works mainly with fine lines to illustrate his world. As a traditional artist, he has worked with various global brands and design agencies. His personal projects that have attracted his online fan following can be found on his website, www.KerbyRosanes.com.

Alasdair Roy is an illustrator and artist. He works at Dark Design Graphics in the United Kingdom. Explore more of his work at www.DarkDesignGraphics.co.uk or contact him at alasdair@DarkDesignGraphics.co.uk.

Gianni A. Sarcone is an author, artist, and inventor. His work explores the wonder and the paradoxes of the mind. He is co-founder of Archimedes' Lab Project, a consulting network of experts specializing in developing creativity. Explore more of his work at www.GianniSarcone.com.

Noah Scalin is a Richmond, Virginia based artist, known for the Webby Award–winning project Skull-A-Day. His fine art has been exhibited in museums and galleries internationally, including multiple solo shows at Krause Gallery in New York City. Explore more of his work on Instagram (@NoahScalin) or at www.NoahScalin.com.

Yuriy Skorohod is a dotwork artist from Minsk, Belarus. Explore more of his work on Facebook (artsc0re.graphics) or www.artscore.by, or contact him at artsc0re.graphics@gmail.com.

Anneline Sophia is a surface pattern designer and illustrator. Explore more of her work at www.AnnelineSophiaDesigns.com or contact her at info@AnnelineSophiaDesigns.com.

Paige Sosebee is an artist based in Houston, Texas. Explore more of her work at www.PaigeSosebee.com.

Xrista Stavrou is a digital design artist based in Nicosia, Cyprus. Her passion for art and fashion led her to draw digital art and transfer them to clothing via silk-screen printing. Explore more of her work at www.XristaStavrou.com.

Monster Steel is your online go-to source for wholesale piercing and tattoo supplies and wholesale body jewelry. In business since 2001, Monster Steel guarantees the best selection of tattoo and piercing products at the best prices. They stock a huge line of the most popular tattoo inks including Eternal, Intenze, and Starbrite, to name a few, and have a loyalty program that lets you earn free credit with each purchase. Whether you are an independent practitioner or own a shop or salon, they stock the supplies needed to run your business. Learn more at www.MonsterSteel.com.

Wilhelm Steiner is an artist based in Vienna, Austria. Explore more of his work at www.deaddreamer.com.

Dino Tomic is a Norway-based tattoo artist. Explore more artwork on Instagram (@DinoTomic) or contact him at dino@atomiccircus-tattoo.com.

Sasha Vinogradova is a Los Angeles–based artist. Explore more work at www.SashaVinogradova.com.

Craig Watkins is a freelance illustrator specializing in character design and unique branding. Working under the name Wotto, he has many clients ranging from industry giants to small, independent brands. Originally born in the United Kingdom, he is currently based in California. Explore more of his work at www.wottoart.com.

Vadim Zhulanov is a Moscow-based artist. Explore more of his work at www.behance.net/artbyvad.

ABOUT THE AUTHOR

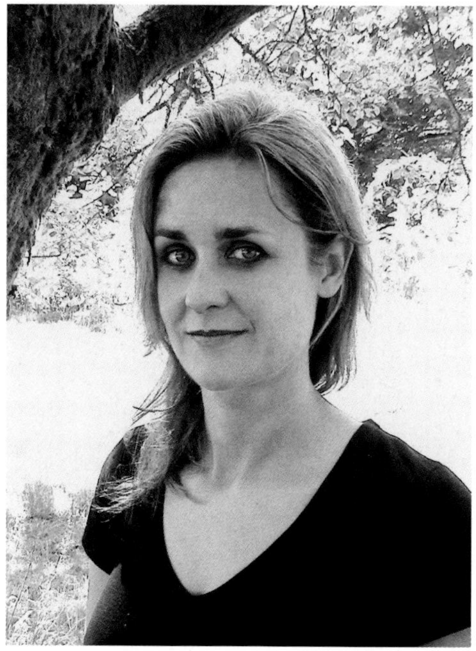

ADELE NOZEDAR is the best-selling author of *The Element Encyclopedia of Signs & Symbols, The Element Encyclopedia of Native Americans, Freaky Dreams, The Secret Language of Birds, The Secret Language of Birds Tarot, The Hedgerow Handbook*, and *The Garden Forager*. She lives in Wales where she writes books and runs the Brecon Beacons Foraging School. Learn more about Adele at www.AdeleNozedar.com.

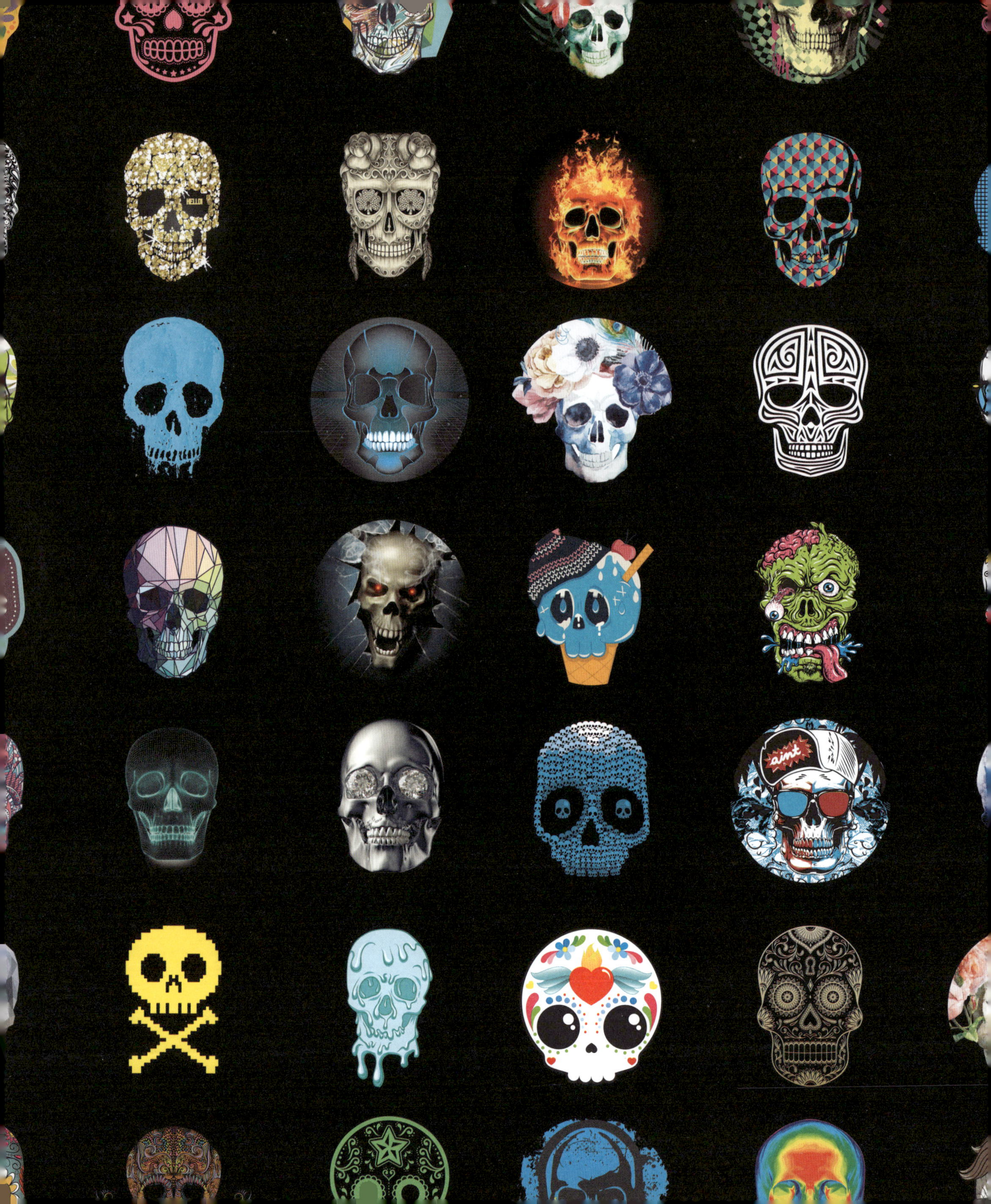
HELLO!
aint